Mahmoud Arghavan, Nicole Hirschfelder,
Luvena Kopp, Katharina Motyl (eds.)
Who Can Speak and Who Is Heard/Hurt?

Culture and Social Practice

Mahmoud Arghavan (Dr.) coordinates the unit for Help for Refugees, Migration, and Integration at the Innere Mission in Munich. He completed his PhD in 2013 at the Free University of Berlin with a dissertation on Iranian American literature. From 2014 to 2016, he worked as an adjunct lecturer at the American Studies department of the University of Tübingen. His research interests include diaspora studies, postcolonial studies, critical race theory, and border studies.

Nicole Hirschfelder (Dr.) is an associate professor at the American studies department at the University of Tübingen and an associate member of the Collaborative Research Center (Sonderforschungsbereich) 923 "Threatened Order – Societies Under Stress". She was a visiting professor at the University of Maryland in 2016 and 2019. Her main areas of scholarship include figurational sociology, inequality, protest, new social movements, and the Civil Rights Movement in the U.S. and Germany.

Luvena Kopp (M.A.) teaches film, literature, and cultural studies at the University of Tübingen. Her research interests include concepts of figurational sociology as developed by Pierre Bourdieu and Norbert Elias, power relations, and the films of Spike Lee.

Katharina Motyl (Dr.) is an assistant professor at the American Studies department of the University of Mannheim. Her research focuses on Arab American literature, global Arab and Muslim perspectives on the 'War on Terror,' the sociocultural history of drugs and addiction in the U.S., the cultural history of failure, African American expressive culture as well as Indigenous cultures and ways of knowing.

Mahmoud Arghavan, Nicole Hirschfelder, Luvena Kopp, Katharina Motyl (eds.)

Who Can Speak and Who Is Heard/Hurt?

Facing Problems of Race, Racism, and Ethnic Diversity in the Humanities in Germany

[transcript]

Printed with support from the Collaborative Research Center 923 "Threatened Order – Societies under Stress" and the Center for Gender and Diversity Research (ZGD), both at Eberhard Karls Universität Tübingen.

Bibliographic information published by the Deutsche Nationalbibliothek
The Deutsche Nationalbibliothek lists this publication in the Deutsche Nationalbibliografie; detailed bibliographic data are available in the Internet at http://dnb.d-nb.de

Cover concept: Kordula Röckenhaus, Bielefeld
Printed by Majuskel Medienproduktion GmbH, Wetzlar
Print-ISBN 978-3-8376-4103-5
PDF-ISBN 978-3-8394-4103-9
https://doi.org/10.14361/9783839441039

Table of Contents

Acknowledgments | 7

Who Can Speak and Who is Heard/Hurt? Facing Problems of Race, Racism, and Ethnic Diversity in the Humanities in Germany: A Survey of the Issues at Stake | 9
Mahmoud Arghavan, Nicole Hirschfelder, and Katharina Motyl

I. RACE IN TRANSLATION: COMPARING RACIAL AND XENOPHOBIC FORMATIONS IN GERMANY AND THE UNITED STATES

'*Ausländer*' – A Racialized Concept? 'Race' as an Analytical Concept in Contemporary German Immigration History
Maria Alexopoulou | 45

Perspective Matters: Racism and Resistance in the Everyday Lives of Youths of Color in Germany
Lili Rebstock | 69

Beyond a Trifling Presence: Afro-Germans and Identity Boundaries in Germany
Anthony Obute | 83

Race and Racism in Translation: "Who Can Speak?" in German Renderings of Literary African American English
Annika Rosbach | 101

II. NORMATIVE WHITENESS IN THE GERMAN HUMANITIES

Post-Racism, Colorblind Individualism & Political Correctness: Contemporary Modes of Materialization in American Studies and German Academia
Courtney Moffett-Bateau and Sebastian Weier | 123

***~~Kanak~~* Academic: Teaching in Enemy Territory**
Ismahan Wayah | 153

The Migrant Scholar of Color as Refugee in the Western Academy
Mahmoud Arghavan | 177

Keeping Academia White: A Case Study
Kai Linke | 195

III. DIVERSITY-CONSCIOUS APPROACHES TO ACADEMIC AND PEDAGOGICAL PRACTICE

On Racism without Race: The Need to Diversify *Germanistik* and the German Academy
Priscilla Layne | 217

"So You Want to Write about American Indians?" Ethical Reflections on Euro-Academia's Research on Indigenous Cultural Narratives
Amina Grunewald | 239

"The Danger of a Single Story": Addressing Contemporary Public Discourse and Protest Movements in American Studies Classrooms in Germany
Saskia Hertlein | 261

IV. SHIFTING PERSPECTIVES: TRANSATLANTIC AND GENRE-CROSSING REFLECTIONS ON WHITE NORMATIVITY

Goethe Meets Baldwin: Notes towards a Comparative Perspective beyond Misappropriation
Derek C. Maus | 285

Notes from the Margin: Academic White Spaces and the Silencing of Scholars of Color
Kimberly Alecia Singletary | 295

Transatlantic Postcolonial (T)Races in the Classroom: From Defoe's Desert Island to Larsen's *Quicksand* and *Black-ish* Suburbia
Elahe Haschemi Yekani | 315

Passing Tone/Note
Marius Henderson | 337

Contributors | 343

Acknowledgments

This project resulted from a workshop in Tübingen in 2016 that was among the first of its kind in American Studies in Germany: it took on the difficult task of self-reflection while talking about racism, power, representation, recognition, and pain – that is, the very issues many of us experience practically and/or work on theoretically as scholars – in a high-pressure academic context that is marked by (racialized) hierarchies and power dynamics. This endeavor proved challenging, at times hurtful, but also crucial and meaningful to all of us.

While scholars usually take pride in the 'analytical distance' they were able to uphold to their subject matter, with regard to this project, we look back with humility: one could say that we as an editing team encountered many issues on a micro-level that this volume seeks to address with regard to society, the university, and the Humanities as a whole. Ultimately, we were able to accomplish our common goal, which was to provide a forum for the voices of the contributors, which will hopefully result in their being heard and being afforded the respect they deserve.

We thus also look back with gratitude and would like to thank a number of people and institutions that have supported, funded, and expressed their belief in this project. First and foremost, we want to express our deep appreciation to the contributors in this volume for their trust, engagement, and patience with each other and us as editors. We also learned much from colleagues who presented at the workshop, but decided not to be part of the book project as well as from the discussions at the workshop – thanks to all presenters, panel chairs, and audience members. The decision to include both the contributions of those who wanted to expand their presentations given at the initial workshop into essays and to invite new colleagues to join this project allowed us to further broaden and diversify the conversation. We are grateful for this added focus.

While projects such as these require lots of idealism, they would not exist without the type of endorsement that ultimately results in a tangible publication, which allows us to reach a wider audience at the very center of the discussion – the academic field. Hence, we would particularly like to thank the Collaborative Research Center 923 "Threatened Order – Societies under Stress" and the Center for Gender and Diversity Research (ZGD), both at Eberhard Karls Universität Tübingen, for their generous financial support.

Astrid Franke has given us a tremendous amount of trust for our work on this project while offering both her intellectual and professional support. We would also like to thank our editor at *transcript*, Annika Linnemann, for her expertise and patience in the course of this publishing process. We deeply appreciate the people who have each done their share to bring this project to fruition: Jasmin Hayn, Enno Küker, Marina Lieb, Vanessa Miller, Nadja Prpa, Denise Soydan, and Michaela Wildermuth, who chiefly helped organize and conduct the workshop, as well as Hanna Bozenhardt, Aileen Priester, Anthony Obute, and particularly Marcel Pichal for their extraordinary motivation, diligence, commitment, and competence throughout the editorial process. We are furthermore indebted to Yasmin Nasrudin for her involvement and input. Finally, we are grateful to those whose love sustained our spirits and thus ultimately this project.

Who Can Speak and Who Is Heard/Hurt? Facing Problems of Race, Racism, and Ethnic Diversity in the Humanities in Germany: A Survey of the Issues at Stake

Mahmoud Arghavan, Nicole Hirschfelder, and Katharina Motyl

Building on Gayatri C. Spivak's seminal essay "Can the Subaltern Speak?" (Spivak 1989), in which she argues that even if those occupying the subject position of the subaltern *can* speak, they will not be heard due to the Eurocentrism underlying power relations, this edited collection offers a forum for analysis of who can speak in the Humanities in Germany with regard to dimensions of race, ethnic diversity, and racism – and of who will be heard. That the contributions comprised in this volume also inquire into who is *hurt* by the social, institutional, professional, and disciplinary logics informing (the opportunity to formulate) current articulations of Humanities research and higher education as well as their reception, highlights the contributors' conviction that there are ethical commitments at stake in discussing these issues. The question whether the subaltern – in this context, Germans of color and migrants (of color) – can speak in the German Humanities, is multidimensional and entails a host of ramifications.

First, there is the question of embodied representation: ethnic minorities and migrants (of color) are hardly represented in Humanist disciplines at German universities in numbers adequate to their proportion in German society at large – not among the student body, and certainly not on the level of scholars/teachers. When contemplating this lack of ethnic diversity, the issue that comes to mind immediately is that compared to other OECD countries, a pupil's educational opportunities in Germany are strongly determined by his or her family's socio-economic background (Kühne/Warnecke 2018). Not only does the German three-tiered system of secondary education put pupils of color and pupils from migrant families[1] at a disadvantage *per se* ("PISA-Studie"); since pupils of color disproportionately hail from economically disadvantaged families (a circumstance itself caused by structural and political reasons), the fact that the German education system's strong emphasis on meritocracy and its early segregation of the pupil population negatively impacts pupils from the lower socioeconomic strata, creates additional obstacles for pupils of color ("PISA-Studie"). Given the exclusionary tendencies of the secondary-education system, it seems logical that the student body at German universities should be less ethnically diverse than German society at large. And if there is but a limited percentage of students of color enrolled at Humanities departments, the lack of ethnic diversity among PhD students and faculty may not come as a surprise, either. But holding the German secondary-education system solely responsible for the lack of ethnic diversity at the Humanities departments of German universities falls short. While reform efforts aimed at increasing the permeability of the German education system are certainly needed (some of which will be explored in the contribution by Ismahan Wayah), from our experience,[2] there *is* a robust (if disproportionate) degree of ethnic diversity among the student body enrolled at Humanities departments.[3]

The question then becomes why so few of those enrolled in Humanities undergraduate studies go on to pursue a PhD and why even fewer decide to pursue a career in academia – that is, why so few Germans of color and migrants (of color) find themselves in positions that grant them the opportunity to speak in Humanist research and as teachers in higher education. "They're simply not interested in pursuing an academic career,"[4] some members of the (predom-

1 We try to avoid using the German concept *Migrationshintergrund*, which is used to categorize people who have migrated to Germany themselves or as hailing from a family who migrated to Germany (however many generations removed). This concept serves to abrogate the right to belong to German society from people of color who were born in Germany or have made a home in this country (cf. the contributions by Anthony Obute and Lili Rebstock).

2 Collectively, the four editors have taught at the American Studies departments at the Universities of Berlin (FU), Mannheim, and Tübingen.

3 Disciplines that strongly connote education ideals associated with the German *Bildungsbürgertum* such as Classics or Medieval German Literature, will probably boast less ethnic diversity among the student body than American Studies.

4 Two tenured professors expressed this view in the discussion group "Race, Ethnicity, and Critical Whiteness" at the workshop "Diversity and/in the GAAS," sponsored by the German Association for American Studies (GAAS) and held in Munich on October 20-21, 2017.

inantly white) tenured professoriate have put forth, pointing out that precarious employment conditions as they currently reign in the German Humanities may deter graduates of color, who often hail from economically challenged families, from going on to obtain a PhD. While an apprehensiveness of facing precarious employment may be part of the issue, it can hardly account for the totality of the problem. High school graduates whose main motivation is a high and steady income would hardly have chosen a Humanist course of study in the first place; people pursuing higher education with the primary aim of obtaining job security and a high salary will opt for studying business, engineering, or the natural sciences. Rather, it appears that the inability of the German Humanities to *retain* a robust percentage of students of color for doctoral studies and, even more so, for an academic career beyond the PhD, may have to do with the university's general and the Humanist disciplines' particular failure to *hear* people of color, causing many of them deep *hurt*. It seems, then, that people of color and migrants are interpellated as "space invaders" (Puwal 2004) in the structurally, culturally, and epistemologically "white space" (Anderson 2015: 10) that is the German Humanist academy. This assertion has various dimensions, which we will briefly outline vis-à-vis the Humanities at large before addressing the situation in German American Studies as a paradigmatic case in more detail.

First of all, racialized power structures in society tend to be reproduced in the university, perhaps particularly in the Humanities,[5] as the contributions by Ismahan Wayah, Mahmoud Arghavan, Kai Linke, Priscilla Layne, and Kimberly Alecia Singletary highlight. For one, there are administrative reasons such as the "similarity attraction paradigm" (cf. Peus 2018: 207): committees hire colleagues 'like themselves,' that is, if a committee is all-white or majority-white, it is very likely that the chosen candidate will be white, too. Moreover, studies have shown that 'the trailblazing scholar' is still predominantly imagined as male and white, and that people of color and/or women who are equally qualified as white, male contenders, are evaluated more negatively by search committees (cf. Peus 2018: 207). Stereotypes about minorities are, of course, deeply engrained in the cultural archive and continuously reproduced by the mainstream media. In light of omnipresent tropes of 'the hypersexual black woman,' 'the oppressed Muslim woman,' or 'the violent black man' circulating in the cultural imaginary, it comes as no surprise that, as the abovementioned

5 One could argue that disciplines that rely on quantification more and/or generate more substantial third-party funding than do the Humanities, are less prone to racial discrimination than the Humanities (cf. Mahmoud Arghavan's contribution in this volume).

studies found, most will not associate the figure of the trailblazing scholar with women and/or people of color.

Arguing that racialized power dynamics are reproduced in the university is not to allege that academic staff and search committees generally discriminate against students and applicants of color in a premeditated fashion. First, in the words of Sara Ahmed, "racism should not be seen as about individuals with bad attitudes (the 'bad apple model'), not because such individuals do not exist (they do) but because such a way of thinking underestimates the scope and scale of racism, thus leaving us without an account of how racism *gets reproduced*" (2012: 44; original emphasis). Second, it is to highlight how much effort is required by most white academics to become aware of their "unconscious bias" against people of color due to, among other things, the cultural narratives that accompanied their socialization, the coloniality of Western knowledge systems that inform these narratives and constitute the philosophical bedrock of the German Humanities, and finally, the force of habit (they are used to predominantly white staff meetings, classrooms, etc.) Becoming aware of one's own "unconscious bias" and actively working towards transforming one's own research, teaching, and administrative practices as well as those of one's department/institution/discipline, requires effort and commitment – there's a reason why it's called "diversity *work*" (cf. Ahmed 2012: 29; our emphasis).

To add insult to injury, students and scholars of color are facing difficulty when it comes to articulating their grievances of racist exclusion and being overlooked, since the hegemonic narrative German society tells itself about itself stipulates that because the use of the term '*Rasse*' ended with the downfall of the Nazi regime in 1945, racism no longer exists in German society (cf. the contribution by Maria Alexopoulou). This situation could be described as a particular type of colorblindness: not seeing racism because the lens that is needed to see it has gone missing (cf. Courtney Moffett-Bateau and Sebastian Weier's contribution). The European consolidation process and the abolition of border controls between EU states has reinforced ties between Germanness/Europeanness and whiteness,[6] while people of color, regardless of their citizenship status, are imagined as perpetual *Ausländer*/Others. In *European Others: Queering Ethnicity in Postnational Europe*, Fatima El-Tayeb argues that Europe is "a colorblind continent in which difference is marked along lines of nationality and ethnicized Others are routinely ascribed a position outside the nation, allowing the permanent externalization and thus silencing of a debate on

6 Since the inception of the so-called 'refugee crisis' in the summer of 2015, some EU states have reintroduced the practice of border controls; for instance, the German federal police is controlling the border of Austria/Germany again.

the legacy of racism and colonialism" (2011: 14). In short, when students and scholars of color seek justice for having been targeted by racism, their cases are all too often dismissed as 'individual' and sometimes 'too petty' stories; they are generally neither heard in the Germany university nor in German society.

Diversity in American Studies in Germany as Paradigmatic Case

Having outlined in broad strokes the dimensions involved in questions of race, ethnic diversity, epistemic coloniality, and racism in the German Humanities at large, we will now turn to the particular case of American Studies in Germany. Pragmatically speaking, this is the professional field of knowledge generation and transmission we know best. But there are other reasons why the discipline of American Studies may serve as paradigmatic case for our analysis of how Humanistic disciplines institutionalized at German universities, professionally organized (in associations, etc.), and embedded in German society deal with race, ethnic diversity, and racism. For one, said issues have received increased scholarly attention in light of the Black Liberation and ethnic revival movements of the 1960s and 1970s, which resulted in the institutionalization of African American/Black Studies and Ethnic Studies programs at U.S. universities. While these research interests have been mirrored in German Americanist scholarship, the mode of activist scholarship that has become the dominant paradigm of U.S.-based American Studies (a development related to the abovementioned social and racial justice movements), is regarded with great skepticism in the German Humanities, which are otherwise eager to adopt American trends in research and teaching, and at least eyed uncomfortably by some scholars in German American Studies. Two distinct philosophical traditions clash here: whereas scholar-activism, rooted in American Pragmatism, considers scholarship worthwhile that is useful to the political struggles of the marginalized, the German Humanities uphold the ideal of scholarship as objective and detached, which is rooted in German Idealism. We will outline the historical, political, and theoretical implications of both philosophical camps in more detail later. *Real-life* repercussions of this clash between two theoretical traditions became public a few years ago in Bremen. While we have decided to briefly outline this case in the following, as discussions about its aftermath also partly informed the workshop on which this volume builds, it is crucial to us to emphasize that what happened in Bremen is by no means endemic to that institution's academic environment, but rather marks such an important example because similar events could have easily taken place in many other universities all over Germany, as well.

In June 2014, colleagues in American Studies and other Humanist disciplines at the University of Bremen filed a grant application for an interdisciplinary Creative Unit entitled "New Black Diaspora Studies: Ethical and Aesthetic Challenges of the 21st Century." In January 2015, both the project, the content of the application, and the envisioned distribution of academic positions were strongly condemned in a 10-page open letter in German and English by signees from Germany and abroad, including the scholar and activist Angela Y. Davis. The main points of criticism were that the applied-for research program was *not* – as had been claimed – an innovative project, since black scholars had been dealing with the abovementioned contents for years without receiving academic recognition, but that it was merely new that black knowledge and culture were now being made the object of research by whites, as the grant application sought to allocate all applied-for positions to white scholars. Black scholars, the signees protested, were systematically excluded from resources, while white academics enriched themselves at the expense of people of color without self-critically reflecting on their (individual and historical) contribution to the preservation of racist structures and mechanisms.[7] This controversy garnered considerable attention in the media (e.g. *Tagesspiegel*, *taz*, *FAZ*). The research group "Black Knowledges" at the University of Bremen, which had co-initiated the controversial grant application, disbanded in February 2015 as a reaction to the open letter, stating that it found the criticism voiced in the open letter justified.[8] As mentioned initially, what lay at the core of this controversy was that the white German scholars who had applied for the grant either seemed oblivious to the tradition of American scholar-activism, or, more likely, 'implemented' it in a way that was firmly rooted in the belief in 'objective' scholarship, to the degree that the question of who can speak about whom with authority (and funding), simply appeared irrelevant.

How does the 'Bremen case' fit into the larger context of German American Studies? Conversations about ethnic diversity, race, and racism have existed in American Studies in Germany for many years. Especially in the past few decades, ethnic minorities, Indigenous peoples, and 'new immigrants' could even be said to have advanced to the *center* of a considerable number of scholarly inquiries. However, these studies have overwhelmingly been conducted by white scholars who were born in Germany, speak German as their first language, and hold German citizenship; that is, ethnic minorities and migrants of

7 For further information, see https://blackstudiesgermany.wordpress.com/statement bremen/).

8 The statement can be accessed here: http://www.fb10.uni-bremen.de/inputs/pdf/BK RG_Aufoesung-Disbanding_deu-engl.pdf/.

color have been the objects, but rarely ever the subjects of Americanist research and theorizing in Germany. While the aforementioned research in such fields as Black Studies and Ethnic Studies in Germany has focused solely on the issues of ethnic diversity, race, and racism *in* U.S. society and on negotiations of said issues in cultural productions, meta-level discussions of ethnic diversity in the field of American Studies in Germany have thus far been absent from scholarship. Moreover, in the Americanist classroom in Germany, educating students about historical and current racist formations in the U.S. is hardly ever accompanied by drawing parallels to racism and xenophobia in the German context. If the critical thinking that a Humanities education is meant to impart is not merely an end in itself, but a means to help create a more democratic society, this seems to be a grave omission in view of the renewed *Salonfähigkeit* of ethnonationalist and racist positions in Germany's political landscape in the context of the so-called 'refugee crisis' of 2015.

The contributions in this volume, more broadly, analyze the mechanisms behind and the implications of the lack of ethnic diversity in the field of American Studies in Germany and in the German Humanities at large. These considerations also necessitate thinking through the workings of race, xenophobia, the education system, and hegemonic understandings of *Germanness* in German society. By highlighting ways in which the academic field in Germany may become more reflective and inclusive of Germany's ethnic diversity, contributions also discuss the role of the ethical and the political in German Humanities research and pedagogy.

To elaborate, it is the edited collection's first aim to highlight how American Studies/the Humanities in Germany will benefit (in terms of epistemic pluralism and social justice) by diversifying itself/themselves. The contributors to the edited collection share the assumption that the Habermasian *Erkenntnisinteresse* and the presuppositions with which Humanities scholars approach their object of inquiry are influenced by the ways of knowing they have acquired by having occupied particular subject positions throughout their biographies. Thus, research results in the fields of Black Studies and Ethnic Studies, but also in American Studies at large, remain selective as long as the ways of knowing of people of color and migrants (of color) are largely excluded from the academy. Moreover, the exclusion of ethnic minorities and migrants (of color) counteracts one of the main ideas behind the institutionalization of Black Studies – that of creating an academic space in which scholars of color and members of other marginalized groups could engage in practices of empowerment by *actively* producing and sharing knowledge about themselves and their (historical, social, political, cultural, and economic) conditions. In light of the current under-representation of racialized and ethnicized minorities in German American Studies and the concomitant overrepresentation of Germans constructed as white, it seems as though this knowledge production about the former has been appropriated, and thus monopolized, by the latter. As editors of the volume (and this probably holds for many contributors, as well), we do not subscribe to the demand that research on the African or Arab diasporas, for example, be entirely left to members of the African or Arab diaspora; but it neither should be – as tends to be the case in Germany – the sole prerogative of whites. Thus, we consider as untenable the situation that white scholars in Germany (two of which are editors of this volume) adopt the theories of W.E.B. Du Bois, Frantz Fanon, and Edward W. Said to analyze cultural productions of the black or Arab diasporas in the U.S., while representatives of the black or Arab diasporas and their ways of knowing are largely excluded from the professional field of American Studies in Germany.

Second, the volume seeks to analyze the mechanisms that exclude people of color and migrants (of color) from a career in German American Studies – and that restrict access for pupils of color to higher education in the Humanities, in the first place. These discussions necessitate considering racial and xenophobic formations in the larger societal context in which the German academic system is embedded. Select contributions compare and contrast the workings of race, racism, and xenophobia in the German and U.S. contexts, analyze hegemonic understandings of Germanness, and theorize the figure of the *Ausländer*. Why, for instance, would a Polish person be called *Ausländer* in Germany, while a Swede would be described as a Swede? Why is Germanness still associated with whiteness, so that Afro-Germans born in Germany regularly receive compliments on their excellent German-language skills?

Third, in highlighting ways in which American Studies in Germany may become more inclusive and conscious of ethnic diversity, the contributions to the edited collection touch upon the role of the ethical and the political in Humanities research and pedagogy in Germany, thus problematizing the conception of scholarship as *objective*. As already alluded to, the model of the scholar-activist, which is quite widespread in U.S. Humanities at the present moment, has not held sway among Humanities scholars in Germany. But is it not more ethical, contributions ask, to explicitly reflect one's own positionality and to make one's political commitments (and potential blindspots) transparent, as the scholar-activist does, rather than to pretend one does not have any?

But before we provide a more detailed survey of the issues at stake concerning race, ethnic diversity, and racism in American Studies and other Humanist disciplines in Germany, a critical engagement with the term 'diversity' and an evaluation of some of the political uses that have been made of diversity in times of neoliberalism, seems warranted.

Diversity: Neoliberal Appropriation vs. Critical Intervention

The insight that according to the logic of neoliberalism, it is economically inefficient to discriminate based on race and other markers of social difference, has become widely accepted among Humanities scholars and social scientists who position themselves in the tradition of critical theory. Walter Benn Michaels writes:

> one of the great discoveries of neoliberalism is that they [racism and sexism] are not very efficient sorting devices, economically speaking. If, for example, you are looking to promote someone as Head of Sales in your company and you are choosing between a straight white male and a black lesbian, and the latter is in fact a better salesperson than the former, racism, sexism and homophobia may tell you to choose the straight white male but capitalism tells you to go with the black lesbian (2008: 34).[9]

More generally, ethnic difference has become commodified *and* depoliticized under the neoliberal paradigm: customers at a Mexican restaurant in Phoenix, Arizona, might consume burritos while talking of the need for tighter border controls, and it's safe to assume that some participants in PEGIDA marches (a political movement in Germany which alleges that Europe is in danger of becoming 'Islamized') stop at Turkish fast-food joints to fill up on *Döner Kebab* on their way home. Not only have corporations realized that discrimination is bad for business; they have come to conceive diversity as an asset to enhance value extraction. Angela Y. Davis sums up: "Diversity is a corporate strategy [...] It's a difference that makes no difference at all" (qtd. in Ting 2015). University administrations, too, have begun to discover that diversity promises profitability – this certainly holds true for private institutions of higher education in the U.S., but increasingly also for German universities, inspite of the fact that

9 While we concur with Michaels in this assessment, we disagree with Michaels' line of reasoning in *The Problem with Diversity: How We Learned to Love Identity and Ignore Inequality*, which fails to take into account that axes of oppression intersect, overlooking, for instance, that communities racialized as non-white are disproportionately affected by the economic inequalities ushered in by neoliberalism. For a detailed critique of *The Trouble with Diversity*, see Priscilla Layne's contribution in this volume.

they are by and large public. Education scientists Gerd Grözinger and Marlene Langholz-Kaiser write that the disproportionate representation of women and ethnic minorities in the German academy not only translates into loss of intellectual potential, but also into loss of role model potential for future generations: "Especially universities that compete internationally – particularly with US-American elite universities, which have a lot of experience in diversity management – can no longer afford this loss" (2018: 200; our translation).

As alluded to in the previous quote, diversity management has been the dominant strategy universities have chosen to commit to the goal of attaining a more socially heterogeneous faculty, student body, and potentially also administrative body: that is, universities have issued diversity mission statements and appointed diversity officers or even created equal-opportunity offices. Sara Ahmed, the author of *On Being Included: Racism and Diversity in Institutional Life* (2012), a book-length study of how institutions of higher education in the U.K. approach (ethnic) diversity, has reached an astounding conclusion. Maintaining that universities' approach to diversity could be paraphrased as "*how not to do things with words*" (2016: 1; original emphasis), she points to what she theorizes as the "non-performativity" of diversity (2016: 1):

> commitments [to diversity] were makeable *because* they were not doable: it seems you can make a commitment because commitments do not commit institutions to a course of action. Commitments might even become a way of not doing something by appearing to do something. Understanding the role or function of institutional commitments was to understand how institutions do not do things with words, or how institutions use words as a way of not doing things. [...] Policies can be adopted, words can be uttered; decisions can be made, without anything really changing. Sometimes we refer to this as the "lip service" model of diversity. To use a word like "non-performativity" is to reveal something about institutional mechanics: how things are reproduced by the very appearance of being transformed (Ahmed 2012: 1-2; original emphasis).

Putting things a little less abstractly, Ahmed explains:

> A typical goal of diversity work is "to institutionalize diversity." [...] However, if institutionalizing diversity is a goal for diversity workers, it does not necessarily mean it is the institution's goal. [...] We could say that practitioners are given the goal of making diversity a goal. [...] An appointment becomes a story of not being given institutional support, as if being "just there" is enough. An appointment of a diversity officer can thus represent the absence of wider support for diversity (Ahmed 2012: 22-23).

In sum, the managerial paradigm with which German universities have operationalized diversity actually depoliticizes issues of ethnic difference, which are, of course, embedded in power dynamics, and thus reproduces racialized power dynamics at the very same time as it nominally celebrates cultural pluralism. As Vanessa Eileen Thompson and Veronika Zablotsky succinctly put it:

> the discourse of 'diversity' objectifies relations of power and stabilizes them through neoliberal inclusion of figures of difference in ways that, as Angela Davis has argued, make no difference […], while simultaneously perpetuating and stabilizing social injustices within the realms of higher education (Thompson/Zablotsky 2016: 79).

A bureaucratic-managerial approach to diversity, then, may render the population represented at a university more heterogeneous; it will not, however, result in the voices of people of color being *heard* in institutional life, in Humanist knowledge production, or in the classroom. As we will argue in the following, if people of color and/or people with migrant experience are to be truly *included* in German universities in general, and in Humanities departments in particular, that is, if they are to feel at "home" in the institution (Mbembe 2015: 5) and if they are to contribute all that they are uniquely positioned to contribute, a number of reforms pertaining to practices of knowledge production, teaching practices, and administrative practices are necessitated. In the words of Thompson and Zablotsky, what is needed is a move away from the model of "neoliberal business management and the individualization of political struggles over social justice" (2016: 91), and towards "a repoliticization of difference as a matter of social justice" (2016: 78), the latter conceived as "a political process of contestation in continued political struggles over processes of deliberation and harmonization with their effects of concealed asymmetrical relations" (2016: 91). The German Humanities stand much to gain from becoming more ethnically inclusive – and we are not speaking of economic profit maximization.

An Intersectional Approach to Ethnic Diversity

That viewing someone predominantly through one singular lens constitutes a reductionist mode of perception and that it is important to consider that a person can be subject to several forms of discrimination at once, is chiefly addressed by scholars by using the term 'intersectionality.' Coined by Kimberlé Crenshaw (1991; 2012), 'intersectionality' captures the fact that different forms of discrimination are not isolated from another but frequently *intersect*, leading to a considerable difference in the lived experience of, for example, a *white* and a *black* woman. Even though both women may be exposed to sexism, it is crucial to understand that sexism may not only look and present differently for each of them but also have very different repercussions in both women's lives, given that a number of other axes of oppression, such as racism, economic marginalization, heteronormativity, etc. must also be taken into account. Transferred to (migrant) scholars and students of color in Germany, an *intersectional* analysis would thus be concerned with the fact that the experience of a black, gay migrant scholar of color may differ substantially from that of other blacks, gays, or migrant scholars, who may each find themselves faced with discrimination but who do not experience the specific *combination* of these facets, which can complicate feelings of solidarity but also recognition for one's pain in a hierarchical environment. While the underlying idea and methods of intersectionality are considered and employed, respectively, in various contributions, discrimination based on race constitutes the focus of this volume. This is not to suggest, however, that discrimination due to sexism, heteropatriarchy, neoliberal exploitation, etc., is undeserving of recognition or was not part of the conversation for (migrant) scholars and students of color in Germany. Rather, this volume seeks to raise awareness for the specific relevance of race and racism in the German Humanities/academy, because these aspects have – compared to discrimination based on gender or class – not received adequate consideration thus far.

Diversifying and Decolonizing Knowledge Production in the German Humanities

As we mentioned initially, the argument that the German Humanities ought to become more inclusive of ethnic diversity is informed by more than ethical and political concerns for social/racial justice (though this is a crucial dimension). Including more faculty, PhD candidates, and students of color will also alter the process of knowledge production at German Humanities departments in fundamental (and, we argue, beneficial) ways. As feminist scholars demonstrated by developing standpoint theories in the 1980s, power relations are analyzed with particular acumen from the vantage point of those oppressed by the power relations in question (cf. Rolin 2009; Harding 2004). Transferred to the thematic context of our volume, this posits that scholarship and classroom discussions pertaining to issues of race, racism, and colonialism will gain additional and vital perspectives by German Humanities departments' *including* (which is, as discussed above, different from merely adding) more researchers and students of color and/or with migration experience. Highlighting the unique perspective scholars of color bring to the study of racialized power dynamics, the colonial residue in the Western cultural imaginary, etc., is not to suggest, however, that scholars of color had better limit their research interests to fields of Humanist inquiry that are concerned with race, ethnicity, and colonialism (cf. the contribution by Amina Grunewald). Rather, as we put it above, the presuppositions with which Humanities scholars approach their object of inquiry are influenced by the ways of knowing they have acquired by having occupied particular subject positions throughout their biographies. (This point is, of course, related to the assumptions informing feminist standpoint theories). Thus, *any* field of inquiry in any Humanist discipline stands to gain from diversifying the ways of knowing represented qua increasing diversity in various ways among scholars, and particularly at the current historical conjuncture with movements such as #BlackLivesMatter and #metoo, which deliberately raise the question of whose *humanity* is truly recognized and respected. Accordingly, the so-called Humanities in Germany *must* critically reevaluate their attachment to the notion of the universal Human subject postulated by Enlightenment philosophies, and broaden their self-definition in order to still live up to their label in the future.

It is the abovementioned link between socialization and epistemic orientation that points to a crucial contiguous dimension that increasing ethnic diversity among Humanities scholars would entail: a move towards decolonizing knowledge, which represents an endeavor that far exceeds diversifying the canon of primary sources, e.g. by amending the canon of American literature with African American, Indigenous, Chican@, etc. voices, and employing (and respecting) different *forms* of rendering knowledge, such as the anecdote that 'theorizes from experience.'[10] (Kimberly A. Singletary and Mahmoud Arghavan employ anecdotal knowledge in their contributions, for example). The fact that decolonial critique questions the very concept of the 'canon,' points to the epistemological, methodological, and political assumptions as well as demands it encompasses. In his lecture "Decolonizing Knowledge and the Question of the Archive," Achille Mbembe argues for the necessity to dismantle the "Eurocentric *epistemic* canon," which "attributes truth only to the Western way of knowledge production" (2015: 9; our emphasis). According to Mbembe:

> Western epistemic traditions are traditions that claim detachment of the known from the knower. [...] They are traditions in which the knowing subject is enclosed in itself and peeks out at a world of objects and produces supposedly objective knowledge of those objects. The knowing subject is thus able, we are told, to know the world without being

10 The method of 'theorizing from experience' was developed by women of color social scientists such as Anna Julia Cooper and popularized by the feminists (of color) associated with standpoint theories (cf. May 2012: 168).

> part of that world and he or she is by all accounts able to produce knowledge that is supposed to be universal and independent of context. [...] This hegemonic notion of knowledge production has generated discursive scientific practices and has set up interpretive frames that make it difficult to think outside of these frames. But this is not all. This hegemonic tradition [...] actively represses anything that actually is articulated, thought and envisioned from outside of these frames (2015: 9-10).

Boaventura de Sousa Santos has conceptualized this last dimension (the Western epistemic tradition's will to power) as *epistemicide* (2014: 149).[11] The utopian *telos* of attempts to decolonize knowledge, that is, to de-center Western epistemologies, diversify syllabi, and transform the university from a "white space" into a "home" for diverse constituencies, is the "pluriversity" (cf. Boidin, Cohen, and Grosfoguel 2012).

Since people of color and/or migrants may have been socialized in cultures that feature epistemologies and cosmologies outside the Western epistemic legacy, scholars of color and/or with migration experience are uniquely positioned to contribute to the project of decolonizing knowledge and the university. However, due to reasons located in Germany's political and philosophical histories, a theoretical and methodological mode that has served influential in other societies' epistemic contexts by challenging the notion of detached, disembodied, and objective scholarship Mbembe identifies as central to hegemonic Eurocentric knowledge production, has thus far remained marginal in the German Humanities: we are, of course, referring to scholar-activism.

Although the exhortation "The philosophers have only interpreted the world in various ways. The point, however, is to change it," was famously penned by a German political economist (Karl Marx), today, activist scholarship is met with considerable skepticism in the German Humanities, while it – albeit not Marxism or Marx himself – is widely accepted in the North American Humanities (and perhaps even mainstream in U.S.-based American Studies).

Scholar-activism is rooted in the philosophical tradition of Pragmatism originating in the U.S., which judges a given proposition's validity based on its

11 Defining epistemicide as "the murder of knowledge" (2014: 149), de Sousa Santos argues: "Unequal exchanges among cultures have always implied the death of the knowledge of the subordinated culture, hence the death of the social groups that possessed it. In the most extreme cases, such as that of European expansion, epistemicide was one of the conditions of genocide. The loss of epistemological confidence that currently afflicts modern science has facilitated the identification of the scope and gravity of the epistemicides perpetrated by hegemonic Eurocentric modernity" (2014: 149).

usefulness. It was simultaneously popularized in U.S.-based American Studies in light of various emancipatory social movements of the second half of the 20th century and the proliferation of identity politics. Academics involved in these political struggles 'on the streets' drove forth the crystallization of such academic fields as African American Studies, Women's Studies, Queer Studies, Indigenous Studies, etc., some of which have become institutionally anchored through the creation of African American Studies departments, Women's Studies departments, etc. Simultaneously, an academic revolt in U.S.-based American Studies took place against the hegemony of New Criticism, which focused on artefacts' aesthetic strategies, but steered clear of discussing artefacts' political implications and was judged by a vanguard of scholars who came to be called 'new historicists' as failing to question social power structures and thus, as colluding with the societal status quo. By delegitimizing aesthetics as a dimension of inquiry, new historicism, then, centrally contributed to ideology critique becoming the dominant method of U.S.-based scholarship on American literary and cultural productions.

In the German Humanities, by contrast, the scholar-activist mode has not gained much acceptance, due to a wariness that arises from the historical awareness of academia's ideological *Gleichschaltung* during the Nazi and SED regimes. After all, Marx' above-cited statement was installed in golden letters in the lobby of the main building of Humboldt University of Berlin (located in what was then East Berlin) in 1953: only those Humanities scholars whose work was deemed as furthering the political goals of the GDR's Marxist-Leninist state doctrine were granted academic positions. Moreover, the lasting impact Enlightenment philosophy has had in the German Humanities has also contributed to the unpopularity of activist scholarship, whereas the U.S.' philosophical legacy of Pragmatism is the *conditio sine qua non* of activist scholarship. Third, every field of knowledge production is conditioned by what Craig Calhoun calls the "hierarchical structure of scientific knowledge" (2008: xviii); to put things a bit provocatively, in German American Studies, research on Henry James or Thomas Pynchon endows a colleague with more symbolic capital than research on, say, African American women writers – especially when research on the latter topic is conducted by a black woman. In the Anglo-American realm, on the other hand, Americanist research that does not practice ideology critique in some form will have to vie mightily for recognition. Unsurprisingly, researchers gravitate towards those topics that promise the most recognition in their respective fields of knowledge production. More generally, activist scholarship is viewed critically, since modern (i.e. Western) science, as indexed in the above quotation by Mbembe, has developed an ideal of knowledge based on detached, objective observation, and activism is widely understood as expressive of individual interests, emotions, or ethical commitments rather than of a broader, more reflective, and more intellectually informed perspective on social issues (cf. Calhoun 2008: xviii).

Of course, there are a number of pitfalls when it comes to activist scholarship. The Pragmatist imperative at the heart of activist scholarship demands that the knowledge Humanities scholars produce be 'useful' to the public; on the one hand, this has entailed research in support of emancipatory struggles, for instance, the African American quest for equality; on the other hand, the assessment of knowledge based on its 'usefulness' has allowed (private) university administrations to interfere with the research agenda conducted at their institutions, as the contributors of the volume *The Entrepreneurial University: Engaging Publics, Intersecting Impacts*, edited by Yvette Taylor (2014), have pointed out. Also, scholar-activism is ill-conceived when researchers assume that activism and scholarship are mutually exchangeable. Scholar-activism neither means that writing about political action is tantamount to engaging in political action; nor does it mean applying already established truths to a scenario of political importance. As Calhoun states:

> [I]t is easy to elide the difference between contributing knowledge and analysis to social movements or other practical efforts and simply sharing in the general tasks of struggle. [...] But if activist scholarship is to contribute all that it really can (and if it is to be appreciated well in either scientific or practical realms), it has to do so through production and mobilization of knowledge (2008: xxi).

In the realm of Americanist activist scholarship, a particular and common pitfall lies in practitioners' disregarding the specific mediality and aesthetic strategies of the cultural productions they analyze. Obviously, however, it makes a difference whether a character in an Afrofuturist novel or a demonstrator at a #BlackLivesMatter march utters a statement.

When minding these potential pitfalls, activist scholarship may also have exciting implications for Humanist disciplines in Germany as a whole; as Calhoun states in view of activist social science:

> The primary purpose of activist scholarship thus may be to address public issues or help specific constituencies. Activist scholarship is one way to make social science useful. But activist scholarship can also make social science better, providing occasions for new knowledge creation, challenges to received wisdom, and new ways of thinking (2008: xxv).

Pedagogical Advantages of Learning in Diverse Classrooms and Methods for Diversity-Conscious Teaching

Education scientists Grözinger and Langholz-Kaiser emphasize two dimensions when it comes to ethnic diversity and the higher-education classroom: first, including more teaching staff of color and/or with migration experience has an empowering impact on students of color/migrant students, because the former serve as role models: "Institutions in liberal societies should be seen as principally accessible to anyone. And the higher the position, the higher the visibility. [...] The symbolic authorization through visibility is crucial. It empowers members of these groups to trust in their capacity to likewise achieve a career of high visibility" (2018: 200; our translation). Second, learning in ethnically diverse classrooms prepares students for a respectful and democratic conviviality in an increasingly diversifying German society and a globalizing world (Grözinger/Langholz-Kaiser 2018: 200).

In terms of subject matters, theories, and methodologies taught in higher education, decolonizing and diversifying syllabi will benefit not only students of color/with migrant experience, but all students, by pluralizing the cultural, political, and epistemic perspectives with which students are encouraged to engage. However, for students of color/migrant students, finding voices represented on the syllabus and acquiring theoretical concepts (such as Eurocentrism) that help them make sense of their own experiences, can prove particularly valuable. Humanities scholars without a designated emphasis in the study of race and ethnicity tend to deflect demands for diversifying and decolonizing syllabi. Priscilla Layne, a professor of German, writes in her contribution to this volume:

> When I raise the issue of diversifying a German Studies curriculum, many of the skeptics echo the same excuses. They say there are not enough texts by People of Color in German. Or they say they do not have the expertise to teach an entire class on "people with a migration background." Or perhaps they claim a unit on "diversity" would not fit into their syllabus. What they get wrong is the assumption that including texts in German by People of Color means you need an entire unit, lesson, or course about diversity or multiculturalism. While this has been the approach taken for decades, it also creates a literary 'ghetto' for authors of Color, keeping them at the margins. It is just as possible to incorporate more texts by People of Color in whatever lesson or syllabus you intend to teach, whether it is about literature and science, Romanticism, or poetry (226).

At the same time, classes pertaining to the study of race and ethnicity in another cultural context (e.g. the United States) often lead students in German Humanities classrooms to engage in what one could describe as 'cathartic fingerpointing': class discussions of police brutality against unarmed African Americans may include passionate denunciations of *American* society's racism, the implication being that contemporary German society is free from such blatant manifestations of racism. Pointing out that police in the state of Bavaria engage in racial profiling; that there is evidence for police brutality against people of color in Germany;[12] that the African and Arab individuals drowning in the Mediterranean Sea on a daily basis are kept out of 'fortress Europe' in the name of German and other E.U. citizens, may encourage students to think more critically of matters pertaining to race, ethnicity, and racism in contemporary German society/Europe. Moreover, in class discussions of U.S. settler imperialism, drawing students' attention to the legacy of German colonialism in Africa may be politically worthwhile.[13]

However, this comparative approach also poses the potential pitfall of students' lack of appreciation for the specificity of the cultural and political phenomena discussed in class. For instance, students should ideally not reach the conclusion that the treatment of the so-called guest workers in Germany in the 1950s was 'just like' the experiences of African Americans in the American South during the Jim Crow era. Saskia Hertlein's contribution in this volume proposes an approach to teaching intercultural subject matters in diverse higher education classrooms in Germany that seeks to sensitize students to related sociopolitical and cultural issues in Germany while doing justice to the specific context of the issue in question.

Orienting one's teaching in an increasingly diverse university classroom towards encouraging students to develop "Critical Diversity Literacy" (and continuously self-reflecting whether one's teaching practice is diversity literate) constitutes a helpful pedagogical methodology. According to sociologist Melissa Steyn:

> "Diversity literacy" can best be characterised as a "reading practice" – a way of perceiving and responding to the social climate and prevalent structures of oppression. The analytical criteria employed to evaluate the presence of diversity literacy include the following: 1) a recognition of the symbolic and material value of hegemonic identities […]; 2) analytic skill at unpacking how these systems of oppression intersect, interlock, co-construct and

12 Cf. the case of Oury Jalloh, which Courtney Moffett-Bateau/Sebastian Weier and Saskia Hertlein briefly discuss in their respective contributions.

13 Thanks go to René Dietrich for bringing this point to our attention.

constitute each other; 3) the definition of oppressive systems such as racism as current social problems rather than a historical legacy; 4) an understanding that social identities are learned and an outcome of social practices; 5) the possession of a diversity grammar and a vocabulary that facilitates a discussion of race, racism, and antiracism, and the parallel concepts employed in the analysis of other forms of oppression; 6) the ability to translate (interpret) coded hegemonic practices; 7) an analysis of the ways that diversity hierarchies and institutionalised oppressions are mediated by class inequality and inflected in specific social contexts; and 8) an engagement with issues of transformation of these oppressive systems towards deepening democracy/social justice in all levels of social organisation (Steyn 2007; qtd. in Steyn 2010: 20).

Those readers seeking to familiarize themselves with methods for diversity-conscious teaching in the higher-education classroom in more detail, may refer to Saskia Hertlein's contribution for further resources as well as to the website of *Zentrum für Kompetenzentwicklung für Diversity Management in Studium und Lehre an Hochschulen* (http://www.komdim.de/). Lili Rebstock's article on the experiences of youths of color in Germany may be of particular interest to (prospective) teachers and those engaged in teacher-training at university.

Institutional Measures to Meaningfully Increase Diversity[14]

Highlighting that universities conceptualize diversity as an issue of management, organizational sociologist Lukas Daubner states that it has become increasingly common for universities to issue mission statements in which they commit themselves "to fostering the perception, recognition, and use of diversity" or to "raising awareness for discrimination and fostering appreciation for difference in any area of life" (2018: 202; our translation). This approach to diversity, of course, is rooted in the neoliberal logic that discrimination is to be avoided because it is economically inefficient. A dominant strategy German universities have used to 'manage diversity,' Daubner critically remarks, has been to add an organizational unit for equal opportunities/diversity to extant structures without changing the latter. Since the "success or failure" of such a unit "depends on

14 The measures introduced here highlight what short-range or medium-range changes may be reached in light of the status quo. That is, while the abolition of the three-tiered secondary education system may be the measure that would most significantly reduce the disparities in educational opportunities, passing this policy is a long-term goal (and, if one assumes a *Realpolitik* stance, politically unrealistic in several German states).

how large its influence is on the operative processes of research, teaching, and administration" (2018: 202; our translation), equal opportunity/diversity officers, committed to their task though they may be, "rarely succeed in influencing the modes of interaction and working, which are often characterized by a large degree of informality" (ibid.; our translation). This dynamics, of course, reflects what we discussed above as Ahmed's theory of the "non-performativity" of diversity (Ahmed 2012: 1). In other words, as institutions, universities will only change in the direction of greater diversity if the administrative steps taken exceed paying lip service to diversity in mission statements and creating positions for diversity officers, and are pursued concomitantly to steps on the part of faculty outlined above towards decolonizing knowledge production as well as diversifying syllabi and fostering critical diversity literacy in the classroom.

What might these administrative steps look like? Claudia Peus, professor of research and science management, underscores that universities need to devise strategies and define clear criteria for recruiting talented individuals at every level (students; non-tenured faculty; tenured professors). If they want to work against the "similarity attraction paradigm" (all-white committees are likely to pick white candidates) and the impact of stereotypes (such as 'the white male research genius'), universities need to devise a selection process that is structured clearly and in detail, since a study of professor recruitment in the Netherlands shows that candidates from underrepresented groups are assessed particularly negatively when there is a lack of clear criteria on which to base selection decisions (for details, cf. Peus 2018: 207). Peus further recommends implementing diversity monitoring in universities' recruitment processes, "e.g. introducing the duty to document the number of applications from underrepresented groups at various levels of the selection process" (2018: 207; our translation). Said measures, Peus argues, are particularly effective when concrete goals are set (e.g. x percent of those candidates invited for job talks should come from communities of color) and when failing to meet these goals will entail consequences, such as proactively approaching candidates of color (ibid.; cf. Peus et al. 2015 for best practices from the German context; cf. Sensoy/Diangelo 2017 for best practices from the U.S. context).

Increasing pupils' of color/migrant pupils' educational opportunities and encouraging gifted pupils to stay course, however, falls outside the terrain of the university. As expressed earlier, a thorough reform of the traditional three-tiered secondary-education system in Germany is structurally best suited to increase the enrolment of students of color/migrant students at university. Until such reforms are enacted, gradual steps towards educational justice for pupils of color/migrant pupils can still be taken: the state of North Rhine-Westphalia has developed an interesting program in this regard; dubbed "talent scouting," the state's Department of Culture and the Sciences has sponsored an initiative since 2011, in which over 70 scouts based at 17 universities/schools of applied sciences visit high schools "to work in unison with teachers to identify motivated adolescents who have not had the opportunity thus far to make the most of their potential" ("NRW-Zentrum für Talentförderung"). Adolescents from non-academic and/or economically challenged families are the particular focus of the program. In individual council sessions, scouts and students determine what educational/professional options there are; scouts continue to be available for the young adults if the latter decide to enroll at university. 12.500 students are involved in the program as of 2018 (ibid.). We would like to close our remarks on increasing the inclusion of students/scholars of color and/or with migrant experience in the German Humanities by pointing to this initiative aimed at increasing the access of students of color to institutions of higher education, since it underscores the holistic approach to diversity-as-inclusion-and-social-justice that we have argued for.

Contributions

In her article, "'*Ausländer*' – A Racialized Concept? 'Race' as an Analytical Concept in Contemporary German Immigration History," Maria Alexopoulou questions the term '*Ausländer*' as a racialized concept and emphasizes 'race' as an analytical concept in contemporary German immigration history. She particularly puts pressure on the fact that a large number of renowned academics deliberately refrain from using the word '*Rasse*' (race) even when writing about the United States in German. Alexopoulou notes that mainstream German historiography almost entirely avoids terms such as 'race,' 'racism,' or '*Volksgemeinschaft*' lest they evoke *Angst* (fear) in white Germans. According to Jacques Derrida's notion of *différance*, Alexopoulou asserts, concepts bear traces of forgotten, suppressed, and silenced meanings. She suggests that this is the case with the binary of '*Ausländer*' and 'German.' Her thesis is that discourses and practices around this binary in the context of migration reproduce racist knowledge that has been transferred historically, albeit it was also transformed or even questioned. While terms and ideas of embodiment of 'the' *Ausländer* changed over time, the negative images associated with these groups have constantly stayed the same. This shows that the term '*Ausländer*' – just like all of its substitutes – are ultimately racialized concepts: they construct a distinct group with particular characteristics, which is set as the Other to 'the Germans.'

Furthermore, intersections such as origin, religion, class, and gender, explain the changing position in the hierarchy of '*Ausländer.*'

In "Perspective Matters: Racism and Resistance in the Everyday Lives of Youths of Color in Germany," Lili Rebstock weaves together interviews with young people of color in Germany and various theoretical insights on *microaggressions*, racism, and intersectionality. She sets out with the contention that since extremely overt or blunt forms of racism have lost moral and social acceptance and have also changed in the course of the past decades, more subtle forms such as *microaggressions* deserve both attention and recognition. Employing Chester Pierce's concept of *microaggressions* and Philomena Essed's concept of *everyday racism*, Rebstock points out that today's lived experience of racism is marked by the staggering, long-term effect of constant demeaning and excluding messages that frequently go unsanctioned, for their harmful impact is either ignored or dismissed once it is addressed. Rebstock argues that in Germany, dominant research about racism mainly represents the perpetrators' views. Thinking about the question of "Who can speak?" with regard to positionality and academia, she asserts that conversations on both race and racism must always take interesecting features such as age, gender, sexuality, class, disability, etc. and power relations into account. This insight is central to her discussion of the perspectives of youths of color on the issue of racism in Germany because their voice is hardly heard. In order to examine the topic of everyday racism and *microaggression*s, Rebstock conducted qualitative research with problem-centered interviews through the *Grounded Theory* approach and the concept of *Co-Construction* as developed by Paul Mecheril (2003). This research design encourages a self-relective approach and transparency in the sense that topics can be evolved together and that Rebstock, as the interviewer, is also involved in sharing her views and feelings with her interviewees. She concludes that *microaggressions* are very hard to detect and to prove and therefore also hard to confront. She suggests that more recognition of the experiences of youths of color, but also support from other persons like peers, teachers, parents as well as educational, medial, and political institutions is needed.

In his article, "Beyond a Trifling Presence: Afro-Germans and Identity Boundaries in Germany," Anthony Obute starts off with the observation that being German and being black does not seem to go together in the mind of most Germans. Whiteness thus functions as a synecdoche, accompanied by the expectation of 'blonde hair and blue eyes' by default in order to qualify Germanness. The imaginary boundaries of German identity also directly confront a remarkable history of the description of the German body. Obute argues that the spirit of citizenship law, tightly constructed around body and color politics, sustain(ed) microaggressive tendencies, as well as the exclusion of black features from German identity and Germany as a space. Hence, Germans of African descent need to situate their belonging within German society. They cannot regard it as a given, as white Germans, by contrast, can. Furthermore, Obute questions the production of black identity through the white gaze and the right of black subjects to define and name themselves. He critiques the tripartite workings of history with regard to black presence in Germany. Emphasizing the importance of race with regard to this discourse, he runs *Critical Race Theory* across the documentary "Afro Germany – Being Black and German" by Jana Pareigis et al. as an example to illustrate said workings. The documentary provides multiple viewpoints and experiences of black Germans across diverse spectra of life. He concludes that the black body emerges from the construction site of the white gaze as the figure of the strange and direct opposite of all that the white body represents. This shows that the production of black identity through the lens of the white gaze relies on self-generated precolonial knowledge of the African space, as the allegedly ill Africa, its space, place and people, were the imagination and production of Europe, too. The future, therefore, remains a void space and a recipient of the fast accelerating past, which imbricates the history of the 'trifling presence' and the exclusion of the black body from the consideration of Germanness.

Annika Rosbach draws attention to translation practices deployed in rendering African American vernacular in the German that have effectually (if unintentionally) reified anti-black racism. Her essay "Race and Racism in Translation: 'Who Can Speak?' in German Renderings of Literary African American English" first foregrounds the aesthetic functions African American vernacular assumes in the narration of Zora Neale Hurston's novel *Their Eyes Were Watching God* (1937) and Alice Walker's novel *The Color Purple* (1982) to subsequently highlight problematic strategies German translators have mobilized to render these literary texts in the German. Rosbach demonstrates that Barbara Henninges' choice to render African American English in the German Rhine-Ruhr dialect in her translation of Hurston's novel, *Und ihre Augen schauten Gott* (1993), represents African Americans as linguistically inferior. In translating Walker's novel, Helga Pfetsch's choice of using German colloquial language (*Umgangssprache*) to approximate African American English in *Die Farbe Lila* (1983), Rosbach argues, limits the main character Celie's voice to indexing a lack of education. Rosbach closes by introducing the translation strategies Hans-Ulrich Möhring has developed in *Vor ihren Augen sahen sie Gott* (2011) as a paradigmatic example of how African American English may be translated in ways that convey the language's aesthetic properties (e.g. its musicality) while eschewing racist stereotypes in portraying African American culture.

"Post-Racism, Colorblind Individualism & Political Correctness: Contemporary Modes of Materialization in American Studies and German Academia," by Courtney Moffett-Bateau and Sebastian Weier, highlights the structures of a specific strategy in racial discourse – that of materialization – which has come to characterize our contemporary post-racist era. The authors define post-racism "as a 'new' period of racism that positions itself in line with mainstream descriptions of colorblind individualism and contemporary language policies that claim we live in a period that is beyond racism and therefore beyond political correctness" (123). Strategies of materialization, then, reinforce the systemic exclusion of blacks by disavowing "race-based structures" and relegating them to the past "in the name of American democratic progress" (123). Moffett-Bateau and Weier highlight examples such as Clint Eastwood's famous "chair-ologue" at the 2012 Republican National Convention, the Dred Scott v. Sanford and State of Missouri v. Celia court cases, the lynchings of black effigies at the University of California, Berkeley in December 2014, and the controversial April 2016 edition of the German academic journal *Forschung und Lehre*, to elucidate the two main forms in racialized discourse: legal and artistic materialization. In so doing, the authors outline the history of the formation of a democratic order – an order that, relying on racialized modes of symbolic violence (Pierre Bourdieu) to exclude blacks from the concept of humanism, "carries an extensive tradition of making people matter differently" (134).

In her article, "~~*Kanak*~~ Academic: Teaching in Enemy Territory," Ismahan Wayah addresses the absence of non-normative bodies of black and people of color (BPoC) from German academic spheres and the concomitant absence of their non-normative knowledge from the white, middle-class, heteronormative system of knowledge production in the German academy. Wayah explains how the exclusion of BPoC bodies begins in the German three-tiered secondary-school system of *Hauptschule*, *Realschule*, and *Gymnasium*, which encourages students from higher socioeconomic status to pursue a university education while preparing children from immigrant families/families of color, who are disproportionately affected by socioeconomic marginalization, for vocational training, leading to low-income job opportunities. Even those BPoC students who make it to the university are expected to mimic a white middle-class *habitus* in order to be considered almost German, but not quite. Germanness, as Wayah argues, in the hegemonic cultural imaginary to this day indexes whiteness and Christianity. Referencing literature and critical thought by and about marginalized people such as migrants and BPoC in Germany, publicly known incidents of racism in the German academy, and her own experiences, Wayah illustrates the odds and obstacles that the racialized, non-normative body of a black female Muslim academic in *hijab*, the "~~Kanak~~ academic," will have to overcome in order to teach in "enemy territory" and to gradually subvert longstanding structural and epistemological regimes of white middle-class heteronormativity in the German academy.

In "The Migrant Scholar of Color as Refugee in the Western Academy," Mahmoud Arghavan proceeds from Giorgio Agamben's remark that the figure of the refugee, who destabilizes the foundations of the nation-state and nationalism, must be the point of reference for thinking the world anew, to tease out parallels between the political-structural positionality of the figure of the refugee (of color) in contemporary Europe and the epistemic-structural positionality of the figure of the migrant scholar of color in the Western Humanist academy. Combining theoretical insights gleaned from personal experience with the theoretical work of Hannah Arendt, Giorgio Agamben, and Zygmunt Baumann, Arghavan concludes that the promise of academic freedom proves elusive for migrant scholars of color upon their arrival in Germany/Europe: while they overwhelmingly left their home countries due to a lack of academic freedom and other indices of political repression, their freedom to research and teach in Germany is often denied in covert ways, as the German Humanist academy is a white-dominated space set on reproducing itself and is generally inimical to epistemic approaches beyond Eurocentrism, which Arghavan, building on Walter Mignolo, calls "epistemic racism." Arghavan discerns certain similarities between this situation and that of refugees (of color), who have fled persecution in their home countries, but continue to face tenuous relations to human rights in Europe/Germany. But, Arghavan contends, just as the figure of the refugee (of color) forces the German society to confront its own racism, the migrant scholar of color forces the German academy to confront its own structural and epistemic Eurocentrism, thus gradually pushing the German Humanities in the direction of decolonization.

Kai Linke discusses the ongoing institutional racism in academia, which excludes people of color through hiring practices and admission policies, in his article "Keeping Academia White: A Case Study." His case of study is an incident that occurred at a public conference at a German university, which led to a series of controversies between students and professors. Through close reading of the controversial statements during the conference and after that, Linke exposes the racist implications and exclusionary consequences of some statements by white people who even perceive themselves as anti-racist but are actually engaged in racist practices. Linke suggests that only by reflecting on the criticisms addressed at white people we shall move away from the illusion of innocent whiteness and towards a more egalitarian society.

Priscilla Layne's article, "On Racism without Race: The Need to Diversify *Germanistik* and the German Academy," deals with the issue of diversity in the field of German Studies (*Germanistik*) in Germany. As an African American scholar working in the field of German Studies and as both a former student and now professor, she has experienced first-hand that there is a lack of diversity in German higher education. She argues that many white academics and students fail to acknowledge their privilege, racializing practices, and racist structures in Germany, which contribute to the lack of diversity within the academy. While this is an issue in American Studies, it is likely an even larger problem in *Germanistik*, as scholars often work within a paradigm of German literature and culture that reflects the homogenous ideal of Herder's *Volk*, and marginalizes the important contributions of minorities, such as those of black German and Turkish German authors. Against this background, Layne emphasizes the importance for white German professors to critically position themselves when working on topics such as *blackfacing*, and proposes that it is important to have more people of color at the university, both in positions of power and as students. She argues that German Studies or *Germanistik*, depending on whether one talks about the U.S. or Germany, tends to see itself as being free from racial problems, aside from recognizing the importance of German Jewish Studies. In order to combat racism in the German academy as a whole, diversifying disciplines like *Germanistik* occupies a central role, Layne asserts. She suggests that one needs to start with *what* is taught and decolonize the canon. This demonstrates that authors of color *belong* to the canon of German literature. With that, German professors should dare to challenge their students and themselves to think more critically about race and think about *how* they can work to deconstruct racial stereotypes in every context. Layne affirms that white Germans have to work towards dismantling white supremacy in the academy and restructure it to create a space of learning in which diverse backgrounds, opinions, experiences, and skills are valued. Only when the universities, departments, and disciplines can commit to deconstructing white privilege, she concludes, real progress can be achieved; progress towards building an academy that better reflects reality.

In "'So You Want to Write about American Indians?' Ethical Reflections on Euro-Academia's Research on Indigenous Cultural Narratives," Amina Grunewald discusses how Indigeneities can be negotiated by Western agents in particular places of education and therefore in spaces of power. She questions what kind of cultural, sociopolitical, and ethical contexts need to be taken into account within a North American decolonizing discourse when dealing with Indigenous peoples, their knowledge and belief systems, and their self-representations. In her article, Indigenous voices such as scholars and artists are given a major authoritative position to represent their agendas in order to let Indigenous culture insiders have their say about problematic Western cultural outsiders' involvements – Western outsiders that often enough have insensitively been trespassing invisible borderlines regarding Indigenous spaces of protection and/or disregarding issues that arise for Indigenous people(s), who are forced to live within Westernized settler nation-states. In this context, Amina Grunewald draws not only from Indigenous authors'/scholars' works and linked methodologies and philosophies, but rather from her personal research stays in Vancouver and Montréal – all dealing with Indigenous issues on the power of (self-)representations and therefore, alterNative political and cultural agendas and visibilities in North American public spaces.

Saskia Hertlein addresses the challenge of teaching contemporary race-related issues in U.S. society in the higher-education classroom in Germany in ways that encourage students to draw parallels to German culture and society while nuancing the contextual specificity of the issue under discussion in "'The Danger of a Single Story': Addressing Contemporary Public Discourse and Protest Movements in American Studies Classrooms in Germany." Taking as her point of departure the recent upsurge of right-wing social and political movements in both the U.S. and Germany, which agitate against racial and religious minorities at the very same time as the U.S. is projected to become minority-white and Germany is becoming decidedly more ethnically diverse, Hertlein builds on literary critic Paula Moya's approach in *The Social Imperative* to outline how close reading can bring the schemas through which individual participants in a classroom discussion engage with issues of race to the fore, and how these schemas may be productively transformed through an open, but respectful discussion of clashing views. Hertlein goes on to highlight how sociologist Melissa Steyn's approach of *Critical Diversity Literacy* can be operationalized as both a method for and a subject of teaching in the Humanist higher-education classroom, and fleshes out why Inter-American Studies approaches and subject matters are particularly suited for inclusion in American Studies syllabi that seek to foster critical thinking regarding racism, colonialism, and border regimes in students.

In "Goethe Meets Baldwin: Notes towards a Comparative Perspective beyond Misappropriation," Derek C. Maus takes an anecdote from his intellectual biography as a point of departure to problematize iterations of the theories of cosmopolitanism and multiculturalism, and subsequently propose ways of 'doing transcultural interpretation' that actively work against misappropriation and Othering. Recalling his enthusiasm for cosmopolitanism as an undergraduate, which he conceived as an antidote to Cold War-fuelled nationalism, Maus details his gradual disenchantment with cosmopolitanist philosophy, whose articulation, he argues, presupposes security afforded by the very nation-state that cosmopolitanism seeks to transcend, and, like multiculturalism, often retains "a quasi-colonialist subjective privilege" (289). Critiquing parallel approaches in the realm of literary criticism, which retain an Othering of the cultural work of authors classified as non-white or postcolonial, Maus proposes frameworks for transcultural interpretation that take their cue from James Baldwin's provocative solution to the outlined dilemma: rather than visit black country (i.e. practice cosmopolitanism), a white person has to 'consent to be black.' Maus subsequently builds on John Pizer's conception of world literature to outline a practice of literary interpretation that takes Baldwin's demand seriously, and "involves remaining receptive to the unfamiliar without prejudging it – either positively or negatively – because of its alterity" so that "the Other ceases to be defined either in opposition to the Self or as a desirable exoticism to be appropriated into it" (293).

Academic white spaces and their (in-)ability to produce knowledge that accurately captures the dynamics of racial, ethnic, and class diversity in 21st century Europe and the U.S. are the focus of Kimberly Alecia Singletary's essay "Notes from the Margin: Academic White Spaces and the Silencing of Scholars of Color." Looking at such spaces – that is, spaces marked by a predominating white presence that relegates people of color to object rather than subject positions when it comes to academic inquiries into racism – entails looking at the "university's position as a white space that silences Other narratives through an adherence to colorblindness and race neutrality, even as the university purports to educate students about racial and social justice" (298). The notion of colorblindness is key here as it allows for particularly subtle forms of black marginalization. "[C]olorblindness," Singletary argues, following Eduardo Bonilla-Silva, "'otherizes softly,' creating opportune moments for covert discrimination and […] a 'racial grammar' [that] normalizes white domination to the point that it remains invisible to many, if not most, people" (306). Singletary augments her investigations of academic white spaces with field notes and diary entries that show how colorblindness undergirds a European problem of "imagining difference," based on a widespread imagination of Europe as completely untouched by 'race' and the effects of colonialism, as well as it informs an American tendency of "choosing ignorance," in other words, an unwillingness to acknowledge, and thereby equalize, the alternative knowledge produced by people of color. In scholars of color such conditions elicit "racial fatigue," as they constantly strive to attain recognition by the white majority, "doing research that benefits their communities while overworking themselves to be seen as equally capable as their white colleagues or avoid[ing] race-related research, opting instead for a colorblind approach that treats whiteness as a race-neutral category at the expense of People of Color" (302). Thus, the question of academic diversity is, as Singletary ultimately argues, not simply one of countering racial discrimination; it is also closely tied to the university's role as an institution of productive "interpersonal and academic relationships" and "a leader in political and social thought" (309).

In "Transatlantic Postcolonial (T)Races in the Classroom: From Defoe's Desert Island to Larsen's *Quicksand* and *Black-ish* Suburbia," Elahe Haschemi Yekani traces race as discursive and aesthetic formation through reconstructing it from its time and space of emergence. In doing so she takes us on an inter-genre journey to revisit three spatiotemporally different texts, that is, Daniel Defoe's *Robinson Crusoe*, Nella Larsen's *Quicksand*, and finally the TV series *Black-ish* through the transnational and intersectional framework of postcolonial epistemology. In *Robinson Crusoe,* Haschemi Yekani investigates the gradual formation of English national identity as white with religious, geographical, racial, and civilizational demarcations. In her reading of *Quicksand*, she interrogates the relationality of race in studying the mixed-race heroine of the novel, Helga Crane, daughter of a white Danish woman and an African American man. In her analysis of *Black-ish,* she reminds us of how much race still matters in 21st century America and that blackness needs to be claimed and remade. Haschemi Yekani proposes a non-binary and non-static framing of categories of difference as a postcolonial contribution to an interdisciplinary understanding of intersectionality.

Marius Henderson rounds off our edited volume with a poetic reflection on the issues under discussion in the volume. His poem is titled "Passing Tone/ Note."

Works Cited

Ahmed, Sara (2012): On Being Included: Racism and Diversity in Institutional Life, Durham and London: Duke University Press.

Ahmed, Sara (2016): "How Not to Do Things with Words." In: Wagadu: A Journal of Transnational Women's and Gender Studies 16/1, pp. 1-10 (http://sites.cortland.edu/wagadu/wp-content/uploads/sites/3/2017/02/v16-how-not-to-do-ahmed.pdf/).

Anderson, Elijah (2015): "The White Space." In: Sociology of Race and Ethnicity 1/1, pp. 10-21.

Auernheimer, Georg (2013): Schieflagen im Bildungssystem: Die Benachteiligung der Migrantenkinder, Wiesbaden: Springer VS.

Batson, Daniel C. (2016): What's Wrong with Morality?: A Social-Psychological Perspective, Oxford: Oxford University Press.

Black Diaspora and Germany Network (ed.) (2018): Black Diaspora and Germany – Deutschland und die Schwarze Diaspora, Münster: edition assemblage.

Boidin, Capucine/Cohen, James/Grosfoguel, Ramón (2012): "Introduction: From University to Pluriversity: A Decolonial Approach to the Present Crisis of Western Universities." In: Human Architecture: Journal of the Sociology of Self-Knowledge 10/1, pp. 1-6.

Bonilla-Silva, Eduardo (2006): Racism Without Racists: Color-blind Racism and the Persistence of Racial Inequality in the United States, Lanham: Rowman & Littlefield.

Calhoun, Craig (2008): "Foreword." In: Charles R. Hale. (ed.), Engaging Contradictions: Theory, Politics, and Methods of Activist Scholarship, Berkeley: University of California Press, pp. xiii-xxvi.

Chin, Rita/Fehrenbach, Heide/Eley, Geoff/Grossmann, Atina (2009): After the Nazi Racial State: Difference and Democracy in Germany and Europe, Ann Arbor: University of Michigan Press.

Crenshaw, Kimberlé (1991): "Mapping the Margins: Intersectionality, Identity Politics, and Violence Against Women of Color." In: Stanford Law Review 43/6, pp. 1241–1299.

Crenshaw, Kimberlé (2012): On Intersectionality: The Essential Writings of Kimberlé Crenshaw, New York: Perseus.

Dabashi, Hamid, and Walter Mignolo (2015): Can Non-Europeans Think?, London: Zed Books.

Daubner, Lukas (2018): "Leitbildprosa reicht nicht. Kann man Diversität in der Universität managen?" In: Forschung & Lehre 25/3, pp. 202-203.

de Sousa Santos, Boaventura (2014): Epistemologies of the South – Justice against Epistemicide, Boulder and London: Paradigm Publishers.

de Sousa Santos, Boaventura (2017): Decolonising the University – The Challenge of Deep Cognitive Justice, Newcastle upon Tyne: Cambridge Scholars Publishing.

Dhawan, Nikita (ed.) (2016): Difference that Makes no Difference: The Non-Performativity of Intersectionality and Diversity (Special Issue), In: Wagadu:

A Journal of Transnational Women's and Gender Studies 16/1 (http://sites.cortland.edu/wagadu/v-16/).

Drews-Sylla, Gesine/Makarska, Renata (eds.) (2015): Neue alte Rassismen? – Differenz und Exklusion in Europa nach 1989, Bielefeld: transcript.

El-Tayeb, Fatima. (2005): "Vorwort." In: Maureen Maisha Eggers/Grada Kilomba/Peggy Piesche/Susan Arndt (eds.), Mythen, Masken und Subjekte: Kritische Weißseinforschung in Deutschland, Münster: Unrast Verlag, pp. 7-9.

El-Tayeb, Fatima. (2011): European Others: Queering Ethnicity in Postnational Europe, Minneapolis: University of Minnesota Press.

El-Tayeb, Fatima (2016): Undeutsch: Die Konstruktion des Anderen in der postmigrantischen Gesellschaft, Bielefeld: transcript.

Fereidooni, Karim (2011): Schule – Migration – Diskriminierung: Ursachen der Benachteiligung von Kindern mit Migrationshintergrund im deutschen Schulwesen, Wiesbaden: Springer VS.

Foucault, Michel (1980 [1972]): Power/Knowledge – Selected Interviews and Other Writings 1972-1977. Colin Gordon (ed.), Colin Gordon/Leo Marshall/ John Mepham/Kate Soper (trans.), New York: Pantheon Books.

Großklaus, Götz (2017): Das Janusgesicht Europas: Zur Kritik des kolonialen Diskurses, Bielefeld: transcript.

Grözinger, Gerd/Langholz-Kaiser, Marlene (2018): "Bewusste Anerkennung von Unterschieden: Diversität in der Wissenschaft." In: Forschung & Lehre 25/3, pp. 198-200.

Gutiérrez y Muhs, Gabriella/Flores Niemann, Yolanda/Gonzalez, Carmen G./Harris, Angela P. (eds.) (2012): Presumed Incompetent: The Intersections of Race and Class for Women in Academia, Boulder: University Press of Colorado.

Hale, Charles R. (ed.) (2008): Engaging Contradictions: Theory, Politics, and Methods of Activist Scholarship, Berkeley: University of California Press.

Harding, Sandra (ed.) (2004): The Feminist Standpoint Theory Reader: Intellectual and Political Controversies, New York: Routledge.

Karen L. Dace (ed.) (2012): Unlikely Allies in the Academy: Women of Color and White Women in Conversation, New York: Routledge.

Kelly, Natasha (2017): "Rassismus betrifft alle Fächer." In: Tagesspiegel August 30 (https://www.tagesspiegel.de/wissen/uni-dozentin-natasha-a-kelly-rassismus-betrifft-alle-faecher/20255456.html).

Kilomba, Grada (2010): Plantation Memories: Episodes of Everyday Racism, Münster: Unrast.

Kilomba, Grada (2015): "Decolonizing Knowledge." Voicerepublic, 62:47 (https://voicerepublic.com/talks/grada-kilomba-decolonizing-knowledge-016ae920-48de-445f-845d-dcfe6926c4b4).

Kühne, Anja/Warnecke, Tilmann (2018): "Schüler im Milieu gefangen." In: Der Tagesspiegel October 23 (https://www.tagesspiegel.de/wissen/oecd-zu-chancengerechtigkeit-schueler-im-milieu-gefangen/23218686.html/).

Kuokkanen, Rauna (2007): Reshaping the University: Responsibility, Indigenous Epistemes, and the Logic of the Gift, Vancouver and Toronto: University of British Columbia Press.

Kuria, Emily Ngubia (2015): Eingeschrieben: Zeichen setzen gegen Rassismus an deutschen Hochschulen, Berlin: w_orten & mehr.

Lennox, Sara (ed.) (2016): Remapping Black Germany: New Perspectives on Afro-German History, Politics and Culture, Amherst: University of Massachusetts Press.

Linde, Frank/Auferkorte-Michaelis, Nicole (2018): "Auf mehreren Ebenen agieren. Diversitätsorientierte Hochschullehre." In: Forschung & Lehre 25/3, pp. 210-212.

May, Vivian M. (2012). Anna Julia Cooper, Visionary Black Feminist. A Critical Introduction, Foreword by Beverly Guy-Sheftall, New York: Routledge.

Mbembe, Achille (2015): "Decolonizing Knowledge and the Question of the Archive." Lecture, Wits Institute for Social and Economic Research (https://wiser.wits.ac.za/system/files/Achille%20Mbembe%20-%20Decolonizing%20Knowledge%20and%20the%20Question%20of%20the%20Archive.pdf).

Michaels, Walter Benn (2006): The Problem with Diversity: How We Learned to Love Identity and Ignore Inequality, New York: Metropolitan Books.

Michaels, Walter Benn (2008): "Against Diversity." In: New Left Review 52, pp. 33-36.

Mignolo, Walter D. (2012): Local Histories/Global Designs. Coloniality, Subaltern Knowledges, and Border Thinking. New preface, Princeton: Princeton University Press.

Mignolo, Walter D./Escobar, Arturo (eds.) (2013): Globalization and the Decolonial Option, New York: Routledge.

Mignolo, Walter D. (2013): "Introduction: Coloniality of Power and De-Colonial Thinking." In: Walter D. Mignolo/Arturo Escobar (eds.), Globalization and the Decolonial Option, New York: Routledge.

Nagel, Mechthild (2016): "Pitfalls of Diversity Management within the Academy." In: Wagadu: A Journal of Transnational Women's and Gender

Studies 16/1, pp. 40-76 (http://sites.cortland.edu/wagadu/wp-content/uploads/sites/3/2017/02/v16-pitfalls-of-diversity-management-nagel.pdf/).

NRW-Zentrum für Talentförderung. https://nrw-talentzentrum.de/talentscouting/grundverstaendnis/.

"PISA-Studie: Deutsche Schüler über dem Durchschnitt." In: Bundeszentrale für politische Bildung December 3, 2013 (https://www.bpb.de/politik/hintergrund-aktuell/174546/pisa-studie/).

Peus, Claudia/Braun, Susanne/Hentschel, Tanja/Frey, Dieter (eds.) (2015): Personalauswahl in der Wissenschaft – Evidenzbasierte Methoden und Impulse für die Praxis, Heidelberg: Springer.

Peus, Claudia (2018): "Fortschritt durch Vielfalt. Ansatzpunkte für das Diversitätsmanagement an Hochschulen." In: Forschung & Lehre 25/3, pp. 206-207.

Puwar, Nirmal (2004): Space Invaders: Race, Gender and Bodies out of Place, Oxford: Berg.

Sensoy, Özlem/Robin Diangelo (2017): "'We Are All for Diversity, but…' – How Faculty Hiring Committees Reproduce Whiteness and Practical Suggestions for How they Can Change." In: Harvard Educational Review 87/4. pp. 557-580.

Sow, Noah (2008): Deutschland Schwarz Weiss – Der alltägliche Rassismus, München: Bertelsmann.

Spivak, Gayatri C. (1988): "Can the Subaltern Speak." In: Cary Nelson/Lawrence Grossberg (eds.), Marxism and the Interpretation of Culture, Urbana and Chicago: University of Illinois Press, pp. 271-313.

Steyn, Melissa (2010): "Critical Diversity Literacy. Diversity Awareness in 12 South African Organisations." In: Melissa Steyn (ed.), Being Different Together: Case Studies on Diversity Interventions in some South African organisations, Cape Town: iNCUDISA, pp. 15-42.

Taylor, Yvette (ed.) (2014): The Entrepreneurial University: Engaging Publics, Intersecting Impacts, Basingstoke: Palgrave Macmillan.

Thompson, Vanessa Eileen/Zablotsky, Veronika (2016): "Rethinking Diversity in Academic Institutions." In: Wagadu: A Journal of Transnational Women's and Gender Studies 16/1, pp. 77-95 (http://sites.cortland.edu/wagadu/wp-content/uploads/sites/3/2017/02/v16-rethinking-diversity-thompson-zablotsky.pdf/).

Ting, Jenevieve (2015): "Angela Davis's Legacy of Collective Solidarity." Ms. Magazine Blog, February 26 (http://msmagazine.com/blog/2015/02/26/angela-daviss-legacy-of-collective-solidarity/).

Vertovec, Steven (ed.) (2014): Routledge International Handbook of Diversity Studies, New York: Routledge.

Zuberi, Tukufu/Bonilla-Silva, Eduardo (eds.) (2008): White Logic, White Methods: Race, Epistemology, and the Social Sciences, Lanham: Rowman and Littlefield.

I. Race in Translation: Comparing Racial and Xenophobic Formations in Germany and the United States

'*Ausländer*' – A Racialized Concept? 'Race' as an Analytical Concept in Contemporary German Immigration History

Maria Alexopoulou

Silencing as Methodology

In an article published in *American Historical Review*, German Professor of American History Manfred Berg and two co-authors describe the problems European – and in his case, particularly German – historians have in using the concept of 'race.'[1] In light of the "insoluble association with Nazi ideology, the very idea of race appears too contaminated for a semantic resurrection," Berg argues (2014: 807). In his view, it seems comprehensible that some German colleagues, despite its analytical usefulness, do not use 'race,' even when writing in German about the United States. According to the authors, "the narrow biological connotation of *Rasse*" (ibid: 803) in Germany does not correspond with the cultural and social meanings *race* has in the U.S. Moreover, there is always the danger of "fall[ing] into the trap of a tainted and cumbersome vocabulary" (ibid: 804), in other words, of mixing *race* as an ascriptive term with 'race' as an analytical concept. Additionally, "[i]n Germany, the term 'race' is never used in connection with immigration from Africa and Muslim countries because branding immigrants as a race would be racist" (ibid: 803). Thus, since the concept 'race' is not used in the German historiography on immigration, it seems that racism is absent in this context, too. Even if it is not totally absent in the discourses, the authors seem to suggest that silencing it is better than polluting oneself with such a word or, even worse, mixing its meanings and

1 The quotation marks signify that 'race' is used as an analytical concept in this case. When italiziced, it is left open whether the respective speaker means a real object or a (social) concept.

using it ascriptively instead of analytically. Besides that, the authors seem to presume that a significant part of readers will understand *Rasse* ascriptively. Why would they believe that? Maybe because they sense that this is exactly the way many people think about this concept – that *Rasse* is still perceived as something biological and real. Non-informed readers, as Berg and his colleagues imply, may re-believe in its existence and the concept may resurrect as an object, particularly when it is mentioned in scientific literature.

Accordingly, mainstream German historiography dealing with the Federal Republic of Germany or newer German immigration history refrains from using the terms 'race' or 'racism.' In Germany, these concepts appeared belatedly in the scientific field in the context of postwar migration, and mainly in cultural studies, critical sociology, or critical migration pedagogy.

In a scholarly debate some years ago, similar doubts were voiced against the term '*Volksgemeinschaft*' as an analytical concept (Wildt 2014). Some younger historians suggested that *Volksgemeinschaft* was a central concept for the Hitler Regime and thus, an important analytical tool for explaining the cohesion of German society during the *Drittes Reich*. However, older historians such as Hans Mommsen refused to use this concept, dismissing it as purely a propaganda formula (Mommsen 2007: 20). Mommsen, Ulrich Herbert, and others even considered its use to be dangerous, arguing that it reified a mere phantasm of the Nazis (Wildt 2011). This argument follows the same anti-logic as the reasons brought up for not using 'race' as an analytical tool in the contemporary German context, even if only studying the fate of this concept after 1945. How plausible is it that all notions about *race* disappeared in a supposed *Stunde Null*? How plausible is it that they had no impact at all on how difference was perceived in Germany after the "Nazi racial state"?[2]

'*Volksgemeinschaft*' and '*Rasse*' seem to evoke *Angst:*[3] naming them and speaking about them could unleash some beasts, neatly kept under the carpet – exactly by silencing them. If they were named, they would come to life again. Or maybe they are there, but by being ignored, they seem harmless. This might be one aspect. Another might be that ignoring racism has some benefits for those defining the discourses, who are, in general, on the privileged side of this system.

2 This term is taken from the subtitle of the book by Chin et al. (2009).

3 *Angst* or even *German Angst* are terms not easily defined, as they are used in different contexts ranging from philosophy to popular culture. But the term implies in all these contexts the quite irrational and hyperbolic character of this anxious mood, in contrast to fear ("*Furcht*"), which is evoked by real dangers.

Interestingly enough, by making this conscious act of ignorance invisible in historiography, too, the very roots of what I call *racial knowledge*, which still structures many aspects of the *Einwanderungsgesellschaft Deutschland* (immigration society), are not cut off, but kept alive. The binarity of '*Ausländer*' (foreigner) and 'German' is one of these aspects; it seems banal or even 'natural,' while, in reality, it is a powerful transmitter and stabilizer of racial knowledge.

Continuities

'The North African' and 'the Italian' seem to have much in common: they are rapists, petty criminals, and 'dangerous fellas.' At least this is the impression if one considers these *topoi* circulating and being known at different times. 'The North African' was marked according to these attributions prior to New Year's Eve 2016.[4] A bar in the German town of Gießen had already banned 'North Africans' in 2012, but it was only two years later, after a report in the local newspaper, the *Gießener Allgemeine Zeitung*, that this was made public. The journalist wanted to know from the staff how they recognized North Africans. By their looks, by their French language, and by their habits, she/he learned. Such persons were not served, and sent out of the bar. What if non-North Africans or foreign students were thrown out, the journalist wanted to know? This was "*Pech*" ("bad luck"), the waiter at the bar responded ("Lokalverbot nur für Nordafrikaner").

Similarly, Italians were not welcome guests in the 1960s, either. In the spring of 1963, bars and dance halls in Schönau, a working-class district in the city of Mannheim,[5] even had signs in their windows saying "no Italians." These signs

4 On New Year's Eve 2015/16, hundreds of young men, likely most with an Arab background, harrassed and in some cases raped girls and women in Cologne and in other major German cities during the festivities, particularly at the *Domplatte* in Cologne. These numerous incidents of sexual violence evoked an ugly campaign against refugees and North African young men, many of whom live in Germany undocumented. Rape was clearly racialized in this discourse, although some women's groups tried to alter the focus on sexual abuse as a problem in general. They had little impact on the medial-political debate. See Weber (2016) for more details.

5 On further aspects of the (more) recent migration history of Mannheim, see my forthcoming book *Mannheim translokal*. My research focus on Mannheim is the reason why cases referring to this city are used in this essay as exemplifications for my hypotheses and theorizations.

revived "unpleasant memories," according to a journalist of the local newspaper *Mannheimer Morgen*, since 25 years before, the same windows had displayed signs stating "no Jews." A week had passed since an Italian *Gastarbeiter* (guest worker) had injured a young German fatally with a knife in a fight.[6] Up to 400 German residents of Schönau protested in at least two silent marches in the streets of their district against Italians and *Ausländer* (foreigners) in general. Italians, having lived for more than a decade in Schönau, were attacked; a Greek, entering a bar, was threatened by German guests with knives. At one of the township meetings, the local authorities were asked to intervene against "these people threatening our *Lebensraum*."[7] Italian guest workers organized in the local section of the *Deutsche Gewerkschaftsbund* (German Trade Union Confederation), pleaded in a newspaper article with the population of Schönau not to pass judgment on all Italians. For years, they had been frequenting the dances organized in Schönau and nothing had happened.[8]

Italian workers were no strangers in Mannheim, anyway. As early as 1907, a local newspaper reported on the "invasion of Italians" as seasonal workers in and around the city, which had occurred for years.[9] The first formal *Anwerbeabkommen* (recruitment agreement) between Germany and Italy was concluded in 1938 – not in 1955, the year that is commonly considered the start of this form of work migration (Sala 2007: 100). Local press coverage of the "Fascist Christmas in Mannheim" being celebrated the third consecutive year at the Italian community house *Casa d'Italia* testifies to the Italian presence in Mannheim in those years.[10] When Italy left the Axis Powers, the 'workers of the befriended nation' turned into *Zwangsarbeiter*. At the end of the war, 18.000 slave and foreign workers were in Mannheim (Kreutz/Wiegand 2008: 203). In 1961, Mannheim counted 8.000 foreign workers; 3.500 of them were Italian.[11]

---

6 Barchet, Mac. "Erregung in Schönau trübt Vernunft." *Mannheimer Morgen* (hereafter MM) 7 May 1963, in: Stadtarchiv Mannheim – Institut für Stadtgeschichte Mannheim (hereafter StadtA MA – ISG), ZGS, S 2/2434.

7 "Schweigemarsch zog durch die Schönau." *MM* 8 Mai 1963, ibid. *Lebensraum* ("living space") was a term used mainly during the Third Reich, meaning the territorial expansion of Germany in Eastern Europe, which aimed at the expansion of the "Aryan *race*" and the enslavement or extinction of the "Slavic *races*."

8 "Gastarbeiter bitten: Kein Kollektivurteil." *MM* no date, ibid.

9 *Generalanzeiger Mannheim* 8 March 1907, ibid.

10 *Neue Mannheimer Zeitung* 10 January 1938, ibid.

11 "Fast viertausend Italiener ohne Treffpunkt." *MM* 11 August 1961, in: StadtA MA – ISG, ZGS, S 2/0040.

Most employers accommodated their guest workers in prefabricated huts or barracks, often at the same spots or plant grounds where slave workers had been forced to live during the Second World War. In 1961, by contrast, the Halbach company rented a building in one of the wealthy districts of Mannheim, Lindenhof, for its Italian guest workers. However, they never moved in. In November 1961, the local newspaper, *Mannheimer Morgen,* reported on protests of "concerned citizens" and councilmen of one of the "preferred residential areas in town." They lamented the "threatening invasion of Lindenhof by Italians," fearing for the security of girls and women in that urban district. Italians drank too much, seduced young and respectable women, were "unsocial elements," and would spoil Lindenhof.[12] They should live in the outskirts of town, one pastor postulated, or in barracks with military rules, the district councilman suggested in a letter to the editor.[13] This statement was rather uncanny if one considers that some 800 meters from the planned guest worker building and 16 years before, there had been a slave worker camp for Polish internees (Schadt/Caroli 1993: 146). The Lindenhof case was even discussed in the city council and Lord Mayor Hans Reschke promised to intervene.[14] The company gave up its plan; the workers were accommodated in the industrial area. Such perceptions of Italians – and more generally, of Southern Europeans and guest workers – as stabbers and womanizers were common in the 1960s and were constantly reproduced in the media, as the 1972 study *Die Gastarbeiter in der Presse* by Manuel Delgado shows.

Even municipal authorities had "no *Ausländer*" signs. In 1974, the newly installed City Commissioner for Foreign Residents in Mannheim, Herbert Lidy, complained at the City Housing Office that the long-established sign "no *Ausländer*" pertaining to a whole section of the notice board with private housing offers, might be offensive to some of the foreign residents. The head of the Housing Office refused to change this practice, as too many landlords definitely did not want "*Ausländer*" as tenants and would otherwise totally withdraw their offers (cf. Alexopoulou 2018).[15] In 1980, the *Ausländerbeirat* (council of foreign residents) of the city of Nuremberg, consisting of elected immigrants, also

12 Preuss, Dieter. "Gedämpftes Willkommen für Italiener." *MM* 29 November 1961, StadtA MA – ISG, ZGS, S 2/2434.

13 "Der Lindenhof und die Italiener," 1 December 1961, ibid.

14 Niederschrift über die öffentlichen Verhandlungen des Gemeinderats, 23 November 1961, p. 448-449, StadtA MA – ISG, Zug. 1/1900, Nr. 344.

15 Aktenvermerk von Herbert Lidy an Bürgermeister Watzinger, 8 January 1974, StadtA MA – ISG, Dezernatsregistratur, Zug. 16/1993, Nr. 76.

decided to petition the city council to prohibit the Housing Office of their town to use "no *Ausländer*" signs. They had no success, either.[16]

Were these incidents merely actions of concerned citizens who feared or were prejudiced against foreigners – in the same manner as hostile reactions against refugees are often explained in the current situation of 2018? Or should they be interpreted as need-based and situative reactivations of knowledge about the migrant Other? It is interesting to note these parallels in the perception of 'the North-African' and 'the Italian.' As the latter were Europeans and well-known as foreign workers for a long time, using the concept of xenophobia to explain such aversion is logically unsound: Italians did not constitute a 'genuinely foreign' group eliciting 'fear.'

Yet, rather than on parallels, I want to focus on continuities in the history of the immigration society Germany, as they elucidate a crucial point, namely the transfer of racial knowledge mirrored in these constructions of Others, of '*Ausländer.*' According to Jacques Derrida's notion of *différance,* concepts bear traces of forgotten, suppressed, and silenced meanings. Additionally, they are interwoven in a nexus of meaning with their alleged opposite. In this case, I suggest the binarity of '*Ausländer*' and 'German.' My hypothesis is that discourses and practices around this binarity in the context of migration reproduce racial knowledge that has been transferred historically, although it was also transformed or even questioned.[17]

'*Ausländer*' at first glance solely denotes a formal juridical status that derives from citizenship. Indeed, citizenship plays a central part in the history of the German immigration society. But citizenship alone does not sufficiently explain this binarity. This binarity becomes clearer if one considers so-called *Ausländerfeindlichkeit*:[18] it was and is not aimed at white foreigners from Sweden, Switzerland, or Australia. There was and still is shared knowledge of who these *Ausländer* are at whom hatred is aimed (Kalpaka/Räthzel 1986: 14-15). Moreover, it is clear that an *Ausländer* or *Migrant* – in the 1990s, the latter term replaced the former as somehow more politically correct, because '*Ausländer*' had virtually turned into a term of abuse – can at any moment be attacked

16 Dokumentation zum Ausländerbeirat bei der Stadt Nürnberg, StadtA MA – ISG, Beaufragter für ausländische Einwohner (hereafter BfaE), Zug. 47/2011, Nr. 45.

17 The author of this article is working on a second-book project dealing with various aspects of this binarity. Most of the theses discussed here are first findings or still tendencies, as this is work in progress.

18 '*Fremdenangst*' is the German term for xenophobia; '*Ausländerfeindlichkeit*' literally means hostility against *Ausländer*. As this term is central to my argument, similar to '*Ausländer,*' I do not translate it.

verbally or physically simply because they are perceived as such. It is also relatively common to think that *Ausländer* or migrants are poorly educated, have low-skilled jobs, and live in closed communities; some might also think that their cultures are premodern, anti-modern, or even barbaric. Today, the term '*Migrant'* evokes the same perceptions that *Ausländer* used to evoke (and still continues to evoke), although a high percentage of those so-called migrants, whom official statistics also refer to as 'persons with migration background,' are German citizens. Here, one can grasp the still prevalent biologistic notion of 'German' as 'Bio-German' and at the same time, the essentialization of its alleged opposite term, '*Migrant*' or '*Ausländer.*'

Over the years, the term itself and the embodiment of 'the *Ausländer*' changed: from the Eastern European seasonal workers of the *Kaiserreich* (German Empire) to the '*fremdvölkische*'[19] worker in the Third Reich to the displaced persons and stateless foreigners after the war and in the early years of the Federal Republic of Germany (FRG), to the '*Neger-Soldaten*' (Black GIs), to the 'Italian stabbers' and '*Spaghetti-Fresser,*' to the '*bildungsferne Anatolier*' (uneducated Anatolians) – a phrase traceable to the highly prestigious historian Hans-Ulrich Wehler – and Muslims, to the Romanians and Bulgarians – mainly Romani people – to North Africans in the present moment. But the negative images associated with these groups stay constantly the same: childish, in need of education, uncivilized, sexually predacious, dangerous. The analogous images of immigrant women reached a wider public at a later time: in need of protection, unenlightened, unemancipated. This shows that '*Ausländer*' and all of its substitutes are racialized concepts: they construct a distinct group with particular characteristics which are cast as Other to 'the Germans.' This process of Othering constitutes the binary and hierarchical relation. As the Othered part of this binarity 'naturally' becomes negatively charged, it is not merely descriptive.

National origin, religion, class, and gender can be regarded as intersecting with the racialized concept '*Ausländer.*' These intersections explain the changing positionalities in the hierarchy of *Ausländer*. These reconfigurations are constantly observable, indeed with different groups, taking place within the respective historical context. However, it should be noted that this hierarchy is not merely a matter of discourses. It is also mirrored in the unequal legal treatment of migrants. In particular, residence permits have a direct effect on the opportunity structures open to migrants and thus, a direct effect on their lives (Hess/Lebuhn 2014: 17). In fact, this hierachy decides who is allowed to be present in Germany at all.

19 This term was common in the Nazi era and still used thereafter. '*Völkisch*' has a clear racial component.

I use the concept 'racial knowledge' deliberately, as I do not view racism in this context as an ideology, i.e. a systematic theory. Similarly, I do not address racism as a faulty individual attitude, as the concepts 'prejudice' or 'resentment' imply, but rather as traces of a racial body of knowledge being inscribed in institutions and structures, and reproduced by them constantly in bureaucracies' knowledge and everyday-life knowledge. I view racism as a power-knowledge complex – in the Foucauldian sense – characterized by imagined hierarchies between allegedly unchangeable, essential, and homogenous groups hypostasized and fixed in social positions. The low social positions of groups are then used as evidence for the inferiority of their origins, cultures, or religions – a process called racialization.

In his book *Geschichte der Ausländerpolitik*, historian Ulrich Herbert gives a very plausible description of a racialization process, although he does not use this term, as it is not part of his terminological repertoire. Initially, Herbert reports how brick factories around the year 1910 proceeded to hire Polish workers, so-called '*Auslandspolen*,' instead of Italian workers, as they turned more pretentious and expensive, as the contemporary complaint went.

> Polish and Ruthenian workers received much smaller wages than Italians and slowly occupied the worst positions held before by Italian brick workers. In the eyes of German authorities, this was proof for the different cultural stages of different *Völker* being "founded in race." "Where there is a presumed need of cheap workforce, the Pole is supplanting the Italian," the trade supervisory board of Baden assessed, because "the Pole likes jobs requiring a minimum of mental work the most. According to his cultural stage, he shows characteristics corresponding to childishly bad habits. He skives off work, has always to be pushed and controlled." Hereby, the circular argument was completed: Since Poles came from a less industrialized country and stood on a "lower cultural stage" as the Italians, they were preferred for particularly simple (and poorly paid) jobs. The fact that they carried out the simplest tasks, proved to the authorities how low their cultural stage was (Herbert 2003: 63, my translation).

The Germans, of course, stood at the top of this hierarchy. As the later head of the *Reichsanstalt für Arbeitsvermittlung* (employment bureau), Friedrich Syrup, put it in 1918, German workers rose to skilled labor when foreigners occupied the lowest ranks. Besides the fact that the national economy had no costs, foreign workers were sent away from Germany as soon as less workforce was needed. At the same time, foreign workers did not spoil the culture of Germans. Because according to their "higher [...] stage of culture" ("*höhere* [...] *Kulturstufe*"; my translation), they detested the rough and hard work foreigners performed with "greatest willingness" (qtd. in Herbert 1995: 214). During the years of forced and slave labor in the course of the World Wars, particularly the Second World War, these racialization processes were brought to their perverse climax.

In the guest-worker system (1955-1973), foreign workers again occupied the most unprestigious and often hardest jobs. Today, particularly newer immigrants, or such with an uncertain residence status or no status at all (undocumented aliens), mostly carry out unskilled labor at the lowest level and thus, Seyla Benhabib argues, even stabilize the economies and labor markets of the Global North (Sierakowski 2015). Who would collect the garbage, work at the cemeteries and slaughterhouses, or take care of the elderly? This exploitation logic has accompanied the work-migration system and related discourses in Germany ever since the *Kaiserreich*, especially in connection with migrants viewed as inferior. Work migration was flanked by a thorough control system for foreigners, who were not allowed to stay permanently in the country – in the case of Poles less than a year. This system bridged the discrepancy of needing additional workforce and the eugenic notion of staying or becoming purely German virulent until 1945. Yet, what happened to these perceptions and beliefs? Did they actually die out with 'unconditional surrender'? Especially if one considers that the legal and structural mechanisms of controlling and administering migrant workers initially stayed the same as before the war and were only transformed in a somewhat coercive manner in the course of the adaptation of the German work-migration system to that of other European nations and now EU partners (Schönwälder 2001: 277-287)?

According to the dominant narrative,[20] Germany welcomed foreign workers, neatly called guest workers, after the war and hosted them in the spirit of international friendship. Their function as *Konjunkturpuffer* (buffer against economic fluctuation) and as helpers for the social promotion of the German working class in forming the new underclass were all desired effects, and at the same time, allegedly to the benefit of the guest workers and their poor countries of origin. They were supposed to leave, anyway, as Germany was not an immigration country. Or rather, it was not in a position to be one, as the official announcements and the master narrative since the 1960s state. However, no one really asked why it was not an immigration country, although the arguments used for the unquestioned mantra "*Deutschland ist kein Einwanderungsland*" ("Germany is not an immigration country") were all but sound. One of them

20 For better readability I will not quote all studies on the general *Ausländerpolitik* or on the history of events. You can find a literature survey in my article Alexopoulou 2016.

was: Germany is overpopulated.[21] In reality, in 1972, the death rate exceeded the birth rate for the first time – a trend only balanced in the next decades by ongoing immigration via family reunion, asylum, and by the high birth rate of immigrants ("Geborene und Gestorbene"). As their birth rate adjusted in the process and the access of asylum seekers was nearly blocked by the so-called *Asylkompromiss*[22] in 1992 and by the Dublin Treaties, the balance tilted again in 2002. One of the reasons why the Christian Democratic Union (CDU) most grumpily endorsed the mass immigration of refugees in 2015 was by virtue of the fact that the population of Germany is eroding and getting older, while the economy needs workforce and the social-security system necessitates contributors. Before the summer of 2015, the Chamber of Industry and Commerce and other major economy actors had already launched campaigns and lobbied for the inclusion of asylum seekers into the labor market ("Berliner Erklärung") – not for humanitarian reasons, but because they desperately needed workforce.

Back in 1973, when the recruitment of migrant workers was banned, the guest workers, who had entered into an immigration process autonomously and had settled down mostly in the industrial areas and cities in which they had lived and worked, were, on the one hand, still needed in the labor market, as their jobs could not be occupied by Germans,[23] despite the ongoing economic crisis. Germany in 1977 ameliorated their resident status, granting many of them the unrestricted right to stay under certain requirements. On the other hand, they were fixed in their social position as *Ausländer* – constituting a group that was denied the right to vote and marked as not really belonging to German society, much less to the German nation. The scheme of *Integration auf Zeit* (temporary integration), which was developed at that time, still presumed that the *Ausländer* would leave someday, although not in the next couple of years.

Ausländer had no right to political participation aside from establishing associations or petitioning others to help them, and so they had no real leverage against any of the policies that affected them. Since the 1960s, non-elected

21 For example in the following statement: Landtag von Baden-Württemberg, Schriftlichte Antwort des Ministeriums für Arbeit, Gesundheit und Sozialordnung auf die Große Anfrage der Fraktion der CDU vom 12.4.1973, Probleme der ausländischen Arbeitnehmer in Baden-Württemberg, Drucksache 6/3280, 21.9.1973, p. 9, in: StadtA MA – ISG, Nachlass Nikitopoulos, 3/1995, Nr. 18.

22 The so-called political compromise on asylum made it possible to change the respective article of the German constitution.

23 According to a statement of the president of the *Bundesanstalt für Arbeit* Josef Stingl in: "Ohne die Gastarbeiter geht es nicht." *MM* 13 April 1976, StadtA MA – ISG, ZGS, S2/0040.

professionals, mostly Germans, represented in a paternalistic way – and the records show exactly that – the alleged interests of *Ausländer*, who by definition did not have civil rights, which again were reserved to German citizens and the German *Volk*. Only German citizens had the right to vote in national and federal elections. Voting rights at the communal level were not bound to citizenship, but restricted to members of the *Volk*. The federal states of Hamburg and Schleswig-Holstein were willing to include the *Ausländer* as a new part of the *Volk* and granted communal voting rights to them. However, the *Bundesverfassungsgericht* (Federal Constitutional Court) ruled in 1989 that this was unconstitutional, as these rights were reserved to members of the German *Volk*.[24] In Mannheim, *Ausländer* did not even have a forum like the *Ausländerbeirat* (council of foreigners), as in other cities such as Stuttgart or Nuremberg. The council of foreigners had no decision-making power, but their members were at least elected by the migrant workers themselves, in very few instances even by asylum seekers. Moreover, in this forum migrant workers could set the agenda themselves, speak for themselves, and petition the municipal council with their concerns. In Mannheim, the first *Migrationsbeirat*, as it was called by then, was only installed in the year 2000.

When one considers the ongoing discourses about migrants, particularly Muslims, who allegedly are incapable of being democrats (*demokratieunfähig*) and, on the other hand, looks at the history of political exclusion of *Ausländer*,[25] one can arrive at a rather contrary conclusion. In reality, German democratic culture is deficient, since it kept a great part of the population – approximately 8 percent at the end of the 1980s[26] – out of the political system, apparently in a deliberate effort to inhibit them from turning into full-fledged immigrants and citizens.

The most powerful institution stabilizing the binarity of 'German' and '*Ausländer*' and safeguarding the self-image of Germany as "*Nichteinwanderungsland*" (Bade/Bommes 2004: 337), was the German citizenship and

---

24 The Court recommended to the government in the same judgment to ease naturalization provisions. You can read the text of the judgment BVerfGE 83, 37 here: http://www.servat.unibe.ch/dfr/bv083037.html/.

25 While foreigners with citizenship in an EU country were granted municipal voting rights in the mid-1990s, the exclusion of Muslims – who in Germany are mainly comprised of migrants with Turkish citizenship and thus, of non EU-members – was and is complete.

26 Tabelle A, 13a, Zeitreihe Bevölkerung nach Nationalität, histat Historische Statistik (24.5.2017) https://histat.gesis.org/histat/de/table/details/70242C10504193247E8378 45EC1361EB.

naturalization law in effect from 1913 to 1999. The German citizenship and naturalization law of 1913 solely rested on the *ius sanguinis*, i.e. right to citizenship by blood, and was directed from the outset against migrant workers. Foreigners considered as "valuable gain for the *Volk*" still had the possibility to acquire German citizenship in an act of discretion (Trevisiol 2006: 200). Yet, as Trevisiol demonstrated in studying the practices of naturalization between 1871 and 1913, there was a shift from economic to racial rationales in the decisions taken, which implies that the self-perception as a *race*, which had to be protected from inferior intermixtures, was prevalent prior to the Nazi time. Trevisiol also points out that the FRG adopted the Third Reich's citizenship and naturalization law without changing one iota; consequently, its thoroughly biologistic and *völkisch* self-definition was transferred, too. His study stops in 1945, but there is much evidence that the praxis thereafter did not change very drastically, at least concerning the naturalization quota, which at the national level fluctuated from 0.25 percent in the 1970s to 0.38 percent in the 1980s, only one third of which had been guest workers (Dornis 2001: 136), an incredibly low figure. Even in the 1990s, with a naturalization quota of 1 percent, Germany ranged below the European average and still possessed a very substantial group of permanent *Ausländer* (Wobbe/Otte 2000: 444).

The mainstream version of this story, now and then, is that the guest workers themselves did not want to become Germans, as almost all of them, being *homines economici* according to the prevalent exploitation logic, wished to return home – or as the Baden-Württemberg Department of Labor called it in an official statement in 1974: guest workers had the "natural will to return."[27] It is true that the vast majority of migrant workers left Germany. However, not all of those who stayed did so by mere coincidence. The microcensus data by the Statistical Office of Baden-Württemberg rather leaves the opposite impression: in April 1978, 45.7 percent of the respondents – family heads with origins in the so-called recruitment states – said that they did not want to return to their home countries. 38 percent answered that they had no plans for a return yet. Out of the 45.7 percent who wanted to stay, 32 percent aspired to acquire German citizenship, as did 12 percent of those who had no concrete return plans. This amounted to a total of 17 percent of all respondents who wanted to acquire German citizenship.[28] Yet, *Ausländer* – in contrast to other privileged immigrant

27 Schriftliche Antwort des Ministeriums für Arbeit, Gesundheit und Sozialordnung auf eine Kleine Anfrage, 24.4.1974, Drucksache 6/4902, StadtA MA-ISG, Dezernats-registratur, Zug. 16/1993, Nr. 35.

28 Die Ausländer 1979, Statistik von Baden-Württemberg, Statistisches Landesamt Baden-Württemberg (Hg.), Stuttgart 1980, p. 91, chart 68.

groups who were *Volksdeutsche*[29] – had no claim to naturalization; they had to meet many provisions, but even if they met all of them, whether they were granted citizenship was still up to the particular officer's or authority's discretion. Until 1978, when a new directive was issued to the local authorities, a secret provision established in 1953 was in effect, instructing them to use their discretional power as far as possible "in the interest of keeping out unfitting elements, which would only afflict the *Volksganze*."[30] The new official provision established in 1977 emphasized that Germany had no interest in naturalizations, as "*Deutschland ist kein Einwanderungsland*."[31] These laws, practices, and discourses remained widely unchanged until 1999 and created '*Ausländer*' as a hereditary status passed on to third- and even fourth-generation migrants.

However, this was not solely a consequence of the naturalization practice. It was also the product of discourses, microaggressions, open discriminations, hate speech, and hate crimes, which created an atmosphere that kept the binarity of 'German' and '*Ausländer*' alive. The advent of more substantial numbers of refugees and asylum seekers from all over the world starting in the 1970s added fuel to the fire (Poutrus 2016: 887). Many of those entering Germany through this regime belonged to the very same groups occupying the lowest levels in the hierarchy of *Ausländer*, who were supposed not to be present at all. In the early 1960s, the Ministers of the Interior of the Federal States and the Federal Department of the Interior decided to completely ban so-called "*Afroasiaten*" ("Afro-Asians") – a cipher for people with dark skin – from work recruitment (Schönwälder 2001: 257-276). In 1963, Baden-Württemberg even deported some thousand Jordanian workers – most of them Palestinians previously living in refugee camps in Jordan – because they fell under the category of "Afro-Asians." In this case, even the exploitation logic did not apply: authorities in Mannheim wanted to keep 'their' Jordanians, as they occupied the lowest and hardest jobs at the commercial ports of Mannheim and Ludwigshafen, jobs even other European guest workers did not want anymore. Additionally, the local

29 In the FRG *Volksdeutsche* referred to individuals and their descendants who had lived inside the German borders of 1937 as Germans; translating the term '*volksdeutsch*' with ethnic German silences the historical context in which this term came up and evolved.

30 Erlass des Bundesinnenministeriums vom 26.6.1953, qtd. in Trevisiol 2006: 209, footnote 40. *Volksganze* cannot simply be translated as 'the people as a whole'; the term has a clear connection to '*völkisch*' thinking.

31 Abschnitt 2.3, Einbürgerungsrichtlinien: https://www.jurion.de/Gesetze/EbRichtl/2/.

police attested that the Jordanians as a group were unsuspicious.[32] However, decision makers acted according to their own racist notions and, anticipating reactions of the German population, kept banning migrants 'too foreign' for Germans to tolerate. In fact, when "Afro-Asians" began to arrive in greater numbers, their anticipation of Germans' attitudes was proven right.

In 1977, when the government of Baden-Württemberg forced Mannheim to house some 250 asylum seekers mainly from North Africa and Southeast Asia in the district of Rheinau, this caused great protest, and not only by the residents of Rheinau and the respective councilmen.[33] Even the Lord Mayor of Mannheim, Ludwig Ratzel, who was known as somewhat pro-immigrant, publicly announced his opposition to this plan, as it would bring more marginal groups to his town. He worked hard behind the scenes to prevent this.[34] Finally, the mostly young men were accommodated at the designated house in the harbor district, totally out of sight and too distant to walk to the inhabited part of Rheinau – all this had been clear to the protesters before. Nevertheless, they had caused agitation with petitions with hundreds of signatures,[35] demonstrations at the city council, local press coverage, and openly racist language. They claimed that Asians and Arabs were threatening their lives,[36] and that Mannheim-Rheinau was not supposed to be the "dumpsite" of Baden.[37]

In 1990, a fire broke out in an accommodation for asylum seekers in the city center. One inhabitant from Nigeria died. The responsible police officer reported at a hearing of the city council about a "terrifying amount of hate" of the German neighbors against the asylum seekers. They were shouting all the time "let the rats burn."[38]

---

32 Preuss, Dieter. "600 Jordanier sollen wieder heimfahren". *MM* 22 October 1964, StadtAMA – ISG, ZGS, S2/40; "Polizei muß Düsenmaschien chartern". *MM* 19 August 1965, ibidem.

33 Egermann, H. "Das ist eine Invasion für die Rheinau". *Rhein Neckar Zeitung* 28 Februar 1977, StadtA MA – ISG, Dezernatsregistratur, Zug. 17/1993, Nr. 119.

34 Most of the relevant documents in this case can be found in: StadtA MA – ISG, Dezernatsregistratur, Zug. 17/1993, Nr. 102 and 119.

35 Petitionen und Unterschriftenlisten, ibid.

36 Egermann, H. "Das ist eine Invasion für die Rheinau". *Rhein Neckar Zeitung* 28 Februar 1977, StadtA MA – ISG, Dezernatsregistratur, Zug. 17/1993, Nr. 119.

37 Councilman Winfried Höhn said that in a radio feature of the SDR. Protokoll des Mitschnitts einer Sendung des Presseamts Mannheim, 2.3.1977, ibid.

38 "Laßt die Ratten verbrennen." *Die Rheinpfalz* 12 April 1990, StadtA MA – ISG, Dezernatsregistratur, Zug. 83/1996, Nr. 209.

Racism as "*Ausländerfeindlichkeit*"

These and similar incidents happening everywhere in Germany, which merely exemplify the atmosphere of hate, not to mention the various forms of microaggressions experienced by *Ausländer*, played a crucial part in stabilizing the binarity of 'German' and '*Ausländer*' in the 1970s and 1980s, particularly in the self-perception of the *Ausländer*. According to one of the originators of critical race theory, the first tenured African American Professor at Harvard Law School, Derrick Bell, one method to dissolve inconsistencies between the master narrative and observed practices is revisionist history. The historical record should be reexamined "replacing comforting majoritarian interpretation of events with ones that square more accurately with minorities' experiences" (Delgado/Stefancic 2012: 24). Ways suggested to detect these inconsistencies are the reevaluation of observed practices, a re-reading of the documents, and counterstorytelling. Some theorists believe that counterstories are not only exemplifications in deconstructing mainstream narratives and unquestioned beliefs, but also provide a new language and opportunities to name wrongs that could not even be spelled out before by those who endured them – a "Cure for Silencing" (Delgado/Stefancic 2012: 49).

Although the former so-called guest workers can tell many stories about discriminations, microaggressions, and being named *Ausländer* hatefully again and again from the onset of their immigration to Germany since the 1960s, they mostly neither call these acts racist nor do they consider themselves as victims of racism.[39] The reason for this seems to be the lack of an appropriate language to address these issues, i.e. the unavailability of the terms 'race' and 'racism,' which were tabooed by German society and historiography in an attempt to portray post-Nazi German society as cured from racism for all times.

A possible approach to countering the master narrative of the newer German immigration history is to analyze these accounts and incidents of so-called *Ausländerfeindlichkeit*, the substitute or cover term in order to avoid naming racism 'racism' – an approach I have been using from the onset of this text. '*Ausländerfeindlichkeit*' contains the racialized group in its name, but at the same time implicitly negates the racism phenomenon. Along with *Fremdenfeindlichkeit* or *Fremdenangst* (xenophobia), these are the accepted labels, even

39 The author has conducted many oral-history interviews with immigrants and has listened to the stories of many former guest workers, being herself an *Ausländerkind* or a second-generation *Ausländer*. Mark Terkessides' study (2004) on second-generation immigrants had similar findings.

for highly visible racist phenomena such as violent attacks on individuals or burning down refugee shelters.

These concepts do not refer to structural and institutional discrimination, cultural essentialism, or Othering processes, but to individual emotions, which are cast as *per se* irrational, but sometimes even understandable. The latest requests of several politicians – from right to left – to take the fears Germans and other Europeans have vis-à-vis refugees and foreigners seriously, are expressions of this approach, which serves to legitimize concentration camps in Libya and other inhumane practices.

In his book *New Racism. Conservatives and the Ideology of the Tribe* (1981), philosopher Martin Barker very convincingly suggests that in Britain, xenophobia as a concept arose in order to mask mainstream Tory racism and to distinguish it from the more obvious racism of the National Front; the National Front, however, also tried to differentiate itself from Hitlerist racism – as Hitler was strongly hated by the British – and called their ideology racialism instead of racism. Barker traces xenophobia as a supposedly natural phenomenon back to the proto-racist philosopher David Hume, and shows how ethology and socio-biology, not being exact sciences, grounded a new form of racism, hidden behind the theory, according to which xenophobia was an anthropological principle. Barker argues that the Tories then adopted this ideology, which, unlike the racism of the National Front, was supposedly not based on prejudices and irrationality but on science. Sociobiology suggests that it is natural to be afraid of strangers; the wish to exclude them is rather natural, particularly in times of economic crisis. Hence, the logic goes, considerate politicians have to take into account the justified fears of concerned citizens.

The term '*Ausländerfeindlichkeit*' emerged in Germany in the late 1970s and was widely used since the 1980s, as the phenomenon *per se* supposedly was '*Fremdenfeindlichkeit*,' while '*Fremdenangst*' (xenophobia) appeared later. *Ausländerfeindlichkeit* allegedly emerged as a reaction to a worsening economic situation, and out of protest against the fact that a part of the guest workers had autonomously turned into immigrants and did not stop bringing more immigrants to Germany in the course of family reunions, thus increasing competition in the labor market. Additionally, the number of asylum seekers was continuously rising. Under the new circumstances of economic crisis and visible immigration processes and in light of the apparent shift of the political discourse shaped by the CDU and Helmut Kohl – who declared the *Ausländerproblem* to be one of the four most important political issues in his first term – openly voicing anti-immigrant sentiments became very common. Hate speech in the media and in manifestos, everyday discrimination, or insults against *Ausländer*, offensive graffiti stating "*Ausländer raus*" ("foreigners out") or "*Alle Türken in die*

GASKAM²ER"[40] ("All Turks into the gas chamber") emerging on the walls of German towns in the 1980s and 1990s, or the denunciation of refugees as fake asylum seekers (*Scheinasylanten*) and welfare scroungers (*Sozialschmarotzer*) (Bade 2015: 5-6) were primarily effects of an eroding economy, according to this narrative. Even the racist pogroms in the 1990s in Hoyerswerda, Rostock-Lichtenhagen, and the killings in Mölln and Solingen, which occurred in light of the unprecedented rise in the number of asylum seekers,[41] were mostly explained as a result of fears of social decline or in terms of a conflict over resources (cf. Frindte et al. 2016). Once again, racism did not appear in the dominant discourses: *Ausländerfeindlichkeit* and *Fremdenfeindlichkeit* were the cover terms which ensured that these phenomena were not seen as connected to the pre-FRG past.[42]

On the one hand, this etiology seems to explain these phenomena. On the other hand, it normalizes the fact that real or imagined economic crises unleash hatred against foreigners. As a consequence *Ausländerfeindlichkeit*, *Fremdenfeindlichkeit*, or *Fremdenangst* seem to be more or less socially accept-

---

40 There are some photographs of such graffiti in the records of the *Ausländerbeauftragte* of the city of Mannheim dating from 1980: StadtA MA – ISG, Beauftragter für ausländische Einwohner, Zug. 34/2003, Nr. 39.

41 Soon after the reunification of Germany that followed the fall of the 'Iron Curtain' and with the beginning of war in Jugoslavia – processes bringing thousands of refugees and asylum seakers to Germany – there was a dramatic increase in hate crimes throughout Germany. The most severe and well-known racist pogroms happened in Eastern Germany. In Hoyerwerda in September 1991 and Rostock-Lichtenhagen in August 1993, a mob of neo-Nazis attacked and besieged refugee shelters and dormitories for Vietnamese workers for days, with the applause of 'concerned citizens' standing by. In Mölln and Solingen, two Western German towns, eight immigrants from Turkey, including children, were killed after arson attacks in 1993.

42 These blatantly racist phenomena have been summarized and studied scientifically in Germany since the 1990s in the discipline of *Rechtsextremismusforschung*, a quite insular area of research, as it externalizes these phenomena to a specific group of extremists – particularly in the public perception of its findings. *Rechts-extremismusforschung* has created a vast amount of data since 1990 and some social scientists like Heitmeyer have for decades dominated the discussion with their theorizations, according to Frindte et al. 'Racism' as analytical tool, an approach suggested by Rommelspacher and Mecheril – exactly because this concept incorporates structural aspects – is mostly rejected as moralistic, reductionist, and essentialist, according to Möller (2016: 136).

able in postcolonial and post-Nazi Germany as long as they are motivated by economic fears (and as long as they do not surface in violent manifestations). This is also true for the symphathetic discourse – particularly in left political circles – about the "*Abgehängte*," the "ones left behind," who are supposedly marching alongside PEGIDA (*Patriotische Europäer gegen die Islamisierung des Abendlandes*/Patriotic Europeans against the Islamization of the Occident) or voting for the AfD (the far-right party *Alternative für Deutschland*/Alternative for Germany) out of fear of a deteriorating personal economic status.

Yet, the following quote reveals a diametrically opposed estimation. It is taken from a draft written by the head of department of the State Ministry of Baden-Württemberg in the context of the formulation of a position paper on *Ausländerpolitik* in 1981: given the emergence "of islands of foreign breeding and foreign cultures in the middle of our towns, which create the impression of *Überfremdung* in our native population," the *Ausländerpolitik* at that point "should not be viewed from the perspective of economic demand for workforce, the capacity of the housing market, and similar criteria, but from the perspective of the psychological resilience of the population living here."[43]

Überfremdung

As mentioned earlier, foreign workers had a long history in Germany, which commenced in the *Kaiserreich* with the establishment of a thoroughly organized system of work migration at the end of the 19th century. Furthermore, in 1944, every fourth worker in Germany was a foreign worker, many of them being quasi-personal slaves in households, shops, or in small farms, living visibly in the centers of the cities, working together with German workers in the same factories. The colonial past of Germany had also left material imprints and traces of racial knowledge, an issue which was silenced until recently.

The simultaneity of being accustomed to foreigners, fixing them as foreign, while fearing them as foreign seems to be connected with the history of German identity, in which the fear of *Überfremdung* has been an ever-returning feature since the end of the 19th century, particularly in relation to migrant workers. Migrants were needed for economic growth, but at the same time were perceived as a threat to German culture and identity, allegedly occupying a "lower cultural level," which resulted from their inferiority as a *Volk* or a *race*, respectively. The

43 Entwurf, 26.2.1981, Positionspapier des Innenministeriums zur Ausländerproblematik, Hauptstaatsarchiv Stuttgart – EA 2/303 Bü 287. *Überfremdung* again is a term that would lose much of its meaning in translation.

deeply racist concept of '*Überfremdung*' was prominent since then, and popped up time and again even in postwar Germany. It was used in the 1950s in reference to the so-called *heimatlose Ausländer*, a group comprised of former DPs[44] who had not had the chance to emigrate elsewhere. In the 1970s, the term recurred in local and national newspapers, in political statements, and in statements of 'concerned citizens' in relation to the guest workers. In 1993, reference to '*Überfremdung*' reached a peak in the so-called asylum debate and was even declared as *Unwort* (most inappropriate word of the year) by an association of German linguists. Since the so-called refugee crisis, '*Überfremdung*' has been used mostly by PEGIDA and AfD officials, but also by regular 'concerned citizens.'

The fear of *Überfremdung* was also a decisive factor that led to the creation of what in the 1970s was widely called ghettos. In Mannheim, these were the same districts that came to be known as *Parallelgesellschaften* (parallel societies) in the late 1990s, in which even today most residents are immigrants, particularly poorer ones, and newly arrived immigrants, i.e. the typical arrival city districts. Germans in the 1960s started leaving the cities for the suburbs. However, this trend was reinforced by the presence of *Ausländer*. In 1977, the *Ausländerbeauftrager* Lidy asserted that Germans left downtown Mannheim due to *Überfremdungsfurcht*.[45] Plenty of Germans did not want to rent apartments to *Ausländer*, as the "no *Ausländer*" signs and the countless stories of pioneer immigrants about their troubles to find decent and affordable housing in those years testify. These individual practices, ranging from open hatred to impersonal refusal, merged with analogous practices of the city administration itself. *Ausländer* were only offered those flats by the city-owned housing corporation in which Germans did not want to live anymore. Flats in the preferred urban districts, by contrast, were reserved for Germans according to a secret decision of higher officials of the administration.[46] It was the first Turkish tenant association in Germany, located in Mannheim, that publicly accused the city administration of discriminating especially against Turkish applicants for apartments – and they were right, as the records prove today.[47]

There is still much work to do in excavating the foundations of the binarity of 'German' und '*Ausländer*,' in analyzing the historical processes of the transfer, transformation, questioning of, and fighting against racial knowledge,

44 The majority of Displaced Persons was comprised of the former inhabitants of concentration camps, slave workers, and foreign workers.

45 Bericht, 11.10.1977, StadtA MA – ISG, Dezernatsregistratur, Zug. 34/2003, Nr. 154.

46 Niederschrift über die Sitzung des Koordinierungsausschusses, 25.9.1978, ibid.

47 Schreiben des Dezernats I an den Oberbürgermeister vom 8.5.1978, ibid.

and against *race* as a category that structures social realities. The critics of 'racism' and 'race' as analytical concepts – particularly in Germany – should acknowledge that racism is not limited to instances of killing people, openly installing an apartheid system, or aggressive reactions to non-Europeans. Racism is contextual, a continuum with many stages and gradual differentiations; 'race' is an evolving and adjusting historic concept (Geulen 2007). It is not acceptable to replace it with other terms, which not only render a phenomenon invisible, but also imply that fear of the foreign is something natural. This attitude contributes to keeping racism alive, and to flanking and stabilizing the neo-racism that thinkers like Stuart Hall and Etienne Balibár detected in Europe in the 1990s.

For as long as groups are Othered on the basis of any difference – be it skin color, name, national origin, religion, culture, sexual orientation, etc. – by those who represent the powerful part of the respective binary and portray themselves as the legitimate and primary holders of the privileges afforded by the democratic, pluralistic, and free global world, there will be racism.

Works Cited

Alexopoulou, Maria (2018): "Producing Ignorance: Racial Knowledge and Immigration in Germany." History of Knowledge Blog, July 25 (https://historyofknowledge.net/category/producing-ignorance/).

Alexopoulou, Maria (2016): "Vom Nationalen zum Lokalen und zurück? Zur Geschichtsschreibung in der Einwanderungsgesellschaft Deutschland." In: Archiv für Sozialgeschichte 56/1, pp. 463-484.

Bade, Klaus J./Bommes, Michael (2004): "Migration und politische Kultur im 'Nichteinwanderungsland'." In: Michael Bommes/Jochen Oltmer (eds.), Sozialhistorische Migrationsforschung, Göttingen: V&R unipress, pp. 437-472.

Bade, Klaus J. (2015): "Zur Karriere abschätziger Begriffe in der deutschen Asylpolitik." In: Aus Politik und Zeitgeschichte 25/1, pp. 3-8.

Barker, Martin (1981): New Racism: Conservatives and the Ideology of the Tribe, London: Junction Books.

Berg, Manfred/Schor, Paul/Soto, Isabel (2014): "The Weight of Words: Writing about Race in the United States and Europe." In: The American Historical Review 119/3, pp. 800-808.

"Berliner Erklärung zur Debatte um verstärkte Einwanderung nach Deutschland des Deutschen Industrie- und Handelskammertags und des Zentralverbands des Deutschen Handwerks." In: Deutscher Industrie- und Handelskammertag

March 26, 2015 (https://www.dihk.de/ressourcen/downloads/berliner-erklaerung-2015.pdf/).

Chin, Rita/Fehrenbach, Heide/Eley, Geoff/Grossmann, Atina (2009): After the Nazi Racial State: Difference and Democracy in Germany and Europe, Ann Arbor: University of Michigan Press.

Delgado, Jésus Manuel (1972): Die Gastarbeiter in der Presse: Eine inhaltsanalytische Studie, Opladen: Leske.

Delgado, Richard/Jean, Stefancic (2012): Critical Race Theory. An Introduction 2nd ed, New York: New York University Press.

Dornis, Christian (2001): Einbürgerung in Deutschland: Ihre Rolle bei der Integration von Zuwanderern und die Verwaltungspraxis im Regionalvergleich, Aachen: Shaker.

Frindte, Wolfgang/Geschke, Daniel/Haußecker, Nicole/Schmidtke, Franziska (2016): "Ein systematisierender Überblick über Entwicklungslinien der Rechtsextremismusforschung von 1990 bis 2013." In: Wolfgang Frindte/Daniel Geschke/Nicole Haußecker/Franziska Schmidtke (eds.), Rechtsextremismus und 'Nationalsozialistischer Untergrund'. Interdisziplinäre Debatten, Befunde und Bilanzen, Wiesbaden: Springer VS, pp. 25-96.

"Geborene und Gestorbene." In: Bundeszentrale für politische Bildung September 26, 2012 (http://www.bpb.de/nachschlagen/zahlen-und-fakten/soziale-situation-in-deutschland/61559/geborene-und-gestorbene/).

Geulen, Christian (2007): Geschichte des Rassismus, München: C.H. Beck.

Herbert, Ulrich (1995): Arbeit, Volkstum, Weltanschauung. Über Fremde und Deutsche im 20. Jahrhundert, Frankfurt: Fischer Taschenbuch Verlag.

Herbert, Ulrich (2003): Geschichte der Ausländerpolitik in Deutschland. Saisonarbeiter, Zwangsarbeiter, Gastarbeiter, Flüchtlinge, Bonn: BpB Bundeszentrale für politische Bildung.

Hess, Sabine/Henrik, Lebuhn (2014): "Politiken der Bürgerschaft. Zur Forschungsdebatte um Migration, Stadt und Citizenship." In: sub/ urban. zeitschrift für kritische stadtforschung 2/3, pp. 11–34.

Kalpaka, Annita/Räthzel, Nora (eds.) (1986): Die Schwierigkeit, nicht rassistisch zu sein, Berlin: Mundo Verlag.

Kreutz, Wilhelm/Hermann, Wiegand (2008): Kleine Geschichte der Stadt Mannheim, Leinfelden-Echterdingen: G. Braun/DRW-Verlag Weinbrenner.

"Lokalverbot nur für Nordafrikaner in Gießener Kneipe." In: Gießener Allgemeine July 10, 2014 (https://www.giessener-allgemeine.de/regional/stadtgiessen/Stadt-Giessen-Lokalverbot-nur-fuer-Nordafrikaner-in-Giessener-Kneipe;art71,93194/).

Melter, Claus/Mecheril, Paul (eds.) (2009): Rassismuskritik, Schwalbach am Taunus: Wochenschau Verlag.

Möller, Kurt (2016): "Rechtsextremismus und pauschalisierende Ablehnungen. Alte Probleme und neue Herausforderungen." In: Wolfgang Frindte/Daniel Geschke/Nicole Haußecker/Franziska Schmidtke (eds.), Rechtsextremismus und 'Nationalsozialistischer Untergrund' – Interdisziplinäre Debatten, Befunde und Bilanzen, Wiesbaden: Springer VS, pp. 131-147.

Mommsen, Hans (2007): "Forschungskontroversen zum Nationalsozialismus." In: Aus Politik und Zeitgeschichte 14-15/1, pp. 14-20.

Poutrus, Patrice G. (2016): "Zuflucht im Nachkriegsdeutschland. Politik und Praxis der Flüchtlingsaufnahme in Bundesrepublik und DDR von den späten 1940er Jahren bsi zur Grundgesetzänderung im vereinten Deutschland von 1993." In: Jochen Oltmer (ed.), Handbuch Staat und Migration in Deutschland seit dem 17. Jahrhundert, Berlin and Bosten: Walter de Gruyter, pp. 853-893.

Rommelspacher, Birgit (1995): Dominanzkultur. Texte zu Fremdheit und Macht, Berlin: Orlanda-Frauenverlag.

Sala, Roberto (2007): "Vom 'Fremdarbeiter' zum 'Gastarbeiter'. Die Anwerbung italienischer Arbeitskräfte für die deutsche Wirtschaft (1938–1973)." In: Vierteljahrshefte für Zeitgeschichte 1/1, pp. 93-120.

Schadt, Jörg/Michael Caroli, (eds.) (1993): Mannheim im Zweiten Weltkrieg, Mannheim: Edition Quadrat.

Schönwälder, Karen (2001): Einwanderung und ethnische Pluralität. Politische Entscheidungen und öffentliche Debatten in Großbritannien und der Bundesrepublik von den 1950er bis zu den 1970er Jahren, Essen: Klartext Verlag.

Sierakowski, Sławomir, (2015): "Benhabib: Nobody Wants to be a Refugee." In: Political Critique September 21 (http://politicalcritique.org/world/eu/2015/benhabib-nobody-wants-to-be-a-refugee/).

Terkessidis, Mark (2004): Die Banalität des Rassismus. Migranten zweiter Generation entwickeln eine neue Perspektive, Bielefeld: transcript.

Trevisiol, Oliver (2006): Die Einbürgerungspraxis im Deutschen Reich: 1871-1945, Göttingen: Vandenhoeck und Ruprecht.

Weber, Beverly M. (2016): "'We must talk about Cologne': Race, Gender, and Reconfigurations of Europe." In: German Politics and Society 34/4, pp. 68-86.

Wildt, Michael (2011): "'Volksgemeinschaft'. Eine Antwort auf Ian Kershaw." In: Zeithistorische Forschungen/Studies in Contemporary History 8/1 (http://www.zeithistorische-forschungen.de/16126041-Wildt-1-2011).

Wildt, Michael, (2014): "Volksgemeinschaft." Docupedia-Zeitgeschichte, June 3 (http://docupedia.de/zg/Volksgemeinschaft).

Wobbe, Theresa/Roland Otte (2000): "Politische Institutionen im gesellschaftlichen Wandel – Einbürgerung in Deutschland zwischen Erwartungen von Migranten und staatlicher Vorgabe." In: Zeitschrift für Soziologie 29/6, pp. 444-462.

Perspective Matters: Racism and Resistance in the Everyday Lives of Youths of Color in Germany

Lili Rebstock

Everyday Racism and Racial Microaggressions

Racism has changed its forms over the past decades. As a result, racism today very often occurs in subtle, hidden, or even unintended forms – which may sometimes be 'invisible' to both black and white people. Studying these alternate forms of racism, pedagogue and psychologist Chester Pierce developed the concept of microaggressions in the 1970s. Pierce stresses the subtle character and danger of these small acts of racism. According to him, "one must not look for the gross and obvious. The subtle, cumulative miniassault is the substance of today's racism" (Pierce 1974: 516). Pierce did extensive research on the devastating effects of racial microaggressions on the physical and psychological health of Afro-Americans. He states that the permanent and cumulating insults in the long term have even more damaging effects than the experience of overt forms of racism. Accordingly, the problem of microaggressions lies, on the one hand, in their cumulative effect and permanence from which People of Color are suffering. On the other hand, microaggressions often cause a deep sense of insecurity, because it is more complicated to react to a situation or a statement when the Person of Color cannot identify it as racist or when the act is possibly not intended to be racist.

There are many ostensibly innocent statements, questions, and acts that, at first sight, may seem harmless. However, when People of Color experience them permanently, they can perceive the communicated demeaning and excluding messages of these acts. One famous example is the question of "Where are you from?", which is then typically followed by the more investigative inquiry of "Where are you really from?" This deceptively harmless question about the origin of a person and not accepting the first answer which usually follows, is a very common form of racial microaggression and has often been discussed in German literature, for example by Mark Terkessidis (2004), Sheila Mysorekar (2007), and Noah Sow (2009). Also the alleged compliment "Your German is very good!", is not 'simply' a compliment, but rather also implies a rejection of belonging to a native German speaker of Color.[1]

Microaggressions are not only transmitted verbally, however. They also occur in non-verbal forms and through non-representation or mis-representation of People of Color in literature, history, and the media. Moreover, spatial and social structures, such as so-called 'white spaces,' or intersections of class and race promote and due to their frequency serve to 'normalize' the social exclusion and subordination of those affected. Derald W. Sue defines microaggressions as "the brief and commonplace daily verbal, behavioral and environmental indignities, whether intentional or unintentional, that communicate hostile, derogatory, or negative racial, gender, sexual-orientation, and religious slights and insults to the target person or group" (Sue 2010: 5). In this definition, Sue also stresses the different features that may be offended by microaggressions and which have to be understood as intersecting.

Another pivotal concept developed by Philomena Essed is the concept of *everyday racism*. Essed also emphasizes the cumulative effect of racism on People of Color: "Everyday racism does not exist in the singular but only as a complex – as interrelated instantiations of racism. Each instantiation of everyday racism has meaning only in relation to the whole complex of relations and practices" (Essed 1991: 52). While a single incident may have no noteworthy meaning, it turns into a racist injury when experienced again and again in everyday life. Miriam,[2] a young Afro-German woman and one of the interviewees in my ongoing PhD project,[3] answered my question addressing people's frequent inquiries about her origin as follows:

---

1 The valid question of how personal motives may be entangled with social and political conditions cannot be tackled at this point. Yet, the sense of entitlement which usually characterizes the tone of people asking this question, can clearly be seen as a result of this entanglement and deserves scholarly scrutiny, for it is typically considered commonplace and adequate by those who ask, but usually hurtful and inappropriate by many addressees of this question.

2 All names of youths are anonymized, i.e. the names mentioned are not the real names of the youths cited.

3 The following quotations are taken from my dissertation study. It has the working title *Alltagsrassismus – Erfahrungen, Wissen und Widerstand von Jugendlichen of Color in Deutschland* and will presumably appear in 2020.

> Sometimes I'm really stressed out, because I tell the story a hundred thousand times and I think how easy life could be if you were really white and you wouldn't have to be visible or wouldn't be visible within the mass of people. And not to have to answer any questions. It's really exhausting to tell something over and over again, for your whole life.

Miriam gets to the heart of the issues mentioned above. She verbalizes the exhaustion she feels about the need to repeatedly answer inquisitive questions, the exhaustion to always be visible in public spaces, knowing this will last forever. She is being questioned not just about her origin, but in an enclosing sense as a person. Being *Black* in Germany, she has to justify her presence verbally as well as physically. While describing her reality she depicts, in contrast, whiteness as an invisible position.

The Perspective of Youths of Color[4] and Experiential Knowledge

The topic of racism can be viewed from different angles and perspectives. In Germany, dominant research about racism mainly represents the perpetrator's – or *white* – views. This means that racism in German society would basically be measured by the attitudes and opinions of, potentially more or less, racist Germans – and not by the statements of the People of Color who could provide information about their perceptions and lived experience of racism (for a detailed discussion of this, see Terkessidis 2004, Essed 1991). Accordingly, very few studies focus on youths, their everyday experiences, and the effects of racism on their lives (cf. Scharathow 2014: 21-22).

As Alana Lentin, referring to Frantz Fanon's writing about the *lived-experience* of Black people, puts it, "Lived-experience is fundamental to the explanation of racism because without listening to those who face the ill-treatment that results from racism, we are in no position to theorize it. The failure to put lived-experience at the center of many explanations of racism is the cause of the inadequacy of many theories of racism" (Lentin 2008: 37).

An actual analysis of racism is therefore not possible without considering the experiences and the knowledge of People of Color. Being interested in the views of Youths of Color and the conditions of their socialization, I focus on their experience with and their knowledge about racism as well as the strategies they develop in order to combat racism in their everyday lives. My premise is that these youths have an experiential knowledge about racism and other forms of

4 By the term '*youths of color*' I refer to youths with different ethnic backgrounds.

discrimination, which can help to conduct profound analysis and support the struggle to counteract racism.

Positionality and Academic Considerations

While dominant knowledge production still hides behind the claims of objectivity and neutrality, other approaches demand for reflexivity and transparency about the scholar's subject position. Especially feminist thinkers questioned the dominant views on history, society, and theory production while developing feminist and standpoint theories (Smith 1987, Hill Collins 1986; 2000, Harding 2004). And although rarely mentioned in the course of conducting academic work, there is no secret about hierarchies and the exclusive character of the academy.

The question of "Who can speak?" brings this exactly into focus, and also my own considerations start from this point: since I never experienced racism myself as a white person, I talk about the experiences of others from a privileged position. This "speaking about" consists of different, intersecting features and power relations, such as age, gender, sexuality, disability, race, and class. Being entangled within these conditions in privileged and subordinate positions, I want to make use of my standpoint in order to contribute to research about social inequalities. Racial conditions affect me as well as my political, social, and familial surrounding and motivate my research focus on racism.

As a result of these considerations and the fact that their voice on this topic is rarely represented, perspectives of Youths of Color on the issue of racism in Germany are central to my discussion. However, this framing poses the potential danger of youths' being paternalized or their stories' being misinterpeted. Research on children and youths always contains this dilemma which is hard to solve, unless we let them radically, that is, without any patronizing filters, speak for themselves – which is also difficult in academic settings with its hierarchies and exclusive discourses. Thus, scholars need to carefully rethink the effects of their work, as Linda Alcoff states: "In order to evaluate attempts to speak for others in particular instances, we need to analyze the probable or actual effects of the words on the discursive and material context" (Alcoff 1991-92: 26).

Dealing with racism and other forms of discrimination bears the difficulty of addressing discrimination in order to analyze it – without reproducing it at the same time. Attributions may well be inscribed and reaffirmed by addressing them repeatedly. Hence, it is appropriate to carefully reflect established categories and to deconstruct them where necessary. Linked to this is the question what effects speaking about discrimination may have on whom. These considerations should have consequences on the way we speak in order not to reproduce insulting speech. For this reason, I will not use derogatory terms and language in order to describe discriminating experience.

To conclude, my perspective is only *one* possible perspective and my interpretations are *one* possible way to convey and to analyze the interviews and the topic of everyday racism in general.

Critical Race Theory

I use Critical Race Theory (CRT) as a theoretical framework of my work. CRT emerged in the United States in the late 1970s and has roots in both critical legal studies and radical feminism (Delgado/Stefancic 2012: 4). By then, mainly scholars of Color questioned neutrality and colorblindness in the legal system and criticized the reproduction of unequal social structures and institutionalized forms of racism in the U.S. Principal figures in the founding of CRT were Derrick Bell, Alan Freeman, and Richard Delgado, other main figures are e.g. Kimberlé Crenshaw and Angela Harris. CRT soon influenced other disciplines such as sociology, history, ethnic studies, gender studies, and education. One central assumption is that the actual, more subtle forms of racism need new approaches and analyses.

Accordingly, studying structural racism and so-called microaggressions became central for the scholars of CRT. Racism as a structural problem is embedded in social interests and norms, which are defined by a white majority in the U.S. Neutrality, objectivity, or meritocracy can, therefore, not abolish racism, but rather strengthen an ideology that sets whiteness as a norm and discriminates against non-white people through established structures. Especially in the field of education, for example, CRT provides a powerful tool to make mechanisms of discrimination visible and to understand them in their interdependency. For the field of education, Daniel Solórzano defined the following five tenets which also serve as a range of common principles on which CRT is based:

1. The centrality of race and racism and their intersectionality with other forms of subordination,
2. A challenge to dominant ideology,
3. The commitment to social justice,
4. The centrality of experiential knowledge, and
5. A transdisciplinary perspective (cf. Solórzano 1998: 122-123)

These tenets form a methodological and political frame and encourage activism to accompany academic work. Besides, another central intention of CRT is to empower communities of Color by naming racist injuries and offering "a liberatory and transformative method for examining racial/ethnic, gender, and class discrimination" (Solórzano et al. 2000: 63).

In the German context, Critical Race Theory so far has rarely been adopted and exists rather as a critical theory of racism, mainly in the field of education, cultural studies, and postcolonial studies (e.g. Melter/Mecheril 2009, Fereidooni/El 2017). There are some special and ambiguous situations regarding the term and concept of race. The German term for race is evaded for historical reasons because of its mere use as a biological category, which serves to generate and justify white supremacy.[5] However, the issue of race is being used in discussing *migration*. Non-white people and people of non-German origin are mostly labeled as people with a migratory *background*. Nevertheless, the label of migration in many ways substitutes and hides other labels such as race, religion, culture, and crime, instead of referring to the movement of people. This, among other things, becomes obvious as certain groups and populations – like Turkish, African, Arab, and Muslim people – are being labeled as migrants more often and in other ways than, e.g., people from European or North American countries. The discourse about migration thus often works as a racializing discourse. This fact and the connection between migration and race have become so deeply intertwined, however, that careful analysis is needed in order to be able to see them separately again and to understand the underlying dynamic between them.

Methodology

In order to examine the topic of everyday racism and microaggressions, I conducted a qualitative research using problem-centered interviews through the Grounded Theory approach (Glaser/Strauss 1967). The interview design is open in the sense that the interviewees can set their own topics and priorities, and it is dialogic in the sense that topics can be developed together and that I as an interviewer am also involved in sharing my views and feelings. By using Grounded Theory principles of theoretical sampling and the circular process of conducting and analyzing interviews, I can use first results from interviews and integrate them in the further research process. For the analyzing process, I combined Grounded Theory and the concept of Co-Construction as developed

5 For a more detailed exploration of this issue, see Maria Alexopoulou's contribution in this volume.

by Paul Mecheril (2003). The concept encourages a self-reflective approach and transparency of the scholar's own parts and thoughts in constructing theoretical analysis from empirical material and from spoken reports.

Altogether, I interviewed 16 youths, including eight boys and eight girls. The youths are between 15 and 22 years old and attend different types of schools. Due to the segmented school system in Germany, I considered all types of secondary schools, including special needs schools. The youths have different ethnic backgrounds, including Turkish, Afro-German, Kurdish, Syrian, Afghan, Roma, Algerian, or Italian. Among the religious backgrounds are Muslim, Christian, and Alevism. Most of the youths were born in Germany and some came to Germany as refugees in their childhood. The youths partially live in big cities or in small towns in East and Southwest Germany. I chose these two regions because of the specific histories of immigration in East and West Germany and the subsequent different social structures concerning diversity. Here, it is important to mention that only approximately 4 percent of immigrants live in the Eastern part of Germany, while almost 96 percent live in the West and constitute a solid part of the society there.[6] I thereby assume that experienced racism may be different depending on the spatial and social living conditions of the youths, in addition to their urban or rural surroundings. By the diverse sample, I try to conduct an intersectional analysis through examining different forms and symptoms of racism, such as anti-Muslim and anti-Black racism as well as racism against Romani people.

Everyday Racism: Experience, Knowledge, and Resistive Strategies

One form of microaggression that many of the interviewees report, are gazes. The youths talk about certain ways they have been looked at and how they try to interpret these looks. These gazes can be conspicuous and clear, but often they are hidden and subtle. The experience of being looked or stared at is mostly connected with strong feelings of uncertainty. Are these looks real or just imagined? What is the meaning of these looks and gazes? And finally, how to react to them, while feeling so unsure about them? Najim, a fifteen-year-old boy who lives in a small town in East Germany, puts it straight:

6 Almost 96 percent of all persons with a history of migration live in the former Western part of Germany including the city of Berlin, which is equivalent to 18,4 million people. In the former Eastern part of Germany, approximately 843.000 people have a history of migration (Statistisches Bundesamt 2017: 39-41).

> I feel strange when I walk through my town. In my town there are not many foreigners, but when you walk there and you are the only one with black hair and darker skin and you are of foreign origin, then you feel different. I mean, you are the only one.

To speak with Fanon, Najim feels "hypervisible" (1967: 109) in a surrounding which is characterized by a lack of diversity. In his observations black hair and darker skin seem to be the marker to define and fix him as someone of foreign origin. Although he was born and grew up in that town, he feels as if he does not belong there. It's not one gaze that makes him feel uncomfortable and strange but the *accumulation* of gazes and other everyday experiences, which reflect how people obviously see him. He elaborates on his experience as follows:

> You are looked at as if you were something below them, something like a subhuman. Not so extremely, but in a way that you perceive it. And generally, the way they talk to me, the tone of their voice and so on, it sounds different.

Najim's statement discloses the subtlety of racism and the racial micro-aggressions which afflicted him through looks and talks with other people. He learned to read the subtextual signs that, over time, instilled feelings of inferiority and contempt in him. Also Dilan, a twenty-year-old student from a bigger town in East Germany, reflects intensively upon looks he perceives:

> These are very small things. You feel that somehow you are looked at in a different way. And you feel that you maybe have your origin somewhere else. Such things. But nobody ever told me you don't belong here. Something strong like this never happened to me.

Dilan brings up his perception about being Othered and the subtle way it is communicated to him. In his experience, he feels being Othered through the looks of other people, and the gazes turn into a bodily sensation. As described above, racism for him does not occur overtly or outspoken, but hidden in looks and in the shape of "small things" which recur. Dilan, as most of the other interviewees, lacks a proper vocabulary for his perception, so the acts remain "small" and devalued in the light of dominant definitions of racism as overt or violent acts. Consequently, he expresses his insecurity, caused by the looks:

> And these are very small things, like now for example, when I walk around here with two friends who as well are not German. I just feel it, that we are looked at in a different way somehow. It may be that it's just imagination, but I, I always feel it, that we are frowned at and that they form a different view of us, although you don't even know the person.

Dilan describes the effects of microaggression as insecurity about his perception, followed by the insecurity about possible reactions and resistance. It takes a long time and effort, emotional and physical distress, as well as resilience to distinguish microaggressions and to develop pertinent responses and coping mechanisms.

Experience always contains an active examination of the situation and some form of reaction. There are short- and long-term strategies which help to resist racist insults and to maintain self-confidence. The strategies are mostly processed individually, since very often racism takes place as an experience that leaves youths in an isolated state. However, youths are also guided by stories and strategies of other persons affected by discrimination. In addition, racism is not static and single-sided, but changes with the resistance that confronts it.[7] Through resistance, youths can sometimes change and shape some parts of their social surroundings and experience self-empowerment. This is a steady process which involves creativity and strength as well as different resources. All youths I interviewed who were confronted with everyday racism develop strategies to protect and defend themselves. They test them, reject them when not useful, improve them – depending on what works in a certain context or situation.

In the following, I will outline some strategies of resistance reported by the youths. Asked how he copes with racist insults, fifteen-year-old Roma Jonuz recounts:

> Some hate me and some don't, some say "hey foreigner, what are you doing here." But I just don't listen. For me, how can I say, as if a piece of paper is talking to me.

Jonuz uses a creative strategy of imagination to protect himself from offensive comments. As someone who came to Germany as a refugee, defending himself against insults that aim to question his belonging possibly is a desperate effort. His strategy of ignoring and at the same time transforming his counterpart into something meaningless seems to be a proper strategy to resist the verbal displacement. Eighteen-year-old Farah tested several strategies to defend herself against racial microaggressions and reflects them in a review:

> When I was younger I never said anything, I swallowed things. But now I speak out because I want them to learn something, I want them to change their minds. I decided to handle it like this and I think this is right.

7 Manuela Bojadžijev exemplifies this in her analysis of migrant struggles in Germany (2008).

Staying silent and "swallowing things" mostly indicates helplessness and powerlessness rather than describing conscious strategies.[8] Oftentimes, youths are stuck in speechlessness when first confronted with racial insults until they develop functioning and stable strategies. According to this logic, Farah's actual strategy is to confront other people with their bias and racist behavior so they will stop insulting her. She aims to change their thinking, and in her reflections she explains her motivation also by a responsibility for other People of Color who may as well be harmed by them if they do not understand their behavior as wrong and the effects it has on others.[9]

Fifteen-year-old Najim first tried a strategy of assimilation because he wants to belong and be accepted within his school and among his peers. Unfortunately, this strategy failed and the experienced racism and exclusion even heightened:

> With Germans I always feel different, no matter what. Even when I give my best, when I try to assimilate as much as possible, I am still considered a foreigner. I can do whatever I want, I will always be a foreigner.

As a consequence of this experience of frustration, Najim withdraws from his old friends and turns to people who share similar experiences and thinking:

> I feel attracted to like-minded people, to those who experience the same, because they understand. I try to distance myself from my old friends, or those who I thought were friends, but they really insulted me.

Here, Najim names a common strategy which Solórzano et al. describe in their studies about students of Color in U.S. colleges as the creation of so-called *counter-spaces* (cf. Solórzano et al. 2000: 70-71). These social counter-spaces provide space to share common experiences and thoughts, and to have an opportunity of withdrawal from a racist/sexist or otherwise hostile environment.

To sum up, the resistive strategies of youths dealing with everyday racism are very diverse and creative. They change over time, adapt to changing forms of racism and ascriptions, and reflect the indignities the youths go through. The experience of everyday racism and microaggressions and their anticipation brings with it a comprehensive alertness/vigilance, as Philomena Essed puts it: "It is the precautions you must take in order to evade possible discrimination and

8 However, silence or assimilation may also constitute a basis for resistance, as described by Christine Riegel (2003: 70).

9 Farah reflects: "Shouldn't I have said something so it won't happen again, because they can still offend others."

the strategies you must develop in order to react adequately and be alert to whatever happens around you" (Essed 1990: 142).

The resistive strategies are always connected to a certain knowledge about racism, which the youths have gained over time. This knowledge helps to understand how racism works and with which strategies one can protect oneself against it. The youths' theories and analyses result from their lived experience and the testing of diverse strategies of resistance. Therefore, they possess a situated knowledge, conditional on their social environment, their school, their age, their sex and gender, their political and medial environment, and their family background, among other features.

Concerning this knowledge, one central result I found is that all of my interviewees see racism as a permanent and lasting phenomenon. They have no illusions concerning the abolition or declining of racism. Racism appears as a constant factor in their everyday lives. Most of them argue that people don't change their minds, which leads to a rather pessimist appraisal, as Laura states:

> I don't believe there will be a solution. There will always be something here in Germany, something they have against us, even if it's just a trifle. They will always find something. And to do something about racism is difficult here in Germany.

Related to this aspect, many youths conclude that current racism is embedded in German history. They relate Germany's colonial and national socialist past to current racist tendencies within society. Besides this, the media play a constitutive role in spreading racism, according to the youths: "I would say the biggest part, how racism occurs or where is this thinking that is in the heads of many Germans, is the media," states Najim. This awareness is in turn connected to the experience of everyday racism, as many interviewed youths feel affected and hurt by racist discourse promoted by media and right-wing parties in Germany.[10]

10 In the interviews, the youths name, among other things, e.g. the writings of Thilo Sarrazin with its anti-Muslim terminology, offensive election posters of right-wing parties, and racist content in social media.

Conclusion

Youths learn about racism mostly from their own experience – in a process which is painful and experienced mostly solitarily. I derive from my data that this happens from the age of approximately eleven to 13, the very years when youths, as it is common in Germany, enter secondary school. At this age, racist knowledge also arises with the students surrounding them. At this time, youths face a growing confrontation with racism together with racist, sexist, and physical stereotyping. The education system does not impart information about racism systematically, which could help to detect everyday racism and its multiple forms. Rather, the preoccupation with racism and related topics depends on the initiative of singular teachers or schools that share a social and political commitment.[11] Given that the majority of teachers continues to be predominantly white, knowledge about the lived experience of being subjected to racism is rare. Apart from that, however, theoretical knowledge about the different forms of racism, especially about its subtle forms, does not exist, either. Accordingly, the youths are rather unaware or unsure about naming their experiences as racism. In many cases, the young people in advance of the interviews stated that they have little or no experience with racism, while in the course of the interview they shared numerous stories of blatant racism or microaggressions, realizing and reflecting only later that they had experienced numerous racist incidents.

As mentioned earlier, microaggressions are very hard to detect and to prove and therefore hard to confront. More recognition of the experience of Youths of Color, but also support from other persons like peers, teachers, parents as well as educational, media, and political institutions is needed. Raising awareness of the subtle forms of racism and thereby supporting empowerment are basic steps to come closer to Chester Pierce's vision:

> It is my fondest hope that the day is not far remote when every black child will recognize and defend promptly and adequately against every offensive micro-aggression. In this way, the toll that is registered after accumulation of such insults should be markedly reduced (Pierce 1970: 280).

11 After all, education about racism mainly rests upon the work of numerous initiatives and trainers who, e.g., cooperate with schools and conduct anti-racism as well as empowerment trainings for children and youths.

Works Cited

Alcoff, Linda (1991-92): "The Problem of Speaking for Others." In: Cultural Critique 20/1, pp. 5-32.

Bojadžijev, Manuela (2008): Die windige Internationale. Rassismus und Kämpfe der Migration, Münster: Verlag Westfälisches Dampfboot.

Delgado, Richard/Stefancic, Jean (2012): Critical Race Theory: An Introduction, New York: New York University Press.

Essed, Philomena (1990): Everyday Racism: Reports from Women of two Cultures, Alameda: Hunter House.

Essed, Philomena (1991): Understanding Everyday Racism: An Interdisciplinary Theory, London: Sage.

Fanon, Frantz (1967): Black Skin, White Masks. Charles Lam Markmann (trans.), New York: Grove Press.

Fereidooni, Karim/El, Meral (eds.) (2017): Rassismuskritik und Widerstandsformen, Wiesbaden: Springer VS.

Glaser, Barney G./Strauss, Anselm (1967): Grounded Theory. The Discovery of Grounded Theory. Strategies for Qualitative Research, Chicago: Aldine.

Harding, Sandra (ed.) (2004): The Feminist Standpoint Theory Reader: Intellectual and Political Controversies, New York and London: Routledge.

Hill Collins, Patricia (1986): "Learning from the Outsider Within: The Sociological Significance of Black Feminist Thought." In: Social Problems 33/6, pp. 14-32.

Hill Collins, Patricia (2000): Black Feminist Thought: Knowledge, Consciousness, and the Politics of Empowerment, New York and London: Routledge.

Lentin, Alana (2008): Racism: A Beginner's Guide, Oxford: Oneworld Publications.

Mecheril, Paul (2003): Prekäre Verhältnisse: Über natio-ethno-kulturelle (Mehrfach-) Zugehörigkeit, Münster: Waxmann.

Melter, Claus/Mecheril, Paul (eds.) (2009): Rassismuskritik. Band 1: Rassismustheorie und -forschung, Schwalbach am Taunus: Wochenschau Verlag.

Mysorekar, Sheila (2007): "Guess my Genes: Von Mischlingen, MiMiMis und Multiracials." In: Kien Nghi Ha/Nicola Lauré al-Samarai/Sheila Mysorekar (eds.), re/visionen: Postkoloniale Perspektiven von People of Colour auf Rassismus, Kulturpolitik und Widerstand in Deutschland, Münster: Unrast Verlag, pp. 161-170.

Pierce, Chester (1970): "Offensive Mechanisms." In: Floyd B. Barbour (ed.), The Black Seventies, Boston: Porter Sargent, pp. 265-282.

Pierce, Chester (1974): "Psychiatric Problems of the Black Minority." In: Silvanio Arieti (ed.), American Handbook of Psychiatry, New York: Basic Books, pp. 512-523.

Riegel, Christine (2003): "Wie junge Migrantinnen mit ethnisiert-vergeschlechtlichen Zuschreibungen umgehen." In: Beiträge zur feministischen Theorie und Praxis 26/63-64, pp. 59-76.

Scharathow, Wiebke (2014): Risiken des Widerstandes. Jugendliche und ihre Rassismuserfahrungen, Bielefeld: transcript.

Smith, Dorothy E. (1987): The Everyday World as Problematic: A Feminist Sociology. Boston: Northeastern University Press.

Solórzano, Daniel G. (1998): "Critical Race Theory, Race and Gender Microaggressions, and the Experience of Chicana and Chicano Scholars." In: International Journal of Qualitative Studies in Education 11/1, pp. 121-136.

Solórzano, Daniel/Ceja, Miguel/Yosso, Tara (2000): "Critical Race Theory, Racial Microaggressions, and Campus Racial Climate: The Experiences of African American College Students." In: Journal of Negro Education 69/1-2, pp. 60-73.

Sow, Noah (2009): Deutschland Schwarz Weiss: Der alltägliche Rassismus, München: Goldmann.

Statistisches Bundesamt (2017): Bevölkerung und Erwerbstätigkeit. Bevölkerung mit Migrationshintergrund. Fachserie 1 Reihe 2.2.

Sue, Derald W. (2010): Microaggressions in Everyday Life: Race, Gender, and Sexual Orientation, Hoboken: Wiley.

Terkessidis, Mark (2004): Die Banalität des Rassismus: Migranten zweiter Generation entwickeln eine neue Perspektive, Bielefeld: transcript.

Beyond a Trifling Presence: Afro-Germans and Identity Boundaries in Germany

Anthony Obute

Introduction

"The perception that Germans are white, fair-haired, and blue-eyed has never been true, and will never be true. It is a myth" (Pareigis et al. 2017). This assertion introduces a 2017 documentary that seeks to expand prevalent identity boundaries in Germany. The documentary is titled *Afro Germany – Being Black and German*, and was aired on the famous German TV channel, *Deutsche Welle* (DW). The statement comes through the voice of an invisible character whose closest reach to manifestation is cordoned off to a Black woman's hair. The facelessness of the speaker is juxtaposed with the constant display of the DW logo of the broadcast giants, firmly placed at the left corner of the screen, running through the approximately 42-minute-long production. The invisibility of the character in contrast to the display of the DW logo connotatively underpins a questionable power construction around the speaker, questions their authority to speak, casts doubt on the reliability of what is said, and logically opens up the abovementioned statement to further interrogation. This very scenario directly situates itself within the central focus of this collection, the question of "Who can speak and who is heard/hurt?"

The ambivalence laced into the introductory scene narrows the power relation at play to the right of speech, and the validity of what is said, possibly dependent on who says what. Stressing the importance of speech in any human society, the French philosopher Jacques Rancière claims that "the supremely political destiny of man is attested by a sign: the possession of logos, that is of speech, which expresses, while the voice simply indicates" (1999: 2). Running the opening assertion reproduced above across the philosophical thoughts of Rancière, the voice could be understood to have merely indicated rather than expressed a reality. As Plato suggests, "The beginning is the most important part of the work,"[1] because it often directly engages the very idea of its existence, and foreshadows its future, which in the given context is revealed in succeeding parts of the documentary.

The mobilized contestation above therefore questions the very suggestion of "being black and German." Given the emotive tone of the voice-over, the speaker attempts to debunk what they term a myth, while connotatively re-evoking the very boundaries that enclose the German identity to "white, fair hair, and blue eyes." The highlighted identity boundary is further affirmed in the succeeding narration, "people always ask me where I am from, like it's not possible to be German and Black?" The question reveals the hitherto unknown speaker as Jana Pareigis, the first female Black news anchor in Germany, who also doubles as the narrator of the said documentary. The oscillating revelation of Jana Pareigis and the question "where do I come from," nips the very identity perceived as alien within German society. Sue et al. believe that questions along the boundaries of "where are you from, where were you born" convey indices of racial microaggression and exclusion in everyday life (2007: 276). Sue et al. conclude that regardless of the consciousness of those who instigate similar questions, they always communicate the message of exoticization and cultural dislocation within a given space.

The scholars' positions gain relevance in this study as these questions cast Jana Pareigis, and by extension, Germans of African descent, within a dislocated space, while highlighting an arguably normalized racial aggression towards minority groups in the society. The absence of "white, fair hair, and blue eyes" features invokes the very idea that questions Afro-Germans' belonging, as only perceived aliens are confronted with questions of situation or validating their identities. For the purpose of clarity, I limit my examination of minority groups in this study to Germans of African descent, whom I also refer to as Afro-Germans or Black Germans.

I therefore contend that in the normative consideration of German identity, Germans of African descent were made to maintain a 'trifling presence' leading to the exclusion of features beyond "white, fair hair, and blue eyes." By 'trifling presence,' I mean that the historical existence of Germans of African descent, predating the citizenship law of 1913, was thought to have constituted no serious value, was ignored at the margins, or suffused as an insignificant part into a larger whole. Against this background, many black Germans probably did not even consider it a possibility to be counted as full-fledged members of German

1 The 4th century BC Greek scholar Plato is largely credited with the saying "the beginning of the work is the most important of any work." Through the saying, Plato stresses how beginnings drive the general outcome of works.

society. The renowned 18th century German philosopher Immanuel Kant had earlier alleged that "Negros had no feelings beyond the trifling" (Yancy 2016: 5-6). Harnisch et al. argue that "being German and being Black just does not seem to go together in the minds of most Germans. In a country where citizenship is based on blood relations rather than birthplace, this seems to hold a certain immanent logic" (1998: 77). This 'immanent logic' proposed by Harnisch et al. is made manifest in the general restriction of thought that only bodies with the features "white, fair hair, and blue eyes" could qualify for German identity. Hence, the logic renders the presence of colored bodies as trifling, absent, or insignificant.

Evidence of this exclusionary imagination of the German identity found national reinforcement in the 1912 political debates leading to the 1913 citizenship law in Germany. It is important to note that at the time, it had become very pressing to strengthen the German identity along racial lines, because of the emerging new human population, partly parented by "white, fair-haired, and blue-eyed" Germans in colonial occupations. At the said debate, the German Secretary of Colonial Affairs amidst long political rhetoric concluded his presentation to the German Parliament thus: "The whole German nation will be grateful to you, if you have no consideration but this: We are Germans, we are white, and we want to stay white" (qtd. in El-Tayeb 2004: 85). On the above submission, I zoom in on the prefiguration of German identity on 'whiteness.' Whiteness herewith functions as a synecdoche, in proxy to the succeeding features of "white, fair hair and blue eyes" by default. Similar political rhetoric produced the 1913 citizenship law that sanctioned the German identity on *ius sanguinis*. Mark Roseman in his work "Racial Discourse, Nazi Violence, and the Limits of the Racial State Model," claims that "there is nothing wrong with attaching 'Racial State' as a descriptive label to the Nazi regime" (2017: 31). The Nazi regime functions, however in contest, as one of the many continuities of the racial-purity inclinations of the 1913 citizenship law.

The opening assertion to the documentary under review, delineating the prevalent boundaries of German identity as "white, fair hair, and blue eyes," directly confronts a remarkable history on the description of the German body. It is arguable that the spirit of the citizenship law, tightly constructed around body and color politics, sustained microaggressive tendencies as well as the exclusion of Black features from the German identity. Hence, the need remains for Germans of African descent to situate their belonging within the larger German society. The obvious absence of the desired German features along the fringes of body and color on Jana Paraigis and her kith and kin informs the microaggression. I further argue that attempts, however, at pushing back this rejection and writing Germans of African descent from the margins into the discourse, has birthed the "Afro-German" identity and corresponding movement. This becomes logical where "colour and culture become the badges of non-Germanism since Germans are supposed to be white, and [an] ideal German is pale with blue eyes and blonde hair, the essential characteristics of whiteness in the mythology of race" (Poikāne-Daumke 2006: 49). This understanding silences the features of colored bodies in the society and their historical presence, thereby presenting a fractured historiography in Germany. Geoff Eley concludes that:

> In weighing the importance of the center and the margins, my general view point is that we need to address both – we need to explore minority viewpoints, open up neglected spaces, and reclaim the marginalized subject and areas of the German past, while also relating them to the explicit and unconscious ways in which our understanding has been centered. Only by that means will the centeredness of German history be properly unsettled and remade (2006: 282).

This study is a response to this call and seeks to expand the 'centeredness' of German identity boundaries and history beyond the closed prefiguration of Germanness on "white, fair hair, and blue eyes." In doing this, I question the production of the Black identity through the white gaze, and the right of speech of the Black subject to define and name themselves. I approach this contribution from the perspective Critical Race Theory, running the latter across the documentary *Afro Germany – Being Black and German*. In the said documentary, Jana Pareigis takes a tour through the German society to unearth the experiences of Black Germans, their cultural dislocation, and daily mobilization of African ancestral roots and German socialization. This documentary provides multiple points of view and experiences of Black Germans across diverse spectra of life. In what follows, this research teases through the temporalities of production and exclusion of the Black identity in Germany, the otherness of the Black body, and the oscillating facets of racism in Germany. In my conclusion, I focus on the mobilization of different forms of speech by the Afro-German community to disrupt the 'center' of German history. I read "Afro Germany," the title of the documentary under review, as a fragmented but safe space, created and peopled by Germans of African descent and Black features.

Temporalities of Exclusion: 'Who Is Afraid of the Black Man?'

> I always had a lot of friends, but I do remember that in kindergarten, we played these funny games like 'Who Is Afraid of the Black Man?' and 'Ten Little Negroes.' I remember that sometimes they ran after me and said 'Jana Africana,' and the problem is that African meant something bad (Pareigis et al. 2017).

Through the synecdoche of Jana Pareigis, the Afro-German community summons a memory that, however, implies a tripartite dimension of history in motion in the German society. By 'history in motion,' I mean a continuity of the past, a past not necessary left behind, but rolled onto the future, in its transformed or re-imagined shapes. On sociopolitical continuities, Afro-German political scientist and activist Joshua Kwesi Aikins avers that:

> A lot of aspects of German colonial history are not widely known. We can't even begin to understand National Socialism without looking at the colonial antecedent, because we find that there are ideological, political, but also personal continuities, linking German colonialism and National Socialism (Pareigis et al. 2017).

Such mobility is, however, aimed at accomplishing the results of the past, in this contest, the occluded imagination of the German identity. The Indian American anthropologist Arjun Appadurai has framed his argument around futurity as a "Cultural fact, produced by ordinary people in the ordinariness of [their] locality" (Appadurai 2013: 16-17). The future, therefore, remains a void space and a recipient of the fast accelerating past, which in the given context imbricates the history of the 'trifling presence' and the exclusion of the Black body from the consideration of Germanness. In the succeeding part of this contribution, I peruse these tripartite workings of history, namely: the presence of the Black body in Germany, body and color politics, and the German gaze on the African space and people.

First, Pareigis embodies the continuity of Black existence in the German society, which dates back to well over four centuries, as she attests: "Black people have been living in Germany for 400 years, and today they number over one million" (Pareigis et al. 2017). Blackshire-Belay, in her work "The African Diaspora in Europe, African Germans Speak Out," contends that

> Julius Caesar brought black legions to Germany, and many never returned. The historical presence of Africans in the courts, universities, monasteries, and bedrooms of the 17th, 18th

and 19th century Europe comes as a surprise only to those scholars pseudo-educated in Europeanized bastions of institutional ethnocentricity (2001: 268).

Situating the common question "Where do you come from?", often posed to Jana Pareigis and Black Germans, in context of the above excerpt, it directly diffuses this historical presence. In other words, it renders the presence of the Afro-German community in these four centuries as a 'trifling presence.' Black presence in Germany is also inextricably linked to at least four historical developments. The German pre-colonial contact with the African continent and the slave trade represent one of the earliest popularized contacts of the Black community with Germany. In the documentary under review, Joshua Kwesi Aikins explains the sculptures at the entrance to the famous *Ermerlerhaus* in Berlin:

> So we had *Ermerlerhaus*, and this here, shows us the trail of tobacco. So you see enslaved African men harvesting tobacco, then enslaved African women packaging the tobacco, then it is soughted and weighed, there is [a] white overseer [...]. So it clearly explains the source of the wealth that was used to build this place, but of course, it also implicates the people who frequented here in the suffering that is also depicted here (Pareigis et al. 2017).

The colonial adventure of Germany in present-day Namibia, Togo, Tanzania, Kenya, and Cameroon among others marked the second significant epoch of Black presence in Germany. Asked why his father migrated from Cameroon to Germany back then, Theodor Wonja Michael, the author of *Black German: The Life of an Afro-German in the Twentieth Century* (2017), simply submits, "Cameroon was a German colony" (Pareigis et al. 2017). Heide Fehrenbach in her essay "Black Occupation Children and the Devolution of the Nazi Racial State" opines that after the defeat of Germany in 1945 by the United States' segregated armed forces, "interracial fraternization between African American GIs and white German women elicited a zealous rage and frequent incidents of verbal and physical abuse by white GIs" (2009: 33). The new wave of postcolonial migration of Africans into Germany will represent the fourth wave of Black presence in Germany. The historical presence of the first three categories of the Black community in Germany and their implied continuity, however, remain the central focus of this essay.

Second, Pareigis' encounter at the kindergarten underscores the prevalent politics of body and color, and the exoticization of the Black body in the society. Through the ordinary experiences of children, in the ordinariness of kindergartens and homes, the games 'Who Is Afraid of the Black Man' and 'Ten Little Negroes' open up the discourse on body and color politics in Germany. Jana Pareigis is projected in the given situation as the Other, embodying the figure of the strange to be feared and possible conquered. The instrumentalization of the game 'Who Is Afraid of the Black Man' at the kindergarten level foregrounds an elementary alteration of collective consciousness of the Black body as an embodiment of the strange, which ought to be feared. I further focus on the antics of body politics along the fringes of white and the Other, the subtle demonization of the black body, and its summary exclusion from Germanness.

Uli Linke in her work "Shame on the Skin: Post-Holocaust Memory of the German Aesthetics of Whiteness" submits that "the symbolics of whiteness and the aestheticization of white skin have played a decisive role in the historical context of German body politics" (2004: 207). Linke argues further that the imagination of white as the enclosed understanding of beauty and the acceptable norm in the German society, extensively influenced the imagination of white features as the possibly only accurate representation of Germanness. Hence, features outside the occluded boundaries of "white, fair hair, and blue eyes" remain the Other within the German society. Linke also enlists the workings of the white body as a location of power, a designation of a political space, a crucial signifier of ethnic difference, and comprises a regime of inequality that is sustained by engagement with normative cultural practices (ibid: 207).

The manifestation of the white body as a location of power and a political space finds demonstration within the purview of Pareigis' experience with the German educational system, implicated in the kindergarten. Here, the white body assumes a somewhat assertive position of power, authority, and political space to impose itself on the Other. In the presence of the attributes of power and authority, it becomes very natural to question all other presences, perceived to be in contest, in the negotiation of shared authority and power. Then it becomes very logical to question the Black presence, or to ask: 'Who Is Afraid of the Black Man?' On the backdrop of the self-imposed power and authority by the white body, a counterimage of the Black body is produced in relation to the imagination of the white body as a place of authority, and as the signifier of racial inequality. The production of the Black image, therefore, proceeds from what George Yancy describes as "White Embodied Gazing" (2016). Yancy posits that "on this score, the white gaze involves the correlative constitution of a racialized field that normalizes the marking of Black bodies through a relationship of white power" (2016: 244).

In the first instance, through the instrument of white power, the white body dissolves into "sociohistorical temporality," enclosing itself within the understanding of white as "daylight, truth, virtue, unraced, and universal." However, it brackets off other aspects of sociohistorical realities within which the white body emerges as the fullness of obsessive racial purity, brutalization, dehumanization, genocide, ethnic cleansing, and the imposed violence upon Black bodies and bodies of color (ibid: 245). On the imposition of psychological violence on the Black body, Theodor Wonja Michael attests to his experiences of "human zoo," "imagine human beings being exhibited like objects, literally exhibited for what they supposedly represented, namely Africa" (Pareigis et al. 2017).

On the flip side of the manifestation of white power, the Black body emerges from the construction site of the white gaze, as the figure of the strange and the direct opposite of all that the white body represents. Hence, the Black body stands in proxy to

> night, doom, darkness, and danger; it is deceptive and devious; it is a site of vice and moral depravity. Hence, the meaning of whiteness, as universal, contains within itself an obfuscated parasitism that reduces the Black body to a wretched particularity (Yancy 2016: 245).

The deducible image of the strange imposed on the Black body could arguably serve as the raison d'être for the game/question, 'Who Is Afraid of the Black Man'? A question that could also be posed on a defeated body or violated body, as an exercise of power and authority. The imagined wretchedness of the Black body through the white gaze, therefore, validates the position of the famous German philosopher Immanuel Kant, who had earlier suggested that "Negroes have no feelings beyond the trifling" (Yancy 2016: 5-6). In clearer terms, the Negro and his feelings constituted nothing of significance, but an absent-presence that is called to be by similar power and authority closely linked with the white body. Tiffany Nicole Florvil concludes that "Germans have formed and adapted their ideas about citizenship and national identity based on their interaction with Afro-diasporic individuals, oscillating from fascination and exoticization to fear and hatred" (2013: 8). Florvil's observation leads this research to another layer of continuity embodied by Jana Pareigis, the imagination of the African world.

Third, the Afro-German demography through the German gaze maintains a continuity of the pre- and postcolonial knowledge of the African continent. This is evident in the documentary under review, as the children sometimes ran after Jana Pareigis, yelling "Jana Africana, and the problem is that Africa meant something bad" (Pareigis et al. 2017). The perceived ill with Africa, its space, place, and people was, however, the imagination and production of Europe. Fanon in his book *The Wretched of the Earth* elucidates:

> At times, this Manicheism goes to its logical conclusion and dehumanizes the native, or to speak plainly, it turns him into an animal. In fact, the terms the settler uses when he mentions the native are zoological terms. He speaks of the yellow man's reptilian motions, of the stink of the native quarter, of breeding swarms, of foulness, of spawn, of gesticulations. When the settler seeks to describe the native fully in exact terms he constantly refers to the bestiary (1963:42).

Fanon captures the workings of the European colonial ideology and the Othering of the African geography and people at home, which was successfully conceived, birthed, and bred at the famous Berlin Conference of 1884. Here, I zoom in on the imagination of African life through the lens of the zoo for three reasons. First, such imagination provides a profitable platform for the reduction of the Black body to the borders of the animal. It thereafter facilitates the common hierarchization between human and animal, and the succeeding binaries of white and black, Europe and Africa, and the body and color politics discussed above. In the presence of the first two points mentioned above, Othering of the Black body follows a natural process. Within this colonial frame of knowledge, the ethnographic exhibition of Blacks in "human zoos" may not need further justification. I have quoted Fanon above to re-invoke the residues of the colonial knowledge, within which the Black body is the Other, hence, its features cannot qualify for the German identity.

The colonial imagination of Africa is, however, advanced beyond the African geography, such that the Black body in diaspora, in this case in Germany, continually loses the right of speech to define or name itself. It is never understood for what it is, except for its assigned status of the 'Other.' West-Pavlov affirms that for Hegel, the renowned German philosopher, the precolonial African society "forms no historical part of the world, it exhibits no movement or development [...] [thus the] invasion is merely the obverse side of assimilation: both bring time to the primitive world, or bring the primitive world into time" (2009: 164). Through the workings and thoughts of Hegel, Africa, the Black body, and its features lack the power to define themselves; instead, a definition is imposed on the body, often delivered through the white gaze.

Mbembe further provides a postcolonial understanding of the African world through the white gaze as the same "very figure of 'the strange.' It is similar to that inaccessible 'Other with a capital O' evoked by Jacques Lacan [...] a headless figure [...]" (2001: 3). Mbembe succinctly recreates the image of a trifling creature, an absent-present one, with no head or significance, hence comparable to the introductory voice-over to the documentary under review. Mbembe further explains;

> First, the African human experience constantly appears in the discourse of our times as an experience that can only be understood through a negative interpretation. Africa is never seen as possessing things and attributes properly part of 'human nature.' Or, when it is, its things and attributes are generally of lesser value, little importance, and poor quality. It is this elementariness and primitiveness that makes Africa the world par excellence of all that is incomplete, mutilated, and unfinished, its history reduced to a series of setbacks of nature in its quest for humankind (ibid:1).

Beyond Fanon's characterization of the colonial imagination, Mbembe's readings of the African world cast the region and its attributes as not properly fitting to human nature, while limiting its understanding to an often negative interpretation. Jana Pareigis, therefore, attests that "the problem is that Africa meant something bad" (Pareigis et al. 2017). The imposed evil on the Black body finds manifestation in the game, "Who Is Afraid of the Black Man." Through the chosen diction of this game, language functions as a continuity channel for framing of the Black body. Toni Morrison argues that language "can powerfully evoke and enforce hidden signs of racial superiority, cultural hegemony, and dismissive 'othering' of people" (1992: x). Morrison simply maintains that language could be a violent tool, structured and used by a majoritarian sect to oppress and demonstrate racist behaviors towards a people. However, the instrumentality of language could also be employed in a direct confrontational phase to demonstrate hate. It could also be used in a less confrontational but silently brutal manner to derogate a people.

In whatever form language is mobilized, it conclusively achieves similar purpose of derogation, expression of ethnocentrism, and the further vilification of the targeted victim. It is on the strength of the secondary operation of language that such German vocabularies as S*chwarzarbeit, Schwarze Liste*, and S*chwarzes Schaf*[2] continue to subtly achieve the negative interpretation of Black attributes. The racist interpretation of these lexicons have remained very vague. Maria Alexopoulou observes and explains that discourse on racism in the German society has been subjected to silencing, as a dismissive technology,

2 These are casual German terms for 'illegal job, blacklist, and black sheep.' These words appear harmless in daily usage, but subtly engage the surrounding questions on the image of the Black subject.

since the collapse of the Nazi regime and the foundation of the Federal Republic of Germany. Alexopoulou claims that because of its "insoluble association with Nazi ideology, the very idea of race appears too contaminated for a semantic resurrection" (Alexopoulou's contribution in this volume: 45). Therefore, the exclusionary experiences of the Afro-German community become a *déjà-vu* of the colonial treatment of the African continent.

I have thus far critiqued the tripartite workings of history in relation to Black presence in Germany. This is, however, predominantly framed through a white gaze, within which the black body emerges from the construction site of white power. The tripartite workings of history discussed above are arguably made possible by an absence of power, authority, and speech of the Black subject to define themselves, claim their space, and impose their identity. Geoff Eley contends that "social historians drew back to the intensive study of the bounded case, in which a particular community, category of workers, or event stood in for the 'whole society' argument" (2006: 184). The trifling presence of the Afro-German emerged from this social situation, within which the German identity was occluded to "white, fair hair, and blue eyes." The intended national cohesion evident in these restrictive national boundaries strikes an uncanny resemblance with Nicole Hirschfelder's understanding of oppression. Hirschfelder suggests that "oppression is a telling term [...] it seems that the subtler the features of oppression are, the more powerful they prove" (2014: 1). The seemingly harmless historical understanding of the German identity as "white, fairhaired, and blue-eyed" became more powerful through its outright exclusion of Black features, or better still, through dislocating Black bodies and bodies of color in Germany. In the face of the given circumstance, it becomes pertinent that the excluded write themselves into history.

Beyond the Margins: Writing Afro-Germans into the Discourse

In 1986, a group of Germans of African descent, under the aegis of *Afrodeutsche Frauen* (ADEFRA), chose to define themselves. Jasim Eding, a co-founder of the group, writes:

> We chose to define ourselves, name ourselves. We called ourselves Black Germans, Afro-German or Afropeans, Blacks in Europe (instead of Europeans) or simply Black. We felt it time to challenge German society that being German doesn't always mean being white and that we also had a long Black history in Germany/Europe. This statement lays the groundwork for the importance of the Black community to define itself rather than be defined by anyone else (Mills 2017).

Regarding the above excerpt, I focus on the evident workings and manifestation of power and authority in the right of speech. The Afro-German community through the proxy of ADEFRA recognizes the insidious operation of silencing and of the institutional understanding of the Black presence in Germany as a trifling one. Hence, the need to write the Afro-German community into the discourse of German identity also becomes pressing. Again, as speech was employed as a demonstration of authority to cordon off the German identity to "white, fair hair, and blue eyes," the Black German community initiates the deconstruction process by naming themselves, as Afro-German. ADEFRA, therefore, answers the all-important question posed by this collection, "Who can speak, who is heard/hurt?" The group acts as a self-activating agent, able to imbue itself with power and authority to make pronouncements, to define itself rather than wait to be defined by others. Through this singular act, the Black German community set a mechanism in motion to reclaim its place in the German history. In the concluding section of this study, I return to the documentary under review to explore the continuity of this self-definition process by Afro-Germans.

Renaming Places and Re-Writing History: Towards an Afro-German Space

"Afro Germany – Being Black and German"

Source: Pareigis et al. 2017

In the documentary under review, Jana Pareigis narrates, "I am here at the 'Street Festival,' aimed at forcing the city to change the name of the street from *Mohrenstraße*." Here, a perlocutionary speech act, stemming from the definition of the Afro-German demography, is activated. The British philosopher of language John Langshaw Austin believes that "saying something will often, or even normally, produce certain consequential effects upon the feeling, thought, or actions of other people: and it may be done with the design, intention or purpose of producing them" (1975: 101). In the "Street Festival," the Afro-German community comes to terms with a common identity, and further utilizes demonstrations to revert the tide of historical continuity. The potency of demonstration cannot be underemphasized, because it naturally summons both conscious and subconscious listeners. In the image presented above, the Afro-German community both literally and figuratively raises its history and people, evident in Anton W. Amo, from the muddy waters of colonial understandings and the white gaze, represented in the name of the street *Mohrenstraße*. Joshua Kwesi Aikins explains the need to reclaim the spaces in the documentary *Afro Germany – Being Black and German*:

The term '*Mohr*' is one of the oldest German words for Black person, but if you look at the root of the term, there [is] [...] a Latin and a Greek root[], *morus* and *marus*, and that means *dark* or *black*. But it also means *stupid*, *heathen*, and *primitive*, and so, we see already in the origin of the word that there is this idea of Black inferiority. But then if we look into the history of the street name, we see that the street was named, or acquired its name in the context of the Brandenburg involvement in the transatlantic enslavement enterprise (Pareigis et al. 2017).

In *Mohrenstaße*, the previously discussed historical continuity of the Afro-German community, through the (pre-/post-)colonial white gaze, hangs revealed. It reinforces the single line of narrative earlier promoted by renowned German scholars like Hegel and Kant, thereby brushing aside narratives of Afro-German scholarship left on the margin. Comprehensive historical research by Marilyn Sephocle on Anton Wilhelm Amo maintains:

Born in Akonu in Ghana, Amo was presented as a gift to the Dukes August Wilhelm and Ludwig Rudolf von Wolfenbuttel by the Dutch West India Company in the year 1707. According to church records: "On the 29th of July 1707, a young Mohr was baptized in the chapel of the Salzthal castle and named Anton Wilhelm Amo" (1992: 182).

In the above submission, and with particular focus on the church's records, the imagination of the Black body through the lens of a 'Mohr' in proxy to Joshua Kwesi Aikins's argument regarding a linguistic connection to stupidity, primitivity, and heathenness of the Black body gains further recognition. However, Anton Wilhelm Amo went on to study in Germany and became a successful scholar of Philosophy, Mathematics, and Medicine at the Universities of Halle, Wittenberg, and Jena (Brentjes 1975). However, the historical succession of Anton Wilhelm Amo remains relatively silent. The "Street Festival," particularly with social pressures on renaming *Mohrenstraße* to *Anton-W-Amo-Straße*, seeks to engrave Afro-German scholarship and also retell the course of postcolonial history of the Black world.

Beyond sociopolitical demonstrations, such as the "Street Festival," aimed at altering very negative and demeaning historical continuity from within, the Afro-German community continues to explore other measures for inclusiveness. Some of these measures include the Black presence in the spaces of entertainment, namely the music industry, journalism, and the football world. Sammy Deluxe, a renowned Afro-German Hip-Hop artist and music producer, observes the acute shortage of black superheroes in Germany and most Western societies. Sammy Deluxe submits in the documentary under review:

> At the time I was reading Harry Potter to my son in the evening at bed time, and one night he said he wished he were white, because then he could be like his friends, then he would paint that thing on his forehead and wear round glasses, then he will [...] look like Harry Potter. That made me realize that there is an acute shortage of dark-skinned super heroes (Pareigis et al. 2017).

The encounter with Sammy Deluxe's son, and his innocently misconstrued understanding of his Black skin, within a white-dominated space, serves as a signifier to the earlier alteration of the Black psyche. The child's understanding of his black Body, enforced by the social construction that Germanness borders 'White, fair hair, and blue eyes,' paints the image of the Black skin as a questionable feature in the German society. In response to this deficit of Black superheroes, Sammy Deluxe produced the hit-track *Superhero* and the video to discontinue the narrative and also to promote Black heroism. Jana Pareigis further strengthens the workings of the media space in promoting the Black voice and presence in Germany with her experience as a journalist. Pareigis suggests,

> When I decided to become a journalist [...] I didn't have any role models. For me, it's really important that Black children, you know, when they switch on the TV, they might see me and say, Oh Great! I can be on the TV and read the news, and I don't have to have a job which fulfills the racist stereotypes. For me that would be great, if I could help bring down barriers, in that sense (Pareigis et al. 2017).

Conclusion

In this article, I have discussed the identity of the Black body in Germany as one that finds itself under constant scrutiny, emanating from the historical occlusion of German identity. I have also examined the production of Black identity through the lens of a dominant white gaze that relies on self-generated pre-colonial knowledge of the African space. The given circumstances have birthed the Afro-German identity and movement, which positively distorts the course of the exclusionary historiography in Germany by writing itself into the center of German discourse. For the constraints of space posed on this research, I did not delve deeper into the internal mobilization of this name and identity by the German demography who continually navigate the double spaces of Black body and German socialization. This could prove a useful line of empirical research for future studies on race relations in Germany.

Works Cited

Anderson, Benedict (1996): Imagined Communities: Reflections on the Origin and Spread of Nationalism, London: Verso.

Appadurai, Arjun (2013): The Future as Cultural Fact: Essays on the Global Condition, London: Verso.

Austin, John Langshaw (1975): How to Do Things with Words. 2nd edition, J. O. Urmson/Marina Sbisà (eds.), Cambridge, MA: Harvard University Press.

Blackshire-Belay, Aisha (2001): "The African Diaspora in Europe." In: Journal of Black Studies 31/3, pp. 264-287.

Brentjes, Burchard (1975): "Anton Wilhelm Amo, First African Philosopher in European Universities." In: Current Anthropology 16/3, pp. 443-444.

Chin, Rita/Fehrenbach, Heide/Eley, Geoff/Grossman, Atina (2009): After the Nazi Racial State: Difference and Democracy in Germany and Europe, Michigan: University of Michigan Press.

Eley, Geoff (2005): A Crooked Line: From Cultural History to the History of Society, Michigan: University of Michigan Press.

Eley, Geoff (2006): "How and Where is the German History Centered?" In: Neil Gregor/Nils Roemer/Mark Rosemann (eds.), German History from the Margins, Bloomington: Indiana University Press, pp. 265-282.

El-Tayeb, Fatima (2004): "'We are Germans, We are Whites, and We Want to Stay White' African Germans and Citizenship in the early 20th Century." In: Birgit Tautz (ed.), *Colors 1800/1900/2000: Signs of Ethnic Differences*, Amsterdam and New York: Rodopi, pp. 185-205.

Fanon, Frantz (1968 [1963]): The Wretched of the Earth. Jean-Paul Sartre (eds.), Constance Farrington (trans.), New York: Grove Weidenfeld.

Fehrenbach, Heide (2009): "Black Occupation Children and the Devolution of the Nazi Racial State." In: Rita Chin/Heide Fehrenbach/Geoff Eley/Atina Grossmann, After the Nazi Racial State: Difference and Democracy in Germany and Europe, Michigan: University of Michigan Press, pp. 30-54.

Florvil, Tiffany Nicole (2013): *Writing Across Differences: Afro-Germans, Gender, and Diaspora, 1970s-1990s*, PhD dissertation, University of South Carolina.

Harnisch, Antje/Stokes, Anne-Marie/Weidauer, Friedemann J. (eds.) (1998): Fringe voices: An Anthology of Minority Writing in the Federal Republic of Germany, Oxford and New York: Berg.

Hirschfelder, Nicole (2014): Oppression As Process: The Case of Bayard Rustin, Heidelberg: Universitätsverlag Winter.

Linke, Uli (2004): "Shame on the Skin: Post-Holocaust Memory and the German Aesthetics of Whiteness." In: Birgit Tautz (ed.), *Colors 1800/1900/2000: Signs of Ethnic Differences*, Amsterdam and New York: Rodopi, pp. 207-236.

Lusane, Clarence (2003): Hitler's Black Victims: The Historical Experiences of Afro-Germans, European Blacks, Africans, and African Americans in the Nazi Era, New York: Routledge.

Mazón, Patricia/Steingröver, Reinhild (2010): Not so Plain As Black and White: Afro-German Culture and History, 1890-2000, Rochester and New York: University of Rochester Press.

Mbembe, Achille (2001): On the Postcolony, Johannesburg: University of Wits Press.

Mills, Nikki (2017): "Writing Ourselves into the Discourse: The Legacies of Audre Lorde and May Ayim." In: *FemGeniuses* June 17 (http://femgeniuses.com/2017/06/18/writing-ourselves-into-the-discourse-the-legacies-of-audre-lorde-and-may-ayim/).

Michael, Theodor Wonja (2017[1988]): Black German: An Afro-German Life in the Twentieth Century. Eve Rosenhaft (trans.), Liverpool: Liverpool University Press.

Morrison, Toni (1992): Playing in the Dark: Whiteness and the Literary Imagination, Cambridge and Massachusetts: Harvard University Press.

Pareigis, Jana/Lenz-Gleissner, Susanne/Ulrich, Adama (2017). "Afro Germany – Being Black and German." Youtube.com, a DW documentary, March 29 (https://www.youtube.com/watch?v=pcfPVj5qR1E).

Poikāne-Daumke, Aija (2006): African Diasporas: Afro-German Literature in the Context of the African American Experience, Münster: LIT Verlag.

Rancière, Jacques (1999 [1995]): Dis-agreement: Politics and Philosophy. Julie Rose (trans.), Minneapolis: University of Minnesota Press.

Roseman, Mark (2017): "Racial Discourse, Nazi Violence, and the Limits of the Racial State Model." In: Devin, O. Pendas/Mark, Roseman/Richard F. Wetzell (eds.), Beyond the Racial State: Rethinking Nazi Germany, Cambridge: Cambridge University Press, pp. 31-57.

Sephocle, Marilyn (1992): "Anton Wilhelm Amo." In: *Journal of Black Studies 23/2*, pp. 182-187.

Sue, Derald Wing/ Capodilupo, Christina M./Torino, Gina C./Bucceri, Jennifer M./Holder, Aisha M.B./Nadal, Kevin L./Esquilin, Marta (2007): Racial Microagression in Everyday Life. In: Implication for Clinical Practice 62/4, pp. 271-286.

West-Pavlov, Russell (2013): Temporalities, London: Routledge.

Yancy, George (2016): "White Embodied Gazing, the Black Body as Disgust, and the Aesthetics of Un-Suturing." In: Sherri Irvin (ed.), Body Aesthetics, Oxford: Oxford University Press, pp. 243-260.

Race and Racism in Translation: "Who Can Speak?" in German Renderings of Literary African American English

Annika Rosbach

Harriet Beecher Stowe's *Uncle Tom's Cabin; Or, Life among the Lowly* was a huge success not only in North America but, amongst other countries, also in Germany (cf. Paul 2005: 127-186). This can be largely attributed to German-language translations that started to appear in 1852, the same year Stowe's novel was first published as a monograph in English. Since then, the German National Library has recorded over 150 translations of *Uncle Tom's Cabin*, most of them abridged children's versions. The first German-language translation directed towards an adult readership was preceded by a note of the translator, who introduces his work by admitting that he had taken some liberty with the English text, adapting it to the taste of the German readership:

The English original [...] is 329 densely printed octavo pages long; and a faithful and complete translation of it would hardly please the German reader; if only because of the many lengthy conversations in provincial dialect and Negro English [...] which to the German reader must be tiring since they introduce him to ideas which can be of no interest to him. [...] It will suffice to give an overall impression [...] a shortened German adaptation in which, however, nothing essential was left out [...] (Ungewitter 1852: 6-7; my translation).[1]

1 The German original text reads as follows: "Das englische Original [...] ist 329 enggedruckte Octavseiten stark; und eine getreue und vollständige Uebersetzung desselben würde dem deutschen Leser schwerlich sehr behagen; schon wegen der vielen und langen, im Provinzialdialekt und Neger-Englisch geführten Gespräche, die [...] für den deutschen Leser ermüdend sind, indem sie ihn in einen Ideenkreis führen, der

This statement is racist not only in terms of its choice of words but with regard to its suggestion that what African American characters had to say in Stowe's novel was not relevant enough to be conveyed in German. Accordingly, in the translation, passages of dialogue written in African American English were silenced either by leaving out direct speech of Black characters altogether, by paraphrasing it in indirect speech, or by rendering it in 'standard' German (*Hochdeutsch*).[2] However, the very dialect and African American English that the cited translator regarded as "nothing essential" (*nichts Wesentliches*) marks one of the most important stylistic devices in Stowe's novel. The degree of non-standard, i.e., African American English[3] or a Southern White dialect, spoken by the novel's White and Black *personae*, for the most part can be read as mirroring the degree of moral rectitude or depravity Stowe wanted them to be associated with, irrespective of racial background.

While openly racist attitudes as expressed by the first *Uncle Tom* translator in 1852 may have altogether vanished from recent introductions to German translations of the novel, 21st century translators of Stowe's novel still rely on the same discriminating strategy, i.e., standardization. The complete erasure of Black voice – or rather, what Stowe as a White author considered to be Black voice – might be dismissed as irrelevant in the case of *Uncle Tom's Cabin* by pointing out that the novel abounds in easily detectable racism, e.g., the depiction of Uncle Tom as intellectually inferior and childlike, which, of course, is not lost in German translation. However, it is the subtlety of racially discriminatory translation strategies beyond single racist words, I wish to argue, that is potentially more dangerous in terms of the conservation of racism in Germany because it goes unnoticed by readers, who, most likely, do not have an English-language copy at hand to read simultaneously and compare with its translation. Moreover, the *whitening* of Black voices in translation – with

ihn durchaus nicht interessieren kann. [...] Genug, wenn ihm der Totaleindruck bleibt, [...] eine[ ] abgekürzte[ ] deutsche[ ] Bearbeitung, in der jedoch nichts Wesentliches ausgelassen ist [...]" (Ungewitter 1852: 6-7).

2 In this article, 'standard' implies that a language is generally regarded as the standard either because it is institutionally defined and codified as such, as is the case with *Hochdeutsch*, or because, like General American, it is the language which is used and accepted by the majority of speakers in formal settings, including literature, and therefore perceived to be the standard.

3 In this article, African American English or Black English is not understood, as it often is in media and academic discourses, as a *lect* of General American but as a language in its own right, i.e., possessing genuine grammatical, syntactic, lexical, and semantic structures, and regional variations (cf. Green 2002).

standardization being only one translation strategy that has been applied, as this paper will show – is particularly damaging to the meaning and reception of novels which are essentially about these voices. Two examples, Zora Neale Hurston's *Their Eyes Were Watching God* and Alice Walker's *The Color Purple*, will be analyzed in this paper. Beforehand, I want to briefly outline the general and specific difficulties in translating literary African American English. This paper ends by pointing out an option for a considerate way of translating African American English into German that seeks to avoid racist connotations, thereby arguing for the importance of the translation of novels that employ Black English as an element to create meaning.

The Difficulty of Translating African American English into German

Rather than constituting the production of a mirror-image of a text in another language, the process of translation serves as what Homi Bhabha calls a "third space" in which linguistic, cultural, spatial, as well as historical specificities, and differences are negotiated. The result of this negotiation is a rewriting, a new, and hybrid text in which the foreign world of the 'original' text is represented, but with a tint of the cultural space the translator is working in and/or the translation is meant for. How far this textual transformation is allowed to go is usually constrained by the translator's ambition to create a text that, in emotional and aesthetic terms, at least potentially will be understood by the target readership as is the original text by the source readership. However much a translator may commit to this functional approach to literary translation, his or her negotiation is never perfectly objective, as translation scholar André Lefevere has pointed out: "All rewritings [i.e. translations], whatever their intention, reflect a certain ideology and a poetics and as such manipulate literature to function in a given society in a given way" (1992: vii). Far from saying that, as the Italian adage *traduttore traditore* (translator, traitor) suggests, literary translators are manipulators by intention, how much of the target culture enters a translation is naturally not only determined by the translator him- or herself but also by actors both within the literary field, i.e., critics, scholarship, institutions, other literature and translations, and beyond this field, i.e., individuals, groups, and institutions with ideological power. According to Lefevere, ideological considerations are the most determinant factors in the translation process: "On every level of the translation process, it can be shown that, if linguistic considerations enter into conflict with considerations of an ideological and/or poetological nature, the latter tend to win out" (ibid: 39). They are, at the same time, the most subconscious elements as they may go unnoticed by both the translator and the readers of the translation.

Literary African American English poses a particularly complex problem to literary translators. The African American language itself is the product of a negotiation of linguistic and cultural difference, and thus functions as a "race memory" (Gates 1987: 183) in which the shared language is an expression of the specific historic experience of its speakers. That is why, Gates argues, it is impossible to translate African American English (ibid: 195). Texts like *The Color Purple* and *Their Eyes Were Watching God* are not universal stories but, as I will elaborate upon later in this article, are based on very specific experiences. They nevertheless are of interest to non-Anglophone Black readers elsewhere, and can, as I will specify in my analyses, serve to draw the attention of White readers, and thus, of an audience these texts were not originally written for, to the problem of racism. Therefore, not to translate American literature that uses Black English is not an option. Translating literary African American English into German, then, means to re-negotiate so as to convey via language use what the original text implies in terms of relations between Blacks and Whites, experiences of racist discrimination, racist ascriptions, self-identification, etc. However, the German language, although rich in dialects and sociolects, lacks a variety equivalent to African American English, i.e., a language connoting its community's historical experience of slavery and racist discrimination. Translators are therefore forced to work with the available German language repertoire, and the translation strategies they choose, as the case studies in this article will show, do not do justice to the original and, even worse, inscribe racist ideology into the texts.

The Racist Potential in Language Use

Language, which is an act that never describes reality in neutral, objective terms, is one of the most subtle, systematic, and thus one of the most powerful vehicles to spread and perpetuate racist ideology. Racist speech acts, which differentiate between an "us" and an "Other" by attributing difference to, among others, biological, cultural, mental, and linguistic features, can take many forms, ranging from overt racist messages communicated directly or indirectly to historically racist words used thoughtlessly and to the refusal to talk about it altogether (Ciçek et al. 2014: 311-312; 321). Such speech acts need not be intentional to cause violence to and feelings of anger, humiliation, or fear in those who are their subject. Translations, as the product of the negotiation process outlined above, are speech acts, as well, and as such can transport structural racism, i.e., normalized and thus legitimized ideology and social practices in a White supremacist society that routinely privilege Whites over People of Color. As my analysis will show, structurally racist linguistic strategies in translation – and, for that matter, anti-Black racism, which this article focuses on – usually go unquestioned by White translators and unrecognized by White readers, thus adding to subtly powerful and dangerous array of racist practices in Germany.

Awareness for present-day subtle linguistic practices of anti-Black racism in Germany is very much constrained by the reductionist tendency to equate racist speech with (neo-)Nazi ideology, thereby negating the omnipresence of racism, for instance, in terms like *Mohr* (*Moor*) or *SchwarzafrikanerIn* (*Black African*), which continue to be used in Germany (cf. Nduka-Agwu/Hornscheidt 2010 and Arndt/Hornscheidt 2004). German publishers' efforts to render children's books politically correct have been met with heavy criticism of parents who argued they would much rather explain to their children the racist connotations of words than have them replaced by less offensive vocabulary. While it may be easy to talk about specific words, the detection of racist translation strategies asks for a close comparison of the translated text with its source. That, however, would require an effort that is unrealistic: for common readers, because it disrupts the joy in reading; for literary translators, because it seems to be too time-consuming, which is why new translations, more often than not, rely on older renderings (Koskinen/Paloposki 2010); for scholars of American Studies, because it seems to be uninteresting, even though the "transnational turn" has generated some interest in the "cultural mobility" of texts (Fisher Fishkin 2012) and in how "an attention to translation can produce crucial insights into just how transnationalism works" (Fagan 2012: 245).[4] This article, then, seeks to contribute to a heightened awareness for racist language use in translation among common readers but especially among literary translators and American Studies scholars by closely comparing Zora Neale Hurston's *Their Eyes Were Watching God* and Alice Walker's *The Color Purple*, respectively, and their German translations with regard to their use of African American English and the strategies applied to render Black voice in German.

4 Studies on the reception outside the U.S. of the oeuvre of American writers in general (e.g. Ishihara 2005; Kutzinski 2006; Fisher Fishkin 2011) and African American writers in particular (e.g. Sollors 1999; Lemke 2009; Paul 2013) attest to this interest.

A "speakerly text": Zora Neale Hurston, *Their Eyes Were Watching God*

The African American language is one of the central structuring elements in Zora Neale Hurston's 1937 novel *Their Eyes Were Watching God,* and as such has been widely discussed.[5] The novel, in a twofold sense, is what Gates has termed a "speakerly text" (Gates 1988: 188): spoken language in the novel is more important than activity, and the text seems to speak directly to a female Black American readership.

The story of female Black protagonist Janie centrally addresses her finding a self-confident, powerful voice as an African American woman, and it is told in a complex mix of narrative voices. First, there is a wealth of dialogue between the characters, who are almost exclusively Black; the passages rendered in dialogue make up about 50 percent of the novel. The language spoken by all Black characters is Black English, featuring more or less the same phonological and grammatical features (Minnick 2004: 124-125). Direct speech, thus, does not serve to differentiate characters according to their age, gender, educational, economic, or geographical background. Rather, the African American language functions as an essential aspect of Black identity in the novel. Secondly, the novel features a heterodiegetic narrator whose language, at least in the beginning, is clearly set apart from that of the characters by his philosophical General American language. Lastly, there are passages of free indirect discourse in which the narrator's standard language merges with Janie's African American English. The constant switching between the two languages reiterates the genuine African American form of *call and response*, which Eva Boesenberg has identified as an important element in the direct-speech passages of the novel's Black community, as well (1999: 82-104). Moreover, that Janie's language is taking over the narrator's reinforces the importance of the protagonist's mother tongue to the telling of her story. Her innermost thoughts, feelings, and perceptions are revealed in Black English, while the objectively visible is left to the standard voice. Also, the more self-confident Janie grows and the more she learns to speak up against her husbands and the Black community, who are equally unsympathetic to her self-determined conduct, the less important the narrator's voice becomes. It is altogether superfluous at the end of the novel: when Janie, looking back on her marriages, compares love to the sea – "Love is lak de sea. It's uh movin' thing, but still and all, it takes its shape from de shore and it's different with every shore" (Hurston 1990: 182) –

5 Cf. Gates 1988; Minnick 2004; Boesenberg 1999; Lemke 2009; Holloway 1987.

she adopts the philosophical, metaphoric manner of expression of the narrator, but – and this is the vital difference – continues to speak African American English. Hurston thereby not only powerfully illustrates the poetic qualities of the African American language, but also implies that it is equal to the standard in terms of eloquence and literariness.

Janie's reflecting on her experiences is not presented in a monologue but in a conversation with her friend Phoeby. Thus, the act of Janie's telling her story is complete only through the presence of an interauditor, an "ideal listener" who listens without evaluating what is being told and, more importantly, who is black, female, and shares the protagonist's experiences: "'Course, talkin' don't amount tuh uh hill uh beans when yuh can't do nothin' else. And listenin' tuh dat kind uh talk is jus' lak openin' yo' mouth and lettin' de moon shine down yo' throat. It's uh known fact, Phoeby, you got tuh *go* there tuh *know* there" (Hurston 1990: 183; original emphasis). Since, as Janie points out, listening without sharing the experience of the speaker is useless, the novel clearly addresses neither White readers nor male Black readers. That the transmission of African American (female) experience rests with a woman is important. In fact, the language of the novel's male characters, rhetorically skillful as it is, rather takes on the form of playing the dozens, wordplay, and flirting or, in the case of Janie's husbands, is an instrument of domination over women and thus lacks the emotional depth of Janie's language. In the end, Janie has not only further developed her personal voice, but has unfolded her identity as an African American woman oppressed by both racism and patriarchy (cf. Kimberlé Crenshaw's concept of "intersectionality"). Therefore, *Their Eyes Were Watching God* does not relate a universal story, both in the sense that it is specifically directed at a Black female audience and told in a language uniquely fit to express this specific experience. A universal story, however, is what has been made of *Their Eyes Were Watching God* in German translation.

Eatonville in the Rhineland? *Their Eyes Were Watching God* in German translation

In a first translation, entitled *Und ihre Augen schauten Gott* (1993), translator Barbara Henninges chose the dialect of the Rhine-Ruhr region to give a German voice to Hurston's African American characters. When, for instance, Janie recounts the moment she became aware of her black skin color in a photo, Henninges has her use the following typical features of the Rhine-Ruhr dialect: "Ich *hab dat* Bild lange *angekuckt un hab gesehn, dat* war mein Kleid *un* meine Frisur, *un* ich *sach*: 'Oje, oje! Ich bin ja gefärbt!'" (Henninges 1993: 19; my emphasis).[6] In the first entry of a glossary attached to the translated text, Henninges justifies her choice of dialect by pointing to the similarities she had found to exist between the Black English of the novel's setting, the town of Eatonville, Florida, and the Rhine-Ruhr dialect, especially with regard to grammar and phonology (ibid: 277). Moreover, Henninges asserts, Black American folklore from the U.S. South sounds familiar to Germans due to its origin in both European and African traditions (ibid: 277). As if trying to counter doubts about her first two arguments in favor of the Rhine-Ruhr dialect being a German equivalent to African American English, Henninges adds that Hurston's novel tells a universal story featuring the key problems and feelings of all of mankind (ibid: 277), thereby providing justification for transplanting the story to whatever place on earth, in this case a western German region.

Linguistically, the similarities of the Rhine-Ruhr region's dialect and African American English go as far as the phonologically and semantically identical use of *dat* for the standard relative pronoun *that* and German *das*, respectively. On the level of reader perception, those non-native to the Rhine-Ruhr region, for whom reading Henninges' translation is slowed down significantly by the constant need to decipher the unfamiliar dialect, will certainly feel denied a proper understanding of the novel, just as non-African Americans might regarding the original. This sense of not belonging is intensified by the very prestigious status of both the Rhine-Ruhr dialect and African American English within their respective speaker group. Studies by Arend Mihm (1985) and Anne Kathrin Becker (2003) on the in-group prestige of the Rhine-Ruhr dialect have found that the dialect is used and highly valued across all social strata and that its speakers strongly associate the dialect with tradition and describe speakers of the local dialect as sociable, warm-hearted, and easygoing. This, of course, bears resemblance to Hurston's understanding of her mother tongue: "In the mouth of the Negro the English language loses its stiffness" (Hurston 2007: 474). However, the high in-group standing of the Rhine-Ruhr dialect is based on its past value as a common, unifying language, symbolizing the work ethos of employees in the regional mining industry, which for a long time was the region's main economic driving force. Henninges' translation and adjunct glossary clearly establish a parallel between the physical and economic hardships of miners in Germany with the racist oppression and exploitation of Black people under slavery, thereby making another claim for universality by

6 Other features typical of the Rhein-Ruhr dialect include the replacement of plosives with fricatives ("[g]-Spirantisierung"), e.g. *ruich* (11) in place of *ruhig*, and the cutting of ending syllables, e.g. [m] instead of *-ben* in "Wo's dat ganze Geld *gebliem* […]" (8; my emphasis).

relativizing the latter to a fate that was supposedly shared by people all over the world. This highly problematic reductionist approach is intensified by the dialect's out-group connotations. Mihm and Becker's studies reveal that among German non-speakers of the Rhine-Ruhr dialect, it is often stigmatized as a flawed, careless way of talking that suggests a low level of education and a low social standing of the speaker (Mihm 1985; Becker 2003). To make it even worse, Henninges adds regionally unmarked misspellings and mispronunciations of standard German words – e.g. "Lebenskomferens" (Henninges 1993: 15) in place of "Lebenskonferenz" (life-conference) or "Kom-Plemente" (ibid: 155) instead of "Komplimente" (compliments) – which must be read as faulty language and thus, strongly imply poor education. These associations, being projected onto a story about Black characters, inevitably conjure up the stereotype of the simple-minded, buffoonish Black that literature authored by Whites abounds with. Even though such a reading may depend on whether or not a German reader is a native speaker of the Rhine-Ruhr dialect, the mere potential of African Americans' being perceived as linguistically inferior reverses Hurston's achievement in demonstrating the discursive value of Black English as a powerful literary language in its own right.

The Voice of the Oppressed: Alice Walker, *The Color Purple*

Walker's 1982 fictional account of a young Black woman, abused emotionally, physically, and sexually since her earliest childhood, who by writing down her story finds strength and courage to raise her voice against her oppressors and to have her voice heard, challenges both racist norms in American society and the patriarchal practices prevalent in parts of the African American community. Written for the most part in African American English, Walker's epistolary novel has been praised for its authentic, non-stereotypical representation of Black speech (cf. Harris 1984), but has been amply criticized for, among other aspects, narrator-protagonist Celie's initial apathetic passivity, its stereotypical representation of Blacks (cf. Harris 1984; Kaplan 1996; Dieke 1999), as well as for its allegedly utopian ending (cf. Kaplan 1996; Hite 1989), when "everybody speaks the same language" (Kaplan 1996: 137). Against this criticism, which detaches language from content, I argue that there is hardly another piece of literature in which African American English is used as a stylistic and functional device which forms a similarly close symbiotic relationship with the content it conveys. This symbiosis is constituted, first, by the importance of Black English for the formation of Celie's "narrative identity" (Ricœur 1992) through writing, thereby counteracting both Celie's alleged passivity and purported Black stereotypes. Secondly, it is constituted in the dialogic engagement of the reader with the text initiated by the alternating use of African American English and General American in the letters Celie and her sister Nettie write to each other.

Through her use of characteristic morphological, syntactical, semantic, and a few phonological features of African American English, Celie indubitably reveals herself as a Black person. Black English functions as a vehicle for her self-definition, self-revelation, and self-development. By verbalizing everyday and traumatic experiences in her letters, she is able to make sense of and come to terms with them. On a linguistic level, this process is most effective because it is rendered in African American English. For instance, when Celie demands that her husband, Mr ------ hand over to her Nettie's letters, he refuses, reasoning that she has no right to make claims against him, because "[y]ou black, you pore, you ugly, you a woman. […] [Y]ou nothing at all" (Walker 2014: 187). In dropping the conjugated forms of the copula *be*, which is a feature characteristic of African American English (cf. Green 2002) and results in a non-distinction between tenses, he implies that Celie's non-being is a state persisting in the past, present, and future. Celie's response repeats her husband's defamatory words but includes present tense forms: "I'm pore, I'm black, I may be ugly and can't cook […]. But I'm here" (Walker 2014: 187). Hereby, Celie weakens his argument as she limits its validity to her present situation only, powerfully claiming her status as a human being irrespective of race, gender, and ability, and at the same time freeing herself from her husband's authority over her. Celie's rhetorically brilliant outburst, of course, does not, on a sociocultural level, put an end to her marginalization as a Black woman, but her explicitly situating this marginalization in the present makes future change ontologically possible. Thus, Walker reminds Black female readers to recognize the value of their identity as African Americans and as women in a racist and patriarchal society, and has Celie advocate for the importance of continuing the fight for equality. Also, since General American does not provide for the resources to establish such multiple meanings through verb constructions – nor does German, which complicates translation – Walker succeeds in demonstrating the richness and literariness of Black English. At the same time, Celie's writing down of what she does not understand, for instance, when she writes "Maybe you [God] can give me a sign letting me know what is happening to me" (Walker 2014: 3), despite having internalized stereotypes that make her believe that she lacks the competency to understand, demonstrates her unconscious will to actively come to grips with what is happening. Thus, the very act of writing is not a passive one at all.

The use of African American English for the greater part of a novel, i.e., not only in dialogue but also in the narrative voice, certainly breaks with White-dominated literary convention. Indeed, for non-speakers of African American English to read and fully understand the language of *The Color Purple* is challenging at first. However, by focusing on a few features easily recognizable for speakers of General American and applying them consistently throughout the novel, Walker provides for easy accustoming to Celie's narrative style. Dialogic engagement of the reader is then realized in two steps: first, through estrangement from conventional literary language through the letters written by Celie and Nettie and, second, through the appropriation of Black English as standard language and language of proximity.

Estrangement occurs when, in the second half of the novel, Celie discovers her sister's letters and cites the first five of them consecutively and uncommented. Interestingly, with growing spatial and temporal distance from home and her past, Nettie's diction changes. The more letters she writes, the fewer features of Black English she uses and the more complex her letters get both in terms of content as well as grammar and syntax. Having just become accustomed to African American English as the standard language for reading *The Color Purple*, readers have to re-familiarize themselves with General American and thus pre-experience the same communicational obstacles Celie does at the end of the novel when she finds that her sister and children "speak a little funny" (Walker 2014: 262), an observation she immediately relativizes by asserting "but us gitting use to it" (ibid: 262). Unlike Celie, for the remainder of the novel the reader is not allowed to get used to the language of *The Color Purple*. The double linguistic estrangement leaves the reader questioning conventional notions of (literary) standard language and language hierarchies. Having distanced her readers from what is regarded as conventional language, Walker, in a second step, draws them into her novel through a language of proximity constituted by both the epistolary form and the appropriation of the traditional orality of African American English, again in the sense of a "speakerly writing." Both letter writing and oral telling establish a sense of privacy and authenticity between communicator and recipient. However, as Felipe Smith rightly points out, the strategy of distance and proximity is more than a means for achieving reader engagement, since in *The Color Purple* the reader functions "as a past, present, future potentiality whose presence and participation fulfill the text" (1999: 111). For readers, plot coherence is created only through their own efforts to piece together information from the sisters' letters, which remain altogether unanswered throughout the novel. Eva Boesenberg's correct conclusion that the sisters meet each other in the reader

(1999: 280) can be extended to the linguistic level, where the meeting takes the form of the symbiotic confluence of two languages, i.e., African American English and General American, in the reader, with the effects described above. This bilingualism in *The Color Purple* opens up new perspectives of thinking and understanding in its readers. This can be read as the linguistic appropriation of what R. Erin Huskey has described as Walker's "gospel ideology" (2009: 98-99): The reader reacts to the text because she is confronted with two distinct languages, then becomes part of the text through its use of a language of proximity, and experiences a personal transformation by reflecting upon standard and non-standard language. Finally, the reader is capable to testify about this experience to others in order to effect social change with regard to race relations.

Language lends itself as a means to achieve this because language in *The Color Purple* crucially influences how Blacks and Whites see each other: "[…] Darlene trying to teach me how to talk. She say US not so hot. A dead country give-away. You say US where most folks say WE, she say, and peoples think you dumb. […] [W]hite folks be amuse" (Walker 2014: 194). In the perception of Whites, who claim sovereignty over defining what is to be considered correct and legitimate language, African American English is associated with stupidity, primitiveness, and laughable buffoonery. The stereotypical equation of Black English with a lack of education and simplemindedness seems to be present in *The Color Purple*, too – after all, Celie, writing in African American English, uses short, oftentimes elliptic sentences, commenting mostly on the domestic sphere, as opposed to the information-dense, complex main/subordinate clause constructions Nettie uses. However, this association is weakened by the fact that although Celie's voice is growing gradually stronger and she is gaining knowledge about herself and the world, her language does not change to General American like Nettie's does. Ultimately, the association is dissolved completely by the liveliness and poetic qualities of Celie's writing at the end of the novel.

Celie dismisses Darlene's attempt at linguistic education by pointing to the importance of the use of one's own language for self-awareness and self-expression: "Every time I say something and the way I say it, she corrects me until I say it some other way. Pretty soon I feel like I can't think. My mind run up on a thought, git confuse, run back and sort of lay down" (Walker 2014: 194). At this point in the novel, the reader can easily sympathize with this sort of confusion caused by shifts in language use. Thus, by the end of the novel, readers – regardless of their cultural background – will likely have reached the same conclusions as Celie has after having discovered her sister's hidden letters, namely "that there *are* hidden, unofficial, even silent stories to be recuperated from behind the original – and false – narratives that obscure them" and that these can be fought back "with her own self-narrations" (Kaplan 1996: 127; original emphasis). By deconstructing stereotypes about Blacks and their language as false narratives and counteracting them by making readable and thus heard a voice silenced by literary convention and by exposing White linguistic and racial norms as defective constructs, Walker aims at breaking down racist marginalization of Blacks since, as Molly Hite has pointedly put it, "a speaking margin cannot be a margin at all [...] and tends to destabilize precisely the hierarchical oppositions that give margin and center clearly demarcated meanings" (1989: 122). Walker "signifies" (cf. Gates 1988) on Hurston in that for Celie, like for Janie, the language of her ancestors is the most effective means to express herself. Walker goes beyond Hurston in that she extends the ideal readership, which Hurston limited to Black women, to potentially every reader, irrespective of racial background and gender. This is also suggested by the ending, which in my estimation does not amount to a utopian leveling of difference. Celie's assertion "us gitting use to it" (Walker 2014: 262) negates the necessity to speak "the same language" (Kaplan 1996: 137), and instead points to the legitimacy of linguistic diversity and the importance of developing strategies to bridge linguistic difference.

The Voice of the Oppressor: *The Color Purple* in German Translation

In the German translation of *The Color Purple*, entitled *Die Farbe Lila* (1984), Walker's carefully constructed dialogicity of languages is reduced to a stereotypical representation of Celie. In the German translation, authored by one of the most renowned German translators for English-language literature, Helga Pfetsch, Celie basically speaks German *Umgangsspache*,[7] i.e. colloquial standard German. Colloquial German, which is widely accepted since it generally does not classify its speakers in social terms, is usually used in more informal, intimate settings (Bichel 1973: 379) and is spoken in hardly distinguishable regional varieties. To create intimacy for the reader is the primary aim of Pfetsch's strategy to have Celie tell her story in transregional and thus easily accessible colloquial German. Her letters can be followed effortlessly thanks to rather easily decipherable phonological rewritings that take the form of speech accelerating markers, e.g., the omission of vowels as in *ner* instead of *einer*

7 For a discussion about the vagueness of the term *Umgangssprache*, see Bichel 1973: 401-403.

(Pfetsch 2005: 23) or the merging of two words into one single word as in *hats* (15) or *könnse* (21).[8] Pfetsch also introduces terms and phrases typical of southern, e.g., *scheints* (ibid: 239),[9] and northern German dialects, e.g., *kodderig* (18),[10] as well as sayings from the Swiss German variety, e.g., "der Schuss geht immer hinten *raus*" (42; my emphasis).[11] These regional nuances, instead of irritating the reader, contribute to linguistic intimacy and identification for readers from all over Germany and the neighboring German-speaking countries.

The function of Celie's language as an important aspect of her identity as an African American woman is entirely lost in German translation. The general neutrality of German colloquial language establishes universality instead of, like Celie's African American English, membership in and identification with a specific group of speakers. The artificiality of Pfetsch's adapting the colloquial to be a *potpourri* of regional colloquial varieties, which no German native speaker would draw on simultaneously in everyday life, undermines the original's propagation of the value of Celie's mother tongue as a literary language. Instead, the choice of colloquial grammar features, especially the non-standard use of the German cases, renders Celie's language childlike, simple, and primitive, and thus characterizes her as poorly educated. Although her lack of formal education is undeniable, descriptive characterization, as I have analyzed, is not the function of non-standard speech in *The Color Purple*.

While this accessibility certainly establishes the basis for readers' emotional involvement in the story, the translation, unlike the original, fails to actively engage the reader. Native speakers of German reading the translation get around the translation task that is part of reading the original text for non-speakers of Black English. Thus, *Die Farbe Lila* thwarts every necessity and chance to bridge differences on a linguistic level through immersing oneself in the foreign tongue and experiencing the foreign world through it. A comfortable read, translations such as *Die Farbe Lila* do not train their readers to think beyond cultural boundaries, which obliterates Walker's grand idea that underlies her novel's linguistic structure. Therefore, the utopianism Walker has been charged with is, indeed, much more present in the German translation in which everyone involved, characters and readers, speak the same language.

---

8 Other colloquial features include the non-standard use of grammatical cases, e.g., *Dativ* constructions with *wegen* (due to), and the modal use of the verb *tun* (do), e.g., "Andre Frauen *täten* sich freun" (30; my emphasis; "Other women would be happy").

9 Instead of *anscheinend* (cf. *Duden* n.p.; both words translate to English *apparently*).

10 For *unwell* (cf. Sick n.p.).

11 In place of German-German "der Schuss geht immer *nach* hinten *los*" (cf. Ammon et al. 2004: 699; my emphasis; German saying for an unintended negative effect).

Completely lacking any dialogue between Celie's language and the literary standard variety, the German translation prevents a deeper level of experiencing the novel in terms of the complex connections it draws between language and identity, and the relationship between what is regarded as the standard versus non-standard language. Language in *Die Farbe Lila* remains without a function for the narrative. Instead, caricature-like and stereotypical, Celie's German voice is reduced to her lack of education, which when read in the context of her fate evokes pity in the reader. Yet, the reader remains a detached spectator who is not prompted to reflect upon what it means to be able and allowed to speak in one's own tongue and to let others speak and listen to them in their own language.

Conclusion

As my exemplary analysis has shown, in German translation, African American literary characters tend to be turned into caricatures in the sense Alice Walker has used the term in her collection of short writings *Living by the Word*:

> For it is language more than anything else that reveals and validates our existence, and if the language we actually speak is denied us, then it is inevitable that the form we are permitted to assume historically will be one of caricature, reflecting someone else's literary or social fantasy (1988: 58).

As caricatures, Celie and Janie are reflections of the fantasies of White translators, who still represent the vast majority of translators working in the language pair English/German.[12] Unwilling to accuse each and every translator of intentional racism, on a structural level, I see two main reasons for the continued use of such highly problematic translations strategies: first, the lack of scholarly studies on and recommendations for the translation of African American English in both the realms of Translation Studies and American Studies, and secondly, the lack of understanding of the nature and effects of racist language on the side of White translators. Translation Studies scholars have described rather general strategies that are supposedly applicable to the rendering of all types of *lects* and all language pairs, and that range from standardization to the replacement of the source language *lect* by a similar *lect* in the target language, the annotated mimicry of certain source *lect* features, or the invention of a *lect*. That there can be no one-size-fits-all translation solution has been rendered clear by the examples analyzed in this paper. While Hurston and

12 To my knowledge, there is not a single work of English-language literature featuring Black characters that has not been translated into German by a white translator.

Walker both employ their mother tongue to develop, affirm, and strengthen their protagonists' awareness of their African American (female) identity and, on a linguistic level, argue for the literariness of Black English (Hurston) and the gain in bridging languages differences and breaking down language hierarchies (Walker), they make very different strategic use of features of Black English, engaging and addressing different audiences. Even though these differences seem to be reflected in the different translation strategies applied in rendering *Their Eyes* and *The Color Purple* into German (standardization and colloquialization, respectively), both translators chose strategies inadequate to capture the meaning of Black English in the original, inscribing racist connotations. What a translational approach that does justice to the original while avoiding racist connotations might look like, has been demonstrated by the author of a second German-language translation of *Their Eyes*.

In an epilogue to *Vor ihren Augen sahen sie Gott* (2011), translator Hans-Ulrich Möhring argues that due to the endemic liveliness, wit, spontaneity, metaphoric richness, and, most importantly, musicality of African American English there is no equivalent to be found in the German language (or any other language, for that matter) (Möhring 2011: 262; 266). That is why, according to Möhring, his only option was to conjure up the sound of the original's Black English with small "Hörhilfen" (ibid: 267), acoustic aids that take the form of African American English sentences inserted into the German-language text. For instance, Janie's grandmother's likening Black women to mules – which points to the intersectional discrimination African American women face – reads as follows in Möhring's translation: "Nach meiner Erfahrung ist die Niggerfrau der Muli der Welt – de nigger woman is de mule uh de world" (ibid: 25). Most of these insertions, as the one quoted here, are translated indirectly by paraphrasing their contents in German either in the same sentence or in one of the preceding or following sentences. Thus, Möhring seeks to negotiate the linguistic and tonal repertoire of Hurston's African American English with that of the German language by placing both languages next to each other, obvious to the reader's eye. This translation strategy compensates the orality of African American English non-reproducible in German and enables readers who understand African American English to experience the sound of the original, thereby building a bridge between the novel's foreign world and that of the reader.

However, one third of the inserted African American English remains untranslated. Interestingly, while half of the insertions that are utterances of Janie's husbands are not translated, this applies to only one quarter of those insertions capturing Janie's voice. Thus, not only is Janie's voice visually highlighted in the German text, but the relevance of her husbands' remarks is relativized, deconstructing their self-claimed intellectual superiority over Janie and women in general. Even though this critique of gender inequality is present in Hurston's text, it is not reflected as obviously on the level of language. Moreover, in one quarter of the untranslated insertions important content is concealed for readers without any knowledge of English. These potential semantic "blank spaces" (cf. Iser's concept of *Leerstelle*; Iser 1976) interrupt textual coherence to the effect of a gradual experience of foreignness: readers who do not understand the African American insertions will feel more alienated from the text than those who do. Regardless of their understanding of the insertions, readers are made aware that the language of the fictional *personae* is not German and that any non-Black reader will remain an outsider to the represented world. This awareness might inspire a critical inquiry into readers' (subconsciously held and practiced) views and actions with regard to race and racism.

Hans-Ulrich Möhring's translation of *Their Eyes* is not only a solution to the question of how to render literary African American English into German that is careful in terms of its approach to race. It is also a brilliant example of what is to be gained from such renderings for readers in general, who might lack the linguistic skills to read the original, and for scholars of Transatlantic and German American Studies, who usually only read the original and might consider translations generally deficient, but also for Blacks in the United States and elsewhere. Beyond mere aesthetics on a linguistic level, the phrases from the original included in the German translation, first, direct common readers' attention to the problem of racial discrimination. Second, they help scholars develop a deeper understanding of the text in terms of its critique of sexism as well as of how race relations in the United States were and are perceived in Germany while, at the same time, uncovering racist practices in Germany. Lastly, and most importantly, translations such as Möhring's give an international forum to authors who are marginalized on the grounds of race, allowing them to speak through their characters and transmitting an assertive understanding of Blackness while avoiding what still too often is common translational practice in Germany: a representation of Blackness as mere literary caricature.

Works Cited

Ammon, Ulrich/Bickel, Hans/Ebner, Jakob (2004): Variantenwörterbuch des Deutschen: die Standardsprache in Österreich, der Schweiz und Deutschland sowie in Liechtenstein, Luxemburg, Ostbelgien und Südtirol, Berlin: De Gruyter.

Arndt, Susan/Hornscheidt, Antje (2004): Afrika und die deutsche Sprache, Münster: Unrast.

Becker, Anne Kathrin (2003): "Ruhrdeutsch – Die Sprache des Ruhrgebiets in einer umfassenden Analyse." PhD dissertation, Universität Freiburg.

Bichel, Ulf (1973): Problem und Begriff der Umgangssprache in der germanistischen Forschung, Tübingen: Niemeyer.

Boesenberg, Eva (1999): Gender – Voice – Vernacular. The Formation of Female Subjectivity in Zora Neale Hurston, Toni Morrison and Alice Walker, Heidelberg: Winter.

Ciçek, Azu/Heinemann, Alisha/Mecheril, Paul (2014): "Warum Rede, die direkt oder indirekt rassistische Unterscheidungen aufruft, verletzen kann." In: Gudrun Hentges/Kristina Nottbohm (eds.), Sprache-Macht-Rassismus, Berlin: Metropol, pp. 309-326.

Crenshaw, Kimberlé (1991): "Mapping the Margins: Intersectionality, Identity Politics, and Violence against Women of Color." In: Stanford Law Review 43/6, pp. 124-199.

Dieke, Ikenna (1999): "Introduction." In: Ikenna Dieke (ed.), Critical Essays on Alice Walker, Westport and London: Greenwood Press.

Duden Online v.s. "Anscheinend." (http://www.duden.de/rechtschreibung/anscheinend).

Fagan, Benjamin (2012): "'Feebler than the Original': Translation and Early Black Transnationalism." In: Udo J. Hebel (ed.), Transnational American Studies, Heidelberg: Winter, pp. 229-48.

Fisher Fishkin, Shelley (2011): "American Literature in Transnational Perspective: The Case of Mark Twain." In: Carolyn Field Levander/Robert S. Levine (eds.), A Companion to American Literary Studies, Hoboken: Wiley, pp. 279-93.

Fisher Fishkin, Shelley (2012): "Mapping Transnational American Studies." In: Udo J. Hebel (ed.), Transnational American Studies, Heidelberg: Winter, pp. 31-74.

Gates, Henry Louis Jr. (1987): Figures in Black. Words, Signs, and the 'Racial' Self, New York and Oxford: Oxford University Press.

Gates, Henry Louis Jr. (1988): The Signifying Monkey. A Theory of Afro-American Literary Criticism, New York and Oxford: Oxford University Press.

Green, Lisa J. (2002): African American English. A Linguistic Introduction, Cambridge: Cambridge University Press.

Harris, Trudier (1984): "On The Color Purple, Stereotypes, and Silence." In: Black American Literature Forum 18/4, pp. 151-61.

Hite, Molly (1989): The Other Side of the Story. Structures and Strategies of Contemporary Feminist Narrative, Ithaca and London: Cornell University Press.

Holloway, Karla F.C. (1987): The Character of the Word. The Texts of Zora Neale Hurston, New York: Greenwood Press.

Hurston, Zora Neale (1990[1937]): Their Eyes Were Watching God, New York: Harper & Row.

Hurston, Zora Neale (1993[1937]): Und ihre Augen schauten Gott. Barbara Henninges (trans.), Zürich: Ammann.

Hurston, Zora Neale (2007[1969]): "Spirituals and Neo-Spirituals." In: Henry Louis Gates Jr./Gene Andrew Jarrett (eds.), The New Negro: Readings on Race, Representation, and African American Culture, 1892-1938, Princeton and Oxford: Oxford University Press, pp. 473-478.

Hurston, Zora Neale (2011[1937]): Vor ihren Augen sahen sie Gott. Hans-Ulrich Möhring (trans.), Gräfelfing: Verlag Silke Weniger.

Huskey, R. Erin (2009): "Witnessing and Testifying: Transformed Language and Selves in The Color Purple." In: Kheven LaGrone (ed.), Alice Walkers The Color Purple, Amsterdam and New York: Rodopi, pp. 95-118.

Iser, Wolfgang (1976): Der Akt des Lesens: Theorie ästhetischer Wirkung, München: Wilhelm Fink.

Ishihara, Tsuyoshi (2005): Mark Twain in Japan: The Cultural Reception of an American Icon, Columbia: University of Missouri Press.

Kaplan, Carla (1996): The Erotics of Talk. Women's Writing and Feminist Paradigms, New York and Oxford: Oxford University Press,

Koskinen, Kaisa/Paloposki, Outi (2010): "Retranslation." In: Yves Gambier (ed.), Handbook of Translation Studies, Amsterdam: Benjamins, pp. 295-98.

Kutzinski, Vera (2006): "'Yo también soy América': Langston Hughes Translated." In: American Literary History 13/3, pp. 550-78.

Lefevere, André (1992): Translation, Rewriting and the Manipulation of Literary Fame, London and New York: Routledge.

Lemke, Sieglinde (2009): The Vernacular Matters of American Literature, New York: Palgrave Macmillan.

Mihm, Arend (1985): "Prestige und Stigma des Substandards. Zur Bewertung des Ruhrdeutschen im Ruhrgebiet." In: Arend Mihm (ed.), Sprache an Rhein und Ruhr. Dialektologische und soziolinguistische Studien zur sprachlichen Situation im Rhein-Ruhr-Gebiet und ihrer Geschichte, Stuttgart: Franz Steiner, pp. 163-93.

Minnick, Lisa Cohen (2004): Dialect and Dichotomy. Literary Representations of African American Speech, Tuscaloosa: University of Alabama Press.

Nduka-Agwu, Adibeli/Hornscheidt, Antje (2010): Rassismus auf gut Deutsch. Ein kritisches Nachschlagewerk zu rassistischen Sprachhandlungen, Frankfurt am Main: Brandes & Apsel.

Paul, Heike (2005): Kulturkontakt und Racial Presences. Afro-Amerikaner und die deutsche Amerika-Literatur, 1815-1914, Heidelberg: Winter.

Paul, Heike (2013): "The German Reception of African American Writers in the Long Nineteenth Century." In: Mischa Honeck/Martin Klimke/Anne Kuhlmann (eds.), Germany and the Black Diaspora. Points of Contact, 1250-1914, New York and Oxford: Berghahn, pp. 115-133.

Ricœur, Paul (1992 [1990]): Oneself as Another, Chicago: University of Chicago Press.

Sick, Bastian (2009): "Nu man bloß nich in' Tüdel geraten!" In: Der Spiegel January 04 (http://www.spiegel.de/kultur/zwiebelfisch/zwiebelfisch-nu-man-bloss-nich-in-tuedel-geraten-a-616380-2.html).

Smith, Felipe (1999): "Alice Walker's Redemptive Art." In: Ikenna Dieke (ed.), Critical Essays on Alice Walker, Westport and London: Greenwood Press, pp. 109-125.

Sollors, Werner (1999): "W.E.B. Du Bois in Nazi Germany, 1936." In: Amerikastudien/American Studies 44/2, pp. 207-22.

Stowe, Harriet Beecher (1852 [1852]): Onkel Tom's Hütte oder Negerleben in den Sklavenstaaten des freien Nordamerika. Dr. Ungewitter (trans.), Wien and Leipzig: Pest.

Stowe, Harriet Beecher (2010): Uncle Tom's Cabin, New York and London: W.W. Norton.

Walker, Alice (1988): Living by the Word. Selected Writings 1973-1987, San Diego: Hartcourt Brace Jovanovich.

Walker, Alice (2005[1983]): Die Farbe Lila. Helga Pfetsch (trans.), Bergisch Gladbach: BLT.

Walker, Alice (2014 [1983]): The Color Purple, London: Phoenix.

II. Normative Whiteness in the German Humanities

Post-Racism, Colorblind Individualism & Political Correctness: Contemporary Modes of Materialization in American Studies and German Academia

Courtney Moffett-Bateau and Sebastian Weier

This paper engages with two moments of American and Americanist political discourse in order to theorize post-racism as a period of racial materialization and racial discrimination specific to our contemporary epoch. Post-racial situations facilitate readings of American events in U.S. histories that negate the existence and effects of more traditional (i.e. phenotype-focused) racism and racial discrimination. We understand the contemporary use of post-racism as a 'new' period of racism that positions itself in line with mainstream descriptions of colorblind individualism and contemporary language policies that claim we live in a period that is beyond racism and therefore beyond the need for political correctness.

In order to arrive at a better understanding of the discursive framing and situating of post-racism, we investigate two contemporary American political events: the 2012 Republican National Convention and 2014 civil protests to police brutality. The aftermaths of these truth claims result in a denial of race-based structures that use narratives of progress to place such structures discursively in the past. Hence, we examine how colorblind individualism and contemporary modes of racial materialization work to systematically disenfranchise, alienate, and disqualify people in the name of American democratic progress.

We try to answer this question by looking at two different modes of racial materialization: the legal and artistic modes. In the mode of legal materialization, we examine how bodies materialize differently within legal frameworks of citizenship, property, and race. The mode in which Dred Scott was forced to articulate a narrative of humanity within the U.S. court of law, for example, mandated a narrative that contested American rights to property. Scott's body interacts with American federal documents, which deny him access to the American legal category of Human. A decision that was later overturned in Plessy v. Ferguson, the case nevertheless emblemizes American democracy by partially extending civil but not equal liberties to populations of people who materialized as black.[1]

Hence, we reflect on the consequences of social progress that allow for American individuals to materialize differently – over time and at different paces – across and in between certain social and legal situations. Later, we consider how artistic modes intervene symbolically as a way to disrupt legal modes of materialization. We contemplate the effigy lynching that was staged in the liberal university setting at UC Berkeley; although they provoke and shock, while temporarily trying to unsettle race-based modes of materialization, they reify past symbols of racial violence into the scene wherein they try to intervene. The efficy lynching disturbs twofold: (1) through shock, it writes violence into a presumably safe liberal space so as to mirror the pain and loss of black life, on the one hand, and the fear associated with symbolic violence, on the other. (2) Functioning as a didactic tool so as to "teach" white students about anti-black symbolic violence, it strikes black students in ways that leave them per usual open and vulnerable to anti-black symbolic violence. Here in this academic

1 Following Richard Rothstein, we have opted to use the lower case when referring to black and white people. In *The Color of Law: A Forgotten History of How Our Government Segregated America*, Rothstein argues: "Because our majority culture has tended to think of African Americans as inferior, the words we've used to describe them, no matter how dignified they seem when first employed, eventually sound like terms of contempt. African Americans react and insist on new terminology, which we eventually accept until it too seems to connote inferiority. So at the beginning of the twentieth century, America's subordinated race was called *colored.* Later, we came to think of it as *Negro*, first with a lowercase and then with a capital N. It was replaced by *black*, a term that has had a seemingly permanent currency. Today *African American* strikes us as most appropriate. In these pages, it's the term I'll use most frequently, but I will sometimes use *black* as well. Occasionally, in describing historical events, I will refer to *Negroes*, intending the same respect that it enjoyed in those earlier periods. The shifting of terminology should not distract us from this underlying truth: We have created a caste system in this country, with African Americans kept exploited and geographically separate by racially explicit government policies. Although most of these policies are now off the books, they have never been remedied and their effects endure" (Rothstein 2017: xvii).

context, we investigate the limits of performance that rely on the bodies of black students to negotiate white rehabilitation.

As contributors to transatlantic discussions on American cultural and political phenomena, we reflect on our position in American Studies scholarship. As scholars based outside the U.S., looking inward is problematic, yet also illuminating. Observing and thereby analyzing American cultural and political phenomena requires us to temporarily inhabit and reproduce languages of exclusion to address, however improperly, the modes in which these phenomena materialize. We end our contribution by looking at the April 2016 edition of the academic professional publication of *Forschung & Lehre* (Research and Teaching) in which a German confectionary was used as a cover to discuss political correctness debates in Germany and the U.S. It would seem that a German iconographic image – that has a painful history for black Germans – presents itself as the supreme democratizing image. Through shock, but not directly through engagement, as none of the articles inside the publication ever react to the problematic history of the image, *Forschung & Lehre* uses a notoriously hurtful image to stimulate a conversation for its white academic audience in an attempt to ease white guilt.[2]

Post-Race and Materialization: How a Black Man Becomes a Chair

During the 2012 Republican National Convention, which would nominate Mitt Romney as the GOP's presidential candidate, actor-director Clint Eastwood spoke to an empty chair that was intended to represent President Barack Obama. In an ad-lib performance, Eastwood bashed the "weak" politics of the Obama administration. Eastwood faulted the Obama administration for the unemployment of over 23 million Americans (which the U.S. Labor Department later corrected as being 12.8 million unemployed people) and insisted that the "true" makers of American industry and culture had to take "back" control of their politics, economy, and country. Eastwood asked the empty chair "How do you handle promises that you've made when you were running for election?": in

2 In 2014, an earlier version of this paper was given as the keynote address by Dr. Sebastian Weier at the student conference "Now You See It, Now You Don't: Interdisciplinary Explorations of Race in 'Postracial America'" at the University of Duisburg-Essen, which was organized by Dr. Dietmar Meinel and Courtney Moffett-Bateau. We would like to thank Professor Dr. Barbara Buchenau from the University of Duisburg-Essen and Luvena Kopp for their feedback.

what the media later called a "chair-ologue," Eastwood endorsed Romney, claiming that it was time a true businessman had the opportunity to work for America. Again, Eastwood addresses the chair: "A stellar businessman. Quote, unquote, a stellar businessman. And I think it's that time. And I think if you just kind of stepped aside and Mr. Romney can kind of take over." This separation between the addressed "you" and the implied first-person plural "we" ensures his Republican audience that "their" political and economic values differ from those of the Obama administration. Proclaiming, in fact, that "*we* own this country," Eastwood continues:

> But I'd just like to say something, ladies and gentlemen. Something that I think is very important. It is that, you, we, we own this country. Thank you. Thank you. Yes, we own it. And it's not you owning it and not politicians owning it. Politicians are employees of ours. And, so, they're just going to come around and beg for votes every few years. It's the same old deal. But I just think that it's important that you realize and that you're the best in the world. And whether you're Democrat or whether you're a Republican or whether you're Libertarian or whatever, you're the best. And we should not ever forget that. And when somebody does not do the job, we got to let 'em go (Eastwood 2012).

Eastwood uses the possessive pronoun "ours" to bring the economic elite values of the Republican Party in conversation with its working-class and middle-class voters. In Eastwood's address, America is envisioned as a business apparatus in which political businessmen like Romney fight for the rights of middle-class, working-class, and unemployed Americans.

Understanding who is implied in the offering of Eastwood's use of "we" and "our" is especially revealing in an age of Trump: Eastwood envisions a fictitious Obama – who is intimidated or threatened by his tough cis-masculinist persona – to rationalize his Republican message as "real." He utilizes the tough-guy persona known from his popular Hollywood films as he delivers this message against a massive backdrop of his profile in a cowboy hat.[3] Considering his

3 Against this backdrop, Eastwood's figure represented "the good ole days" when white masculinities were the heroes of their families, hard on crime, and tough in war. Consider the 'variety' of Eastwood's many film personas; *The Outlaw Josey Wales*, an American Western, which Eastwood starred in and directed, was based on the novel *The Rebel Outlaw: Josey Wales* (1972), written by Asa Earl Carter under the alias Forrest Carter Clinton, who was a Ku Klux Klan leader and pro-segregationist. The film is set in the American Civil War after Union soldiers killed Josey Wales' family. In a bloody, violent exchange of events, Wales joins a pro-Confederate guerilla group in order to avenge his family and restore order. After the film's release, many saw

placement as an upper-class man, with an over $ 375 million net worth, Eastwood's appeal to his American audience functions here rhetorically rather than practically. The dominant wealthy speakers of the Republican Party unite here in dialectical relationship with its prominently working-class and middle-class white audience in rhetorical ways that bind them together. In more ways are they deeply divided – as their cultural, economic, and symbolic power can never match that of the representatives of their Republican Party. Eastwood cements his Republican message in a rhetoric of American exceptionalism, which, to the amusement of his audience, justifies the displacement of the Obama administration and restores (white) racial order.

This scenario of course only works without the actual presence of Obama. Eastwood calls Obama's masculinity into question and sits "Obama" down in a chair as if he were a child to be scolded. Unable to respond, the chair is a passive, silent object of Eastwood's Republican message. At times the actor pretends that the chair is able to respond: "What do you mean shut up?" Eastwood interjects this question while making a criticism of Obama's inability to close Guantánamo Bay. As a rhetorical devise, shutting the Obama persona up is often used when the actor seems to have lost his train of thought or needs to reestablish his logic. He ends his speech by referring back to Obama: "And when somebody does not do the job, we got to let them go." After which he takes his index finger and draws it across his own neck in a single motion. He bends his head down to chuckle to himself as the crowd cheers.

In the aftermath of Eastwood's Convention speech, chairs were lynched in Virginia and Texas. What is sometimes described as effigy lynching, did not, however, start with that speech. Six years earlier, in another famous case, rope nooses were hung from a tree in a high school courtyard in Jena, Louisiana, after black students had sat under a tree that was a well-known hangout place for white students. Although these nooses were later claimed to have been intended as a prank directed towards members of the school's rodeo team, tensions occurred that led to the burning of the school's main building, the beating of a white student, and the much-debated prosecution of the so-called Jena Six.[4]

Eastwood as a fascist. Later films, such as *Million Dollar Baby* and the critically acclaimed 2008 *Gran Torino*, allowed Eastwood to channel (but also redeem) his tough-guy, working-class, cis-male persona in feminist and multicultural settings.

4 See the *New Yorker* article by Steve Coll, "Disparities: The Jena Six." Coll states: "Just over a year ago, during a high-school assembly in Jena, Louisiana, a black student asked the school's white principal if it would be all right to sit under an oak tree outside, an oasis of shade known as the 'white tree,' because only Caucasian students congregated there. The principal said that the young man could sit where he

Extracting a linear explanation from the abovementioned situation deserves a deeper individual investigation of both internal and external forces that contribute to effigy lynching. In at least two cases after Eastwood's speech, signs with the printed words "Nobama" were found attached to these noosed chairs. In a 21st century political spectacle these performances of hanging empty chairs from trees align themselves with special rituals of American white supremacy. Inspired by Pierre Bourdieu, this article supposes a direct relationship between political power and its representations, and thereby questions how the history of lynching newly reimagines itself in 21st century American discourse.

Following Bourdieu, symbolic violence is largely focused on creating a "habitus" (1979: 78). It works as a form of dressage through which power seeks to determine its subjects' behavior *a priori*. It prepares the moral ground for physically violent repression – should resistance arise nevertheless. An example of this would be the attempt to prevent anti-racist struggle by both portraying it as imprudent (acting unwittingly against the "facts") or evil (acting against the "facts" in spite of knowing better) and by legitimizing anti-black violence through its constant repetition in the symbolic order. The mutilation of black bodies may be constantly reproduced so as to make it seem "natural" and inculcate in black people the futility of resistance, while also preparing for that mutilation to be actualized. In looking at the post-racist situations that surround artistic and legal materialization – here in the form of effigy lynching – this article analyzes how changes in the political and symbolic order interact to reshape racial domination. Furthermore, we interrogate the capacity of the symbolic order to impact the social order, as well as the necessity of the former to react to permutations in the latter.

---

liked. Later, the student and some African-American friends walked over to the oak and chatted with some white schoolmates. The next day, somebody fixed two nooses to the tree's branches." Coll goes on to comment: "Although some of the evidence in the Jena case is murky, a cumulative verdict of racial double standards lies beyond reasonable doubt. Between Reconstruction and the end of the Second World War, more than two hundred and fifty people in Louisiana, the great majority of them African-Americans, were lynched [...] Last December, at the school, a black student coldcocked a white student, Justin Barker, knocking him briefly unconscious; other black students allegedly kicked the victim while he lay on the ground. Barker was treated for cuts and bruises at a hospital and released a few hours later. The police arrested six black students, aged fourteen to eighteen, and Walters charged them with attempted second-degree murder and a conspiracy count; if convicted, they faced up to seventy-five years in prison" (Coll 2007).

The question becomes in which ways and to what extent the curators of this effigy lynching inhabit the "we" in Eastwood's speech. The nooses found in the Jena Six event cannot only exist in a narrative vacuum that is somehow dislocated from the history of symbolic violence – materializing as a 'simple' prank towards other school members. Nooses are not only engrained within the American imagination to visually connote death, through suicide, but also bear a history of making meaning practices in which (white) extrajudicial justice extracts symbolic capital from (black) death as a means of restoring (white) social and political uniformity.[5]

Simply put, materialization describes how something becomes matter; however, it does not simply describe how something turns into a thing from nothing, but rather asks how specific forms or constellations of matter (say: a chair) relate to non-material forms or constellations (say: an idea and/or a social construction). Consider the following questions:

1. How does a material thing become imbued, charged, or cathected with a non-material thing, that is: How does matter become meaningful?
2. How does that process constitute and transform that non-material thing?

Analyzing materialization, then, means asking: how does a chair become racialized as a black man and/or how does a black man become a chair? Indeed, the chairs hanging from trees materialize as black via Eastwood's chair-ologue, just as the nooses hanging from the schoolyard tree evoke a history of black death and white supremacy. These two events cannot be credibly framed only in relationship to the rodeo team, but must also be read in the *longue durée* of anti-black racism and lynching.

5 According to Pierre Bourdieu, "Symbolic capital enables forms of domination which imply dependence on those who can be dominated by it, since it only exists through the esteem, recognition, belief, credit, confidence of others, and can only be perpetuated so long as it succeeds in obtaining belief in its existence" (Bourdieu 2000: 166). Also see the Equal Justice Initiative's report "Lynching in America: Confronting the Legacy of Racial Terror." EJI researchers "documented 4075 racial terror lynchings of African Americans in Alabama, Arkansas, Florida, Georgia, Kentucky, Louisiana, Mississippi, North Carolina, South Carolina, Tennessee, Texas, and Virginia between 1877 and 1950 – at least 800 more lynchings of black people in these states than previously reported in the most comprehensive work done on lynching to date" ("Lynching in America").

Furthermore, the appeal to innocent 'play' – feigning ignorance in a situation that is clearly engrained in a history of symbolic violence – is not a new reaction of "good"-intentioned white people. It is an evocative response of people in relative power attempting to uphold their position within relationships of dominance. It is a maneuver that Eastwood routinely reverts back to in his speech – as seen in his insecure or bashful body language, his hesitation to use insults when speaking "back" to the Obama-as-chair "I can't say that," or his many other insinuations – and serves to underline the ostensible defensiveness in Eastwood's speech. Therefore, we understand the effigy lynching in the aftermath of Eastwood's speech as a materialization of anti-black violence as valid whether it be 'innocent' play or intentional "anti-Obama" political commentary; and furthermore contend that Eastwood's words – through televised spectacle – envision a binary of power that establishes the chair/Obama as passive and/or mute recipient to Eastwood's position as authoritative speaker.

We situate this event within a narrative of racial materialization as the visual signals – i.e. the noose or the chair as referent to Obama, the nation's first black president – pervasively illuminate a familiar code of symbols in which black existence and/or uplift is reprimanded through white extrajudicial symbolic violence. Earlier modes of racial materialization endorsed modes of racism that allowed white extrajudicial violence to designate full or ultimate power. The new 'post-racism' mode of contemporary materialization denies that the white speaker – because he/she is supposedly powerless, i.e. experiences relative power in a white supremacist capitalist patriarchy – is capable of reproducing these race-based structures. Both the Eastwood RNC event and the effigy lynching that transpired afterward, somehow claim a free-floating action that wishes to exist on its own. It also seems that many opponents of political correctness would struggle to see these two events as inherently racist.

Legal Materialization and the Colorblind Individualism of American Humanism

In this newest moment of American post-racism, narratives of racial materialization claim to have diverged from previous "traditional" modes – yet reify these same modes nevertheless. These narratives insist upon a new American individualism that needs to be colorblind or to "not see race" in order to speak. In our first example, Eastwood imagines himself as authoritative speaker with a mute referent. Eastwood's message cannot be obstructed by Obama's message or presence because he is imagined as *not being* (there) or *not existing*. In the effigy lynching that transpires afterwards, Obama's *not being* is imagined through

(1) the *erasure* of his presence through the materialization of the chair and (2) the symbolic gesture of "his" death being simulated through the lynching of the chair. If, indeed, symbolic violence was traditionally a tool of white self-assertion, it would seem that the *denigration* – via racial materialization – of the chair enables a new form of racialization. Eastwood's socio-symbolic power allows him to materialize desires of white supremacy as a race-based white working-class utopia.

In 1857, when Dred Scott sued for his freedom in the Supreme Court case Scott v. Sanford, Scott was not a mute referent of the law; however, the law still silenced him. This case shows how important a person's race was in order to know if that person could legally be enslaved, or claim protection by certain laws or access to specific trades that were reserved exclusively for whites. Hence, humanism was a category within American chattel slavery that was reserved for non-black people.[6] We understand American humanism as an essential category to reflect upon legal materialization for its socio-symbolic power to organize moments in American history. New forms of white supremacy might suspend the visual capital of black death in hopes of recuperating power. It condenses changes in the nation's modes of racialization into epochal shifts. From the Declaration of Independence (1776), the Seneca Falls Convention (1848), and the Emancipation Proclamation (1863), to landmark court cases such as Plessy v. Ferguson (1896) and Brown v. Board of Education (1954) or the Civil Rights Legislation of the 1960s and the aforementioned election of President Obama, changes in the materialization of race – the morphing of a socio-genetic idea into a concrete material presence – have always been *emplotted* as inevitable moments in American democratic progress.[7] Under

6 "From the discovery that at the origin of law there is nothing other than arbitrariness and usurpation, that it is impossible to found law in reason and right, and that the Constitution – no doubt what resembles, in the political order, a Cartesian primary foundation – is merely a founding fiction designed to disguise the act of lawless violence which is the basis of the establishment of law, Pascal draws a typically Machiavellian conclusion: since the people cannot be made to understand the liberatory truth about the social order (*veritatem qua liberetur*), because that truth could only threaten or ruin that order, the people must be 'deceived,' not allowed to see 'the fact of usurpation,' the inaugural violence in which law is rooted, by 'making it appear as authoritative, eternal'" (Bourdieu 2000: 168; original emphasis). In this way, the symbolic structures which very knowingly impose restraint upon certain (but not all) bodies within U.S. slave systems, are narrated as logical, if not natural.

7 Building on Hayden White's *Metahistory* (1973), here we choose to reflect on key events in American history which continue to be *emplotted* into rehearsed or canonical

American progress, the materialization of race understands itself as part of a sound American society progressing towards its true self. The representatives of American progress depend on a sociopolitical narrative in which the cultural values of his/her subject position are best represented.[8]

American national history. We borrow White's concept of emplotment to describe the process by which historians narratively focus on certain events in time to create an overall series, representational of a coherent national narrative. According to White, "a given historian is forced to emplot the whole set of stories making up his narrative in one comprehensive or *archetypal* story form" (1973: 8; original emphasis). We use the term so as to reflect how moments of American exceptionalism and American individualism are routinely used retroactively to associate social progress and moments of democratic advancement with white cis-male heroicism. White's *Metahistory*, which offers a history of historical consciousness in 19th century Europe, states: "[…] Anglo-American philosophers have produced a massive body of literature on the epistemological status and cultural function of historical thinking, a literature, which, taken as a whole, justifies serious doubts about history's status as either a rigorous science or a genuine art. The effect of these two lines of inquiry has been to create the impression that the historical consciousness on which Western civilization views its relationship not only to cultures and civilizations preceding it, but also to those contemporary with it in time and contiguous with it in space. In short, it is possible to view historical consciousness as a specifically Western prejudice by which the presumed superiority of modern, industrial society can be retroactively substantiated" (2). Also see Paul Ricœur, *Time and Narrative* (1990): "The productive imagination at work in the metaphorical process is thus our competence for producing new logical species by predicative assimilation, in spite of the resistance of our current categorizations of language. The plot of a narrative is comparable to this predictive assimilation. It 'grasps together' and integrates into one whole and complete story multiple and scattered events, thereby schematizing the intelligible signification attached to the narrative taken as a whole" (x).

8 Cf. Wilbur C. Rich's argument: "To be fair to the framers, as political scientist Charles Beard, in *An Economic Interpretation of the Constitution of the United States*, and Robert Dahl, in *A Preface to Democratic Theory and How Democratic Is the American Constitution*? pointed out, they were not trying to build a civil society in which all residents regardless of race, gender, country of origin and income level would be protected by the new government order. Wealthy white men were a minority that needed protection from the horde or the mob who might tyrannize them if given a chance. They felt that the nation did not owe blacks and the indigenous people anything. This is why these issues were not spelled out in the new Constitution. The idea that women and landless white males should be able to vote or to have their vote

Rather than symbolic representations that made visible prior shifts in the fabric of the social, these historical moments were and are narrated as results of the action of heroic individuals or groups of individuals. They stand as proof of the progress of an inherently sound American society, rather than revolutionary events that re-made history in spite of its teleology of white supremacy. The Slavery Code of the District of Columbia of 1862 explicitly stated that to *kill* a slave was *not murder*, but that it did qualify as a legal case of damaged goods that had to be refunded to the owner. The following decision was issued by Chief Justice Roger B. Taney in the Scott v. Sandford case:

> The unhappy black race were separated from the white by indelible marks, and laws long before established, and were never thought of or spoken of except as property, and when the claims of the owner or the profit of the trader were supposed to need protection (Dred Scott v. Sandford, 1856: 60 U.S. 411).

In articulating his case for black humanity, and thereby a reasonable request for self-autonomy within American democracy, the Dred Scott case exemplifies American democracy, although social progress for African Americans is denied: It showcases the positive power of American democracy because it allows for resistance to take part as a welcomed, if not natural, part of its process towards an inclusive nation. This, however, does not make it an example of a post-racial situation but of legal materialization. It functions historically as a 'natural' precursor to the 1896 landmark "separate but equal" case of Plessy v. Ferguson, in which democratic progress is finally, if not rationally and naturally, realized.

Although laws such as Tennessee's "One-Drop Rule" (which considered any person black who had an ancestor of sub-Saharan descent, no matter how remote) or Virginia's similar "Racial Integrity Act" were not enacted until 1910 and 1924, respectively, legal arrangements for constructing racial identity and the legal status tied to it already existed in the antebellum period of the United States. These legal arrangements were enshrined in terms such as "quadroon" or "octoroon" (people who had one black grand-parent or great-grand-parent,

---

be of equal weight to that of the landed gentry never crossed the minds of the attendees. The framers were designing a set of rules that fit their political and economic times. Like all generations of Americans, their political decisions defended what they understood as their interest. This grandiose plan called for the elevation of men with British ancestry and Protestant backgrounds to the apex of society. These men would dominate the politics and economy of the new nation. The wealthy minority would rule forever. Government would serve their interests" (2015: 12-13).

respectively) and in court cases such as Scott v. Sandford. Phenotype[9] as a legitimate allowance of race-based structures has assumed the hegemonic weight of inextricably linking what is physically visible with the internal value, type, or capability so as to characterize, indicate, or *symbolize* any given thing or being.

This first step of legal materialization defines *what* blackness means and then shows what consequences it has both for the person who materializes as black and the white person interacting with that supposed blackness. In "Seduction and the Ruses of Power," Saidiya Hartman uses the example of an enslaved woman named Celia to reflect on the language of rape in American common law. In the 1885 court case State of Missouri v. Celia, the defendant, Celia, is tried for killing Mr. Newsome, a white man, who during her four-year enslavement repeatedly raped her. In her testimony, Celia reveals a pattern of forcible rape that she intended to free herself from: "She said the old man (Newsome, the deceased) had been having sexual intercourse with her. That he had told her he was coming down to her cabin that night. She told him not to come and if he came she would hurt him" (cf. Hartman 1996: 537). Her argument of self-defense was rejected in the American court of law and she was sentenced to death by hanging for the killing of her master. Hartman states:

> In 19th-century common law, rape was defined as the forcible carnal knowledge of a female against her will and without her consent. Yet the actual or attempted rape of an enslaved woman was an offense neither recognized nor legislated by law. Rape was not simply unimaginable because of purported black lasciviousness, but its repression was essential to the displacement of white culpability that characterized both the recognition of black humanity in slave law and the designation of the black subject as the ordinary locus of transgression and offense (ibid: 537).

Legal protection from sexual violence was not afforded to black women within these intersecting histories of American racism, sexism, and capitalism. The statutes of the law did not view black women as Human and sought to protect cis-male white property owners' moral and legal culpability. This example of legal materialization reveals how social conventions of middle-class *white* femininity – which would enforce punitive rules for rape – were 'justifiably' un-available to cis-women who legally materialize as black (Cf. Butler 1993).

The history of legal materialization carries an extensive tradition of making people matter differently. Materialization allows for a definition of race and gender to be enshrined in law that includes such paradigmatic statements as:

9 From the Greek *phainein*, meaning "to show," and *typos*, meaning "type," or from the German word *Phaenotypus*.

Being black means not having the status of being a legal person or a citizen and therefore none of the rights and protections such a status would imply. *Being black* means not being human, but a commodity and legally susceptible to (sexual) violence. "Hypodescent" stabilizes racial categories not for the sake of etymology, but in order to give legal security to categories that were and still are central to the functioning of American society and economy.

Post-racist moments in democratic progress rely on the partial extension of civil but not equal liberties to Americans who materialize legally as black. Within this legal mode of materialization, democratic progress is able to uphold systems of lawful domination through the creation of a new Jim Crow caste system in which race legally materializes anew, as it did after the end of the Civil War, when the social, political, and economic function of race-based slavery was transferred onto convict labor through a legal system geared to punish people of color.[10] It is within this contemporary epoch of post-racism, which is defined by both its distance to former systems of legal and structural racism and its endorsement of new materializations of injustice and oppression, that our diagnosis of a new transatlantic post-race era informs itself. It is one that does not witness an abolition but rather a permutation of familiar epistemic modes of racial materialization.

As has followed, the rise to power of the 44th President has indicated that individual success and upward social mobility no longer know any limits within the frontiers of the United States, and thus have finally overcome the identity frontiers of race. Obama's election and reelection effectively ennobled the country, emboldened its political system and its model of creating and measuring the citizen-subject's liberty and success through and on the level of the supposedly free market of American values and vanities as well as global capitalism. Yet this trope of what has now evolved into a colorblind indi-

10 See the Equal Justice Initiative film "Slavery to Mass Incarceration." EJI reports that "African Americans are 6x more likely to be sentenced to prison for the same crime as a white person. One in three black males born today can expect to spend time in prison during his lifetime" ("Slavery to Mass Incarceration"). They then go on to state: "The Equal Justice Initiative believes that racial bias remains a serious problem and is a direct and lasting legacy of American slavery and our failure to deal with this history of racial injustice. The Equal Justice Initiative seeks to foster an honest conversation about the legacy of slavery, about mass incarceration and racial inequality, and how it still effects millions of people today. We can confront and overcome bias and discrimination. Please join in this conversation so that we can move forward together" ("Slavery to Mass Incarceration").

vidualism is nevertheless still articulated in traditional modes of legal materialization.

Artistic Materialization

Legal modes of materialization have a violent history and deep investments in U.S. law enforcement. These events almost always claim an ahistorical reading of the event, as U.S. law enforcement almost always claim to act as colorblind extensions of democratic fair and legal treatment.[11] Understanding this aspect of legal materialization against what President Herbert Hoover called the "rugged individual" – who obeys no higher authority and knows no structural or cultural disadvantage or repression, but creates His life Himself out of the labor of His own hands – allows us to read U.S. law enforcement against an extensive American history of 'fair' extrajudicial killing. Condensed into the American Dream at the turn of the last century, this rugged American individual knows that the possibilities of success are endless if one "simply works hard enough" and has the will, desire, and discipline to "pull oneself up by the bootstraps."[12]

---

11 According to Diana Yates, "A new study reveals that police recruits and experienced officers are more likely than others to subscribe to colorblind racial beliefs – the notion that they – and people in general – see no differences among people from different racial groups and treat everyone the same. The findings appear in the journal *Race and Social Problems*" (Yates 2016).

12 Referring to World War I in his 1928 speech "Principles and Ideals of the United States Government," the self-made millionaire and American president Herbert Hoover stated: "We were challenged with a […] choice between the American system of rugged individualism and a European philosophy of diametrically opposed doctrines of paternalism and state socialism. The acceptance of these ideas would have meant the destruction of self-government through centralization […] [and] the undermining of the individual initiative and enterprise through which our people have grown to unparalleled greatness […] I would like to state to you the effect that [...] [an interference] of government in business would have upon our system of self-government and our economic system. That effect would reach to the daily life of every man and woman. It would impair the very basis of liberty and freedom." He continues: "The greatness of America has grown out of a political and social system and a method of [a lack of governmental] control of economic forces distinctly its own – our American system which has carried this great experiment in human welfare farther than ever before in history. [...] And I again repeat that the departure from our American system [...] will jeopardize the very liberty and freedom of our people, and

The American individual is subjected to no king or master; he is neither estranged from His labor nor His body and He is responsible to no one except a God of His own choosing.[13]

In early December 2014, national protests took place in response to police brutality and the killing of black men and women in major cities across the U.S., including Chicago, New York, Washington, D.C., Boston, Minneapolis, Atlanta, and Oakland. Gwen Carr, the mother of Eric Garner, led a "Justice for All" march from downtown Washington to the U.S. capitol together with Civil Rights activist and *CNN* guest correspondent Reverend Al Sharpton. In Oakland, Wanda Johnson, the mother of Oscar Grant (1986-2009), held a demonstration at the Fruitvale BART (Bay Area Rapid Transit) Station, where her son had been killed by BART Police officer Johannes Mehserle while Grant had already been apprehended in handcuffs. In response, a group of queer black artists staged an effigy lynching on the UC Berkeley campus. From the fatal situations of Emmett Till (1941-1955) to Tamir Rice (2002-2014), one can see how phenotype – that which is *physically* visible intersected with the assumed *behavior* – ritually justifies the fatal outcome of the two boys.[14] In almost all instances, black death becomes legally justifiable because the accused is perceived as a threat to white

will destroy equality of opportunity not only to ourselves, but to our children" (Hoover 1928).

13 For capitalization, see Calvin Warren: "The 'white, western-god-man' (or the 'American god') that [J. Kameron] Carter describes bears resemblance to what Sylvia Wynter would call 'Man' (2003, 322): both are philosophical-theological apparatuses of anti-blackness, and they function to colonize essential spheres of existence ('Man' colonizes human and the 'white, western-god-man' colonizes God). The 'white, western-god-man' and 'Man' index a process of extreme epistemological and metaphysical violence, and this violence serves as the foundation of Western society and its politics" (Warren 2015: 23).

14 See the findings of a study commissioned by the *American Psychological Association*: "Black boys as young as 10 may not be viewed in the same light of childhood innocence as their white peers, but are instead more likely to be mistaken as older, be perceived as guilty and face police violence if accused of a crime, according to new research published by the American Psychological Association. 'Children in most societies are considered to be in a distinct group with characteristics such as innocence and the need for protection. Our research found that black boys can be seen as responsible for their actions at an age when white boys still benefit from the assumption that children are essentially innocent,' said author Phillip Atiba Goff, PhD, of the University of California, Los Angeles. The study was published online in APA's *Journal of Personality and Social Psychology*" (Goff et al. 2014).

security. Emmett Till is not physically seen or valued as a young *boy* "flirting" with a white girl, but is punished as being a *man* who *had* sexually assaulted an innocent girl.[15] Tamir Rice is not envisioned as a twelve-year-old boy playing in the park with a toy, but as a threatening *man* on the verge of using an assault weapon. In both instances, their death is justifiable based on phenotype: both are envisioned as older and larger than they actually were – as was the case with Michael Brown and Officer Darren Wilson, who, although both men were of similarly large stature, claimed Brown looked like a "demon" (McCoy 2014).

Here we turn to artistic modes of materialization to reflect on the effigy lynching on the UC Berkeley campus. We understand this contemporary mode of grassroots activism engaged by a group of queer black artists as aiming to act as a counter-materialization in that they seek to reveal the continuity of a race-making material infrastructure in the United States in spite of the many situations of post-racism that seek to veil it. Hence, we see artistic modes of materialization as interventions into 'common sense' perceptions of race. Artistic interventions illuminate the crisis that is necessary for change and engage with this discomfort so as to unsettle rituals that disproportionately impact young people of color.

On Saturday, December 13, 2014, *The Guardian* reported that "Effigies of black men [were] found hanging by nooses at UC Berkeley campus" ("Effigies of Black Men"). The hanging images shocked the campus. The life-size photos of lynching victims had been superimposed on cardboard and were found hanging from Sather Gate on the UC Berkeley campus. Across one of the images the words "I can't breathe" were printed in bold white letters. These were Eric Garner's last words before he died at the hands of NYPD Officer Daniel Pantaleo, who was caught on tape using an illegal chokehold.

The effigy lynching at UC Berkeley was part of a broader intervention by a black art activist group. Initially, it was unclear whether the effigy lynchings

15 In January 2017, Carolyn Bryant Donham revealed to Timothy B. Tyson, who was to write the book *The Blood of Emmett Till*, that her claims against Till were false. *The New York Times* reported: "In it, he wrote that she said of her long-ago allegations that Emmett grabbed her and was menacing and sexually crude toward her, 'that part is not true.' [...] As a matter of narrow justice, it makes little difference; true or not, her claims did not justify any serious penalty, much less death. The two white men who were accused of murdering Emmett in 1955 – and later admitted it in a Look Magazine interview – were acquitted that year by an all-white, all-male jury, and so could not be retried. They and others suspected of involvement in the killing died long ago" (Pérez-Peña 2017).

were in support or in protest of the #BlackLivesMatter march that had taken place earlier that day. In their artist statement, they wrote:

> We are a Bay Area collective of queer black and PoC artists responsible for the images of historical lynching posted to several location[s] in Berkeley and Oakland. These images connect past events to present ones – referencing endemic faultlines of hatred and persecution that are and should be deeply unsettling to the American consciousness (qtd. in Frank 2014).

Printed in all black across another effigy lynching stood the hashtag "#I CAN'T BREATHE." Directly beneath it, inches away from where the subject's toes dangled lifeless underneath her dress stood the words printed in red: "Laura Nelson Year 1911." The effigy lynching was intended to participate in national efforts to increase the visibility of police brutality that had gained momentum after the shooting of Trayvon Martin and Michael Brown. These moves seek to anchor the artwork in digital as well as print movements of anti-racist activism that reference the radical media work completed by black feminist and queer bloggers (cf. Dinsman et al. 2016).

At first glance, the artistic works fit well into the institutional history informing the university habitus and/or social field because of Berkeley's radical activist campus history: from the Free Speech, Anti-War, and Women's Movements to the Student Nonviolent Coordinating Committee (SNCC) and later the Black Panther Party.[16] Artistic modes intervene into legal modes of racial materialization that naturalize cultural rituals of anti-black violence. Through shock, the hanging images intervene symbolically into the liberal college setting.

Upon finding the effigy lynchings, Chancellor Nicholas B. Dirks and Executive Vice Chancellor and Provost Claude Steel called them "deeply

---

16 According to UC Berkeley spokeswoman Claire Holmes, "UC Berkeley has a long history of student activism, from the Free Speech Movement to the campaign for divestment from apartheid South Africa. The Free Speech Movement, when students organized for the right to campaign for political causes on campus, is now lauded by the student guides on campus tours. Less mentioned is the period of protest against the Vietnam War... Many anti-war demonstrators were also involved in the Third World Liberation Front, an organization that sought to start various departments dedicated to the study of ethnic minorities, such as a department of Black studies. The movement began at San Francisco State University in 1968 and was taken up at UC Berkeley in 1969" (Holmes 2017).

disturbing" and urged the creators of the images to come forward in a letter released on Sunday, December 14, 2014:

> The African American community has historically faced the terrorism of lynchings used in an attempt to suppress and control. While we do not know the intent of the effigies, the impact that it has had on our campus community is undeniable. No individual or group has claimed responsibility. If you were responsible for this, we invite you to come forward [...] We must all be vigilant to ensure that we are creating a campus environment that allows for the free exchange of ideas and doesn't frighten or intimidate people. Our campus climate reminds us that we still have [a] great deal of work to do to make this campus a welcoming place for all members [...] (UC Berkeley Public Affairs 2014).

UC Berkeley spokeswoman Claire Holmes told *Berkeleyside* "It's unclear what the intent was [...] It could be that these are related to police violence and the protests that are going on in Ferguson. It could be racially motivated as well" (Dinkelspiel/Raguso 2014).

The *LA Times* reported that many black students were shaken by the effigy lynching, provoking the response: "Oh, my God, another racist attack against the black community" (Abcarian 2014). Just two years before, as part of its Halloween décor the Berkeley chapter of Theta Delta Chi fraternity had hung a "zombie body" from a rope from the second-story window. It had been stuffed with straw and "had a dark gray head and was dressed in jeans and a white shirt." At the time, the former Black Student Union chair Salih Muhammad told *Onyx Express Magazine*: "This is another incident that convinces me and my community that this campus and society is not post-racial, and racis[m] is alive and well" (cf. Rondoni 2012).

In the mode of artistic materialization, the effigy lynching writes the pain and loss of black life into the presumed safe space of the neoliberal campus. Yet, the fear associated with symbolic violence is left disproportionately among black students, leaving them per usual open and vulnerable to anti-black symbolic violence. For black students, the violence of counter-materialization is a lesser evil than post-racist race-making, however, both effects socially alienate, disenfranchise, and actively inscribe black death into the university habitus – which, in contrast to the university system, offers two ways out: the staging of anti-black symbolic violence allows the university to both restore its (temporarily unsettled) symbolic order and to simultaneously celebrate its open-mindedness (through engaging with critique and/or anti-black symbolic violence).

Taking into consideration the artists' intension allows the administration of UC Berkeley to restore the effigy lynching from its apolitical narrative vacuum and continue its institutional narrative of being a modern institution with a radical past. In Sacvan Bercovitch's observations regarding the Myth and Symbol School, Bercovitch noted that the power of American ideology – which we take here as American democracy – is its ability to welcome dissent as a defining, if not compulsory, feature of its being. Hence, the power of American ideology is its capacity to incorporate dissensus as fundamentally American. "To criticize myth is to 'appreciate' it from within, to explicate it 'intrinsically,' in its own 'organic' terms. To criticize a piece of ideology is to 'see through' it, to 'expose' its historical functions, necessarily from an 'extrinsic,' and usually from a 'hostile' perspective" (Bercovitch 1986: 638). Not only does the incorporation of dissensus re-stabilize the liberal attitude of the liberal campus, but here we see how welcoming the critique into the very definition of what constitutes the American institution (of higher education), allows for that institution to narrate itself as modern, liberal, and inclusive.

The societal dialogue within the social field/habitus of the American university system offers further insight into the structural organisms that house, make up, and sustain the liberal "well-meaning" symbolic order of modern institutions of higher education. Dissensus – here in the form of an artistic mode of materialization – produces permutations of (rather than abruption from) earlier modes of legal materialization. In both scenarios – as a potential negative threat and/or positive artistic expression – the university habitus or social field controls how the event is narrated into its institutional history and symbolic order.

Here we turn our attention to the artists' intention to gain a better understanding of *how* their artistic message – and its function(s) – materializes within the university habitus. During the brief period when neither students nor the UC Berkeley administration knew how to narrate the effigy lynching within their campus habitus, the university administration poses the question of the artistic intention to its campus community so as to decipher whether the staging of black death in the university campus (1) *advocates* the inscribing of symbolic violence into the 'safe' liberal university space with the hope to suppress struggle and legitimize anti-black violence via a hate crime, or (2) functions within its university habitus so as to *shock* and ultimately rehabilitate white liberal students (by making them aware of their implicit bias). The effigy lynching has a "larger" democratizing effect: it allows the university to be liberal and negotiate its American values.

Following Bourdieu, it would be impossible for the artist to avoid the errors of an anti-black system which would allow the artist message to transcend race. There will be permutations that claim social progress; however, they will always resurrect new power dynamics that keep its previous infrastructure intact.[17] The artistic mode of materialization shows an attempt to extract a historical event from its utopian vacuum and temporarily place it within an alternative national history. Even within an alternative mode of materialization, namely the disruptive artistic mode, the denial of race-based structures continues to persist. It shows the limits of performativity: inasmuch as it intervenes symbolically, the artwork does so by simulating anti-black violence, which through new modes of artistic engagement reproduces similar enactments of violence that many black students have already experienced.

The artists attempt to intervene into a symbolic system that normalizes or presents black death as a natural or logical precursor to the American symbolic order. It interjects the loss of black life into white cultural memory and begs for an alternative treatment of black bodies as living, human bodies. As Calvin Warren argues in his "Black Nihilism and the Politics of Hope," the continuance of the American political order depends on the hope for a better future. Hence, what Warren terms the "not-yet-social-order" depends on the mythic construction of American hope to justify the continuation of the pain and suffering associated with that symbolic order.

> The Political and anti-blackness are inseparable and mutually constitutive. The utopian vision of a "not-yet-social order" that purges anti-blackness from its core provides a promise without relief – its only answer to the immediacy of black suffering is to keep struggling. The logic of struggle, then, perpetuates black suffering by placing relief in an unattainable future, a future that offers nothing more than an exploitative reproduction of its own means of existence. Struggle, action, work, and labor are caught in a political metaphysics that depends on black-death (Warren 2015: 233).

17 "Symbolic violence is the coercion which is set up only through the consent that the dominated cannot fail to give to the dominator (and therefore to the domination) when their understanding of the situation and relation can only use instruments of knowledge that they have in common with the dominator, which, being merely the incorporated form of the structure of the relation of domination, make this relation appear as natural; or, in other words, when the schemes they implement in order to perceive and evaluate themselves or to perceive and evaluate the dominators (high/low, male/female, white/black, etc.) are the product of the incorporation of the (thus naturalized) classifications of which their social being is the product" (Bourdieu 2000: 170).

Here, what Warren terms "the logic of struggle" collides with Herbert Hoover's rugged individualism to claim not to see or to be beyond race while simultaneously turning a blind eye to how bodies materialize differently and are forced to perform under race-based legal systems. In this way, anti-blackness is an integral, unavoidable, if not compulsory component of the American democratic order. Artistic and legal modes of racial materialization intersect in important ways; they inscribe anti-black violence as a prerequisite into the American symbolic and democratic institutional order. They use anti-blackness to make themselves part of Berkeley's "radical activist" campus history – one which celebrates free speech and artistic freedom – to assert their desire or "hope" for American democracy.

Modern institutions of higher education embrace this paradoxical situation (of being able to welcome dissensus as confirmation in all forms) as a defining mark of their democratic order. In our hope for a progressive transnational American future, we undoubtedly commit ourselves to speaking and acting within a language and thought system that is based on American and global anti-black racism. It represents a new social order where the simulation of *black death* is needed in order to restore white rehabilitation. In both scenarios, where the administration restores the symbolic order either from redemption or by just cause, anti-black violence occurs.

Post-Racism in Research and Teaching in Germany: the *Forschung & Lehre* Political Correctness Debate

We extend the initial prompt of this collection of essays to reflect on the discursive power of the image to ask who can speak and who is *hurt* when negotiating power. As junior transatlantic scholars of American Studies, we are living, working, and creating within a German academia that largely does not posit race, especially the German term *Rasse*, as a German national problem and/or valid conceptual reference. Due to its connotation with the history of the holocaust, these terms are used with caution.

German approaches to American Studies use the English-language category "race" and race-based theories to describe an American phenomenon, but rarely and selectively apply them to German events. In *Doing Race: 21 Essays for the 21st Century,* Paula Moya and Hazel Markus make the following observations:

> As is clear from the definitions of race and ethnicity, the process involved in doing them overlap in important ways. In addition, the terms 'race' and 'ethnicity' are sometimes used interchangeably. In the United States, for example, a person's race and a person's

ethnicity can both be claimed as a source of pride and identity. This is not the case, however, in Europe, where the word 'race' is almost never used (Markus/Moya 2010: 22).

Markus and Moya proceed to observe that "In Europe and elsewhere, ethnicity is commonly used to refer to cultural, linguistic, religious, and/or geographical 'others.' As a result, ethnicity often functions in Europe the way race usually works in the United States" (ibid: 22). As Markus and Moya contend, modes of materialization function differently in Germany. For example, unlike in the U.S., it is illegal and considered taboo to complete race-based or ethnicity-based census reports.

Although race in a statistical sense is largely still invisible, understanding where legal modes of materialization publicly perform can at times be quite obvious: it happens when the citizenship of non-white Germans is put into question. It also happens in more overt ways, like in the 2005 killing of Oury Jalloh in Dessau. Jalloh, a Sierra Leonean asylum seeker, was found tied to a mattress in his single cell after he had burned to death, assumedly set afire by the correctional staff. There exists a strong tendency to discredit knowledge systems that attempt to give credit to race-based or ethnicity-based knowledge infrastructures in Germany. Far more common and therefore 'believable' is the appeal to see these situations as individual, isolated events in time. We ask ourselves how a German confectionary materializes as black in German-language discourse if there is no visible infrastructure that would allow us to historically narrate it as such? We believe that at its core, systems of race can only persist because German discourse denies the existence of race-based structures in the first place.

Center stage on the cover of the April 2016 issue of the German academic magazine *Forschung & Lehre* was a picture of the German confectionary "*Schokokuss*" (chocolate kiss) with a small shadow. The cover is in a stark white with the words "Political Correctness" printed underneath. This issue was particularly controversial considering the disturbing history the dessert bears within German language discourse – as they typically were referred to as "*Mohrenköpfe*" (moor heads) or "*Negerküsse*" (Negro/n***er kisses). Since its publication, a petition was started at Change.org in which the initiators stated that the cover expressed a certain "nostalgia for a past that was free of politically correct speech" (Working Group, "Political Correctness in Germany"). The image stands alone and is never critically addressed in any of the six articles – besides a brief rhetorical question in which Elisabeth Wehling asks "Ist es nicht lächerlich, dass wir uns den ‚Negerkuss', dieses deutsche kulinarische Kulturgut, verbieten?" (2016: 300),[18] in order to enter her own argument in defense of political correctness. Wehling is the only female contributor, while all others are white men.

The cover display of a *Schokokuss* with the title "Political Correctness" performs a post-racial materialization akin to the chair lynchings in that it does not mobilize bodies of color, yet activates an extremely provocative register of racialized violence which is then left to a series of white men to defend in relation to German democratic values. We contend that German academia and German Americanist discourse perform modes of post-racialization no different than American political discourse. Like the latter, the former relies on a limited yet self-universalizing humanist and white "we" and its attendant catalogues of equality, life, liberty, and the right to own property in order to read identity politics out of democratic political discourse, thus concealing the function of such instances as race and gender in the formation of academic regimes of labor and capital.

Similar to Eastwood's use of "we" in his speech, the essays situate political correctness within a dialectic between "the" disenfranchised (*die Randgesellschaft*) and mainstream society. "Dazu gehören etwa Menschen, die Frauen sind, die keine weiße Haut haben, die geistig oder körperlich verletzt sind" (Wehling 2016: 300).[19] This framing of political discourse reduces complexity and situates a pseudo-generalized difference *outside* of the *internal* politics of the debate. Not only do the authors position political correctness *within* American humanism and thus *outside* German culture and its knowledge formations, they at times frame it as a threat to democracy due to its perceived undermining of freedom of speech and thought: "Ist die Political Correctness zu einem sehr ernstzunehmenden und die Demokratie mit Rede- und Gedankenfreiheit bedrohenden Feind geworden?", Volker Kronenberg asks in his contribution (2016: 297).[20] Alexander Grau staunchly argues against political correctness by portraying it as geared towards "Gedankenmanipulation und der Schaffung eines neuen Menschen" (Grau 2016: 294).[21] This publication fails to address the prominent issue that *within German-language academic discourse*

18 "Isn't it ludicrous that we have to forbid this German culinary cultural artifact?" (our translation).

19 "This includes people who [identify as or] are women, those who do not have white skin, [and/or] those who are hurt physically or mentally [i.e. differently abled]" (our translation).

20 "Has Political Correctness become a serious enemy of a democracy based on freedom of thought and expression?" (our translation).

21 "[M]anipulation of thought and the creation of a new man" (our translation).

there lacks a language to talk about race-based or ethnicity-based structures. By continuing to posit race and racism as a U.S. phenomenon, German-language academic discourse claims the existence of a colorblind academic space while nevertheless relying on race-based structures to enter into conversations on political correctness.

German modes of racial materialization are bound to the existence of race-based or ethnicity-based structures. Similar to the legal modes of racial materialization in American national contexts, the staging of a mute referent is used to negotiate and save white culpability. As an intervention, the image fails to engage thoughtfully with anti-racist discourse, but instead reproduces a unilateral or paradigmatic defense of whiteness. Those who will be hurt by this image are of no concern. So much so that no reasonable explanation is used to anchor the image within the German research on race and ethnicity that *Forschung & Lehre* claims to engage. Without its specific language and race-based history in German-language discourse, the image could not materialize race. From thereon, the confectionary is allowed to materialize as black so as to negotiate white neoliberal language policies; only when it is convenient for the discussion of German-language policy is its material valuable; however, when it is hurtful – and the speakers must assume responsibility – it bears no real symbolic meaning and therefore becomes dispensable.

Whereas American modes of materialization claim that we live in a new period that is beyond racism and therefore beyond political correctness, German modes of materialization function slightly differently in that they fail to acknowledge the history of ethnicity-based or race-based discrimination of non-white people. German academia has yet to fully integrate anti-racist discourse in its teaching and research strategies. It still lives on the margins. Here the argument against political correctness is quite different: it says because racism has (1) either not transpired like it has in the U.S., or (2) never truly existed for non-white people, political correctness is as much unnecessary as it is an assault on our right to a democratic self-image.

Of interest for this article is how the issue of *Forschung & Lehre* performs a series of epistemological acts in which German academia demonstrates its involvement in current regimes of transatlantic labor and capital. Debates concerning political correctness become reduced to the freedom to use violent language. In order to defend this right, the conversations in *Forschung & Lehre* relocate matters of political correctness to the United States. Failing to recognize the social inequalities found in German society, political correctness becomes both an index of racism and post-race at the same time. It is read both as a remedy and an erasure of existing problematic race relations, serving both as an unnecessary cure to racism and as a tool to signal the overcoming or absence of racism.

Conclusion

The denial of race-based systems is salient to their further materialization. In asking who can speak and who is heard or hurt, the debate on post-race and its enmeshment with labor and capital regimes such as class, demonstrates the necessity to engage questions of who is human and who is a (citizen-)subject – i.e. whose voice is heard or can be heard – and how the definitions of these concepts foreclose certain horizons of political action. American humanism has relied on "we"-narratives of equality-as-humans regardless of race, gender, or class in order to veil the differences these terms mark and the regimes of labor and capital they sustain. At the intersection of these social, political, and economic dispositifs, race continues to be enmeshed with the formation of class.

Post-race has to be understood as marking contemporary permutations of race and its role in the reproduction of labor and capital relations under post-capitalist deindustrialization and digitalization, rather than signaling a post-race after race, racism, and racialization. Without the need to invalidate race-based systems, a crucial aspect of materialization would not exist. The hegemonic shift from rugged American individualism to colorblind individualism is not only necessary for the further materialization of race-based structures, but evocative of a whole generation of scholarship that will need further attention. Under this current epoch of post-racism, race continues to materialize into American and German life. Artistic interventions are necessary, but not fully adequate to tackle the political and social realities of post-racism. Finally, it is time for German research to finally take seriously the impact of anti-racist work. Otherwise, it runs the risk of experiencing a true crisis within democracy instead of the current perceived threat that political correctness debates are being used to simulate.

By moving from the cruel spectacle of mob lynching at the turn of the 20th century to contemporary effigy lynching, the article followed the transformations of such racial materialization from the common-sense race model closely connected to phenotype and the presence of black bodies, to contemporary examples of post-race and the post-racialization of chairs and desserts. As race, racialization, and racism are no longer tied to the immediate presence of black phenotypes and bodies, just as contemporary (post-)capitalism no longer relies on the immediate presence of the body of the worker (whose function is often outsourced to the so-called third world), so race transforms into post-race. Within this newest era of post-racism, we are witnessing the lethal violence and structural discrimination of people of color on the one hand; on the other, we see the pervasive hegemonic move to refute or erase race in order to consolidate the ideology of American humanism and the regimes of labor and capital it supports.

Works Cited

Abcarian, Robin (2014): "Berkeley effigies' powerful racial message was hurt by unclear intent." In: L.A. Times December 19 (http://www.latimes.com/local/abcarian/la-me-1219-abcarian-berkeley-effigies-20141219-column.html).

Bercovitch, Sacvan. "The Problem of Ideology in American Literary History," Critical Inquiry 12.4 (1986): 631-653.

Bourdieu, Pierre (1979): "Symbolic Power" In: Critique of Anthropology 4/13-14, pp. 77-85.

Bourdieu, Pierre (2000): Pascalian Meditations, Stanford: Stanford University Press.

Butler, Judith (1993): Bodies that Matter: On the Discursive Limits of 'Sex', New York: Routledge.

Chan, Wilfred (2015): "Columbia Becomes First U.S. University to Divest from Prisons." In: CNN June 24, (http://www.cnn.com/2015/06/23/us/columbia-university-prison-divest/).

Coll, Steve (2007): "Disparities: The Jenna Six." In: The New Yorker October 8 (https://www.newyorker.com/magazine/2007/10/08/disparities).

Davis, Angela Y./Cassandra, Shaylor (2001): "Race, Gender, and the Prison Industrial Complex: California and Beyond." In: Meridians 2/1, pp. 1-25.

Dinkelspiel, Frances/Raguso, Emilie (2014). "Community Responds to Noose Effigies Found at Cal." In Berkeleyside December 13 (https://www.berkeleyside.com/2014/12/13/effigy-hung-from-sather-gate-before-berkeley-protest-march).

Dinsman, Melissa (2016): "The Digital in the Humanities: An Interview with Jessica Marie Johnson." In: Los Angeles Review of Books July 23 (https://lareviewofbooks.org/article/digital-humanities-interview-jessica-marie-johnson/).

Eastwood, Clint (2012): "Transcript: Clint Eastwood's Convention Remarks." In: NPR August 30 (http://www.npr.org/2012/08/30/160358091/transcript-clint-eastwoods-convention-remarks).

"Effigies of Black Men Found Hanging by Nooses at UC Berkeley Campus." In: The Guardian December 13 (https://www.theguardian.com/us-news/2014/dec/13/effigies-black-men-hanging-nooses-berkeley-campus).

Ellison, Zusha/Walter, Shoshana (2011): "Latest BART Shooting Prompts New Discussion of Reforms." In: The New York Times July 16 (https://www.nytimes.com/2011/07/17/us/17bcbart.html).

Fears, Darryl (2014): "Thousands join 'Justice for All' march." In: The Washington Post December 13 (https://www.washingtonpost.com/national/health-science/sharpton-to-lead-justice-for-all-march-in-dc/2014/12/13/36ce8a68-824f-11e4-9f38-95a187e4c1f7_story.html?noredirect=on&utm_term=.5895a0e77400).

Frank, Priscilla (2014). "Artist Collective Takes Responsibility For UC Berkeley Lynching Effigies." In HuffPost December 15 (https://www.huffingtonpost.com/2014/12/15/uc-berkeley-lynching-effi_n_6328434.html?guccounter=1).

Goff, Atiba, et al. (2014). "Black Boys Viewed as Older, less Innocent than Whites, Research Finds." In: American Psychological Association March 6 (https://www.apa.org/news/press/releases/2014/03/black-boys-older.aspx).

Grau, Alexander (2016): "Resultat einer Weltfluchtagenda: Eine Kritik der Political Correctness." In: Forschung und Lehre 4/16, pp. 294-296.

Hartman, Saidiya (1996): "Seduction and the Ruses of Power." In: Callaloo 19/2, Emerging Women Writers: A Special Issue, pp. 537-560.

Holmes, Lilian (2017). "Looking into UC Berkeley's History of Activism." In The Daily Californian April 9 (http://www.dailycal.org/2017/04/09/looking-into-uc-berkeleys-history-activism/).

Hoover, Herbert (1928). "'Rugged Individualism' Campaign Speech." Digital History (http://www.digitalhistory.uh.edu/disp_textbook.cfm?smtid=3&psid=1334).

Kronenberg, Volker (2016): "Verschärft (un)korrekt: Amerika und die Präsidentschaftswahlen." In: Forschung und Lehre 7/17, pp. 296-298.

Lorde, Audre (2007): "The Master's Tools Will Never Dismantle the Master's House." In: 1984. Sister Outsider: Essays and Speeches, Berkeley: Crossing Press, pp. 110-114.

"Lynching in America: Confronting the Legacy of Racial Terror", June 3, 2017 (http://eji.org/reports/lynching-in-america).

Maher, Bill (2011): "Real Time with Bill Maher on HBO October." October 9, 2011 (https://www.youtube.com/watch?v=tZed_FpQxwI).

Markus, Hazel/Moya, Paula (2010): Doing Race: 21 Essays for the 21st Century, New York: W.W. Norton.

McCoy, Terrance (2014): "Darren Wilson explains why he killed Michael Brown." In: The Washington Post, November 25 (https://www.washingtonpost.com/news/morning-mix/wp/2014/11/25/why-darren-wilson-said-he-killed-michael-brown/?utm_term=.c3458dc0ec73).

Pérez-Peña, Richard (2017): "Woman Linked to 1955 Emmett Till Murder Tells Historian Her Claims Were False." In: New York Times, January 27 (https://www.nytimes.com/2017/01/27/us/emmett-till-lynching-carolyn-bryant-donham.html).

Rayman, Graham (2016): "NYPD cop Daniel Pantaleo's pay rises to $120G while working desk duty since killing Eric Garner." In: New York Daily News, September 12 (https://www.nydailynews.com/new-york/chokehold-salary-rises-killing-eric-garner-article-1.2789313).

Rich, Wilbur (2015): The Post-Racial Society is Here: Recognition, Critics and the Nation-State, New York: Routledge.

Ricouer, Paul (1990): Time and Narrative, Chicago: University of Chicago Press.

Rondoni, Ally (2012): "Fraternity's Halloween decoration prompts outcry" In: Dailycal October 31 (http://www.dailycal.org/2012/10/30/fraternitys-halloween-decoration-prompts-outcry/).

Rothstein, Richard (2017): The Color of Law: A Forgotten History of How Our Government Segregated America, New York: W.W. Norton.

Scott v. Sandford, 60 U.S. 393 (1856).

Sexton, Jared (2011): "The Social Life of Social Death: On Afro-Pessimism and Black Optimism." In: Intensions 5/1, pp. 1-47.

"Slavery to Mass Incarceration", July 6, 2016 (http://eji.org/videos/slavery-to-mass-incarceration).

UC Berkeley Public Affairs (2014). "UPDATE: UC Berkeley Administrators Issue Statement on Saturday, Dec. 13 Incidents." In Berkeley News December 14 (https://news.berkeley.edu/2014/12/14/uc-berkeley-administrators-issue-statement-on-saturday-dec-13-incidents/).

Warren, Calvin (2015): "Black Nihilism and the Politics of Hope." In: CR: The New Centennial Review 15/1, pp. 215-248.

Wehling, Elisabeth (2016): "Alles, nur bitte keine 'Political Correctness': Wie Sprache unser Handeln beeinflusst und politische Ziele untergräbt." In: Forschung und Lehre 4/16, pp. 300-302.

West, Cornel/Buschendorf, Christa (2014): "Moral Fire - Martin Luther King Jr." In: Cornel West/Christa Buschendorf (eds.), Black Prophetic Fire, Boston and Massachusettes: Beacon Press, pp. 65-88.

White, Hayden (1973): Metahistory: The Historical Imagination in Nineteenth-Century Europe, Baltimore: John Hopkins University Press.

Working Group, "Political Correctness in Germany". "Political Correctness in der deutschen Wissenschaft." Change.org (https://www.change.org/p/forschung-lehre-political-correctness-in-der-deutschen-wissenschaft).

Wynter, Sylvia (2003): "Unsettling the Coloniality of Being/Power/Truth/ Freedom: Towards the Human, After Man, Its Overrepresentation – An Argument." In: CR: The New Centennial Review 3/3, pp. 257-337.

Yates, Diana (2016): "Study: Police More Likely than Others to Say They are Blind to Racial Differences." In: Illinois News Bureau May 16 (https://news.illinois.edu/blog/view/6367/362597).

~~*Kanak*~~ Academic: Teaching in Enemy Territory

Ismahan Wayah

> *I feel most colored when I am thrown against a sharp white background.*
> *(Zora Neale Hurston, "How It Feels to Be Colored Me")*

Introduction

Much has changed in recent decades in the humanities and social sciences, where marginalized groups have advanced to the center of a considerable number of scholarly inquiries. There is a growing awareness of the need to oppose white normativity in Western educational systems, accompanied by an increasing demand to decolonize the university. Nonetheless, academia continues to be embedded in racist and colonial practices and thinking, which are reflected in research methodologies, university curricula, and teaching methods. Many German academics in British, American, and Postcolonial Studies are well-versed in critical race theory, Black movements in North America and the United Kingdom, slavery, (post-)colonialism and intersectionality, but seldom can they name the milestone works by People of Color in German-speaking contexts, e.g. *Farbe bekennen* (1986), *Kanak Attak* (1998), *Mythen, Masken und Subjekte* (2005), or *Re/Visionen* (2007). Furthermore, it is not common to link, compare, and discuss discourses around race and racism in the U.S. or the U.K. with those in Germany. This disregard can be attributed to the prevalent underlying notion that post-World War II Germany is a race-less (*rassenlos*) country.[1] It is presumed that since race (*Rasse*) does not exist in the German

1 With 'race-less' (*rassenlos*) I want to draw attention to the fact that in Germany the term '*Rasse*' (race) is rarely used, even in discussions around racism. The reason is that, in contrast to 'race' in the Anglophone context, the term '*Rasse*' has not undergone sociohistorical transformations of meaning. '*Rasse*' is still primarily

context anymore, racism cannot exist, either; however, this post-racial discourse perpetuates the hegemony of whiteness (Roig 2017: 671). In this ostensibly race-less German society, Black and People of Color (BPoC) academics and educators are expected to erase their bodies and uphold 'universal,' 'neutral,' and 'objective' truths, as if knowledge were not also tied to the body (hooks 1994: 139).

This article emphasizes the embodiment of knowledge and interrogates the cultural work of non-normative bodies. More precisely, it poses the questions: What does it mean to research and teach as a Black Muslima in a hetero-normative, predominantly white, and middle-class space? How can BPoC students and academics occupy a subject position within an education system that has traditionally reproduced the exclusion of non-white, female, queer, and working-class bodies as subjects, and which still keeps these bodies at bay by implicitly and explicitly perpetuating the notion that non-normative bodies are at best objects of study? How do BPoC students and academics work around and within social power structures that are already set? This article is concerned with the ambivalence of agency for deviant bodies in violently normative academic spaces. I argue that the academy is one of the most significant pillars of imperial Europe; within this white, middle-class, heteronormative power/knowledge system, hence, there is structurally and discursively no space for BPoC bodies and thought. Furthermore, I propose '~~*Kanak*~~ academic' as a concept *sous rature* (under erasure) to open up subversive conversations around racist, classist, heteropatriarchal, and ableist discourses and practices in the German-speaking context.

---

understood as a biologistic terminology and a residue of Germany's Nazi history, although the 'science of races' was already established in the 17th century (Cremer 2008: 7). Those that reject the term '*Rasse*' often argue that its usage seems to suggest the existence of different distinct biological races. In the humanities and social sciences in Germany, there is a tendency to use the English term 'race' instead of German '*Rasse*' to emphasize the social construction of race and its real-life consequences for racialized people. Yet, Susan Arndt warns that this is an easy way out, as it implies that the term 'race' is less steeped in violence, colonialism, and genocides than '*Rasse.*' Furthermore, it creates a distancing from Germany as an object of study (Arndt 2005: 343; cf. Maria Alexopoulou's contribution in this volume).

Structural Racism within the German Education System

Why do BPoC students and academics feel alienated from academia? One way to answer this question is to list the various personal, social, and economic factors that impact educational opportunities in Germany. For one, it has been clear since the first *Program for International Student Assessment* (PISA) study in 2000 that educational success in Germany depends on the socioeconomic status of the student. Not much has been done to address structural inequality, with the result that Germany continues to be among the lowest-ranked of all OECD countries when it comes to social equity in education (OECD 2013; IFS 2012). The reluctance to transform the educational system in its entirety is manifested, in particular, in the unwillingness to abolish the three-tiered secondary-school system of *Hauptschule*, *Realschule*, and *Gymnasium* established in the 19th century by Otto von Bismarck (Phillips 1995: 4; Berg et al. 1991: 129). This is in spite of the fact that international studies have shown that in countries with large socioeconomic differences between distinct school forms, such as in Germany, the decisive obstacle to educational success is the social background of the student. This is why these studies caution that selecting students based on their academic achievement to attend either *Gymnasium*, *Realschule*, or *Hauptschule* strengthens the connection between socioeconomic status and performance. This mode of selection, they argue, "tends to accelerate the progress of those who have already gained the best start in life from their parents" (OECD 2016: 2).

One of the central drawbacks of the three-tiered secondary-education system is that it limits university-entry qualifications to one school form, namely the *Gymnasium*. Students of higher socioeconomic status are much more likely to be transferred to a *Gymnasium*, while students from socioeconomically disadvantaged backgrounds – in particular the children of immigrant parents and parents of Color – disproportionately end up in *Hauptschule* or *Realschule*, which prepare students for vocational training rather than an academic profession. Again, various studies have demonstrated that BPoC children – specifically of Turkish, Italian, Kurdish, Arab, or African descent – on average attend a *Hauptschule* twice as often as white German children of the same social class, and three times as often end up in a *Förderschule* (school for students with learning disabilities) (Autorengruppe Bildungsberichterstattung 2012: 85; Weishaupt/Kemper 2009: 99). Daniel Gyamerah's study illustrates that it is rather difficult for students of African descent to attain education that surpasses graduating from *Hauptschule* (2015: 42). Moreover, Kerstin Merz-Atalik's study shows that teachers tend to send students with a Muslim background to

Förderschule (Amirpur 2011). Thus, it is very likely that students who are both Black and Muslim, as a doubly marginalized group, end up in the lowest school forms. These students are condemned to a cycle of marginalization because education at a *Hauptschule* or a *Förderschule* mostly leads students to unemployment or, at best, low-income job opportunities.

According to an OECD study on Germany, only 9 percent of disadvantaged students succeed in beating the socioeconomic odds against them and exceed expectations in performance (OECD: 2015: 5). This comes as no surprise, since many poor, working-class, and/or BPoC students report having been discouraged by their educators from pursuing a *Gymnasium* education and an academic career. Aladin El-Mafaalani demonstrates in his qualitative study of social mobility in Germany that those who succeed in climbing the socioeconomic ladder have often not been supported by their educators, but rather by people outside school, such as a sports coach, a part-time employer, or a random acquaintance who took a chance on them (2014: 27).[2] El-Mafaalani also points out that educational success, for the participants in his study, often meant shedding working-class and immigrant values and attitudes in order to be able to adopt and perform a white German, middle-class *habitus* (2014: 23-26). This manifests itself in embracing specific cultural practices (e.g. museum visits, concert-going) and specific preferences in literature and music, deemed 'highbrow' art. However, as Pierre Bourdieu's studies have shown, having 'good taste' is not 'naturally acquired' but socially produced and thus, closely linked to educational level and social origin (Bourdieu 1984: 1). The deeply seated notion that studying at university is only for a select group of people, namely white middle-class students, is so pervasive that poor or working-class students are reluctant to enroll at university even if they have the required qualifications. According to Bertelsmann's education report, 77 of 100 students whose parents are university graduates enrolled in a university compared to 13 of 100 students whose parents graduated from a *Hauptschule* (Autorengruppe Bildungsbericht 2012: 125). In many ways, the three-tiered secondary-school system is successful in preserving

2 When considering social mobility, it is also worth looking into the number of students who have to change to a lower school form. One study found that in Baden-Württemberg, for every student that advances to a higher school form (*Hauptschule* to *Realschule* or *Gymnasium*), two students decline. Most worrying is Niedersachsen, where for one social climber, there are ten decliners, and Berlin, where for one social climber, there are seven decliners (Institut für Schulentwicklungsforschung 2012; cf. Schmoll 2012). This study does not take race or migration status into account, however, which in turn begs the question of how many of these might be BPoC students.

white class interests by reproducing the existing racialized social classes. In other words, what is at stake here is the hegemonic power of the existing white German ruling class.

I have started by laying out the problems of the secondary-education system to highlight the fact that when discussing the lack of a diverse student body and faculty at institutions of higher education, we first have to acknowledge the many barriers that BPoC and socioeconomically disadvantaged students and academics have to overcome. For those of us who are first-generation immigrant, working-class, and/or BPoC students and academics in Germany, the road to attending or working at a university has often proven to be very long and rocky, indeed. So how can we think about transforming a system that so many of us have been taught to believe and trust in? This question becomes even more pressing when considering how many BPoC students and academics are taught to respect, admire, and never challenge the legitimacy of the very educational system that oppresses them. BPoCs are often encouraged to uphold the notion of meritocracy, as our presence in academic spaces seems to suggest that race, class, or gender are indeed only incidental to educational success. However, Raymond Williams astutely notes that the image of meritocracy as a ladder available for everyone to climb is a bourgeois idea of society, for it is a device which can only be used individually. He further argues that the promotion of social mobility as self-interest legitimizes inequality and weakens communities by requiring marginalized people to be in competition with each other (1958: 331).

What Does It Mean to Be German? Black – Muslim – German

Taking into account social and economic structural obstacles faced by BPoC students is undoubtedly necessary; however, Robbie Shilliam observes that there is a tendency for university administration and staff to focus exclusively on external factors, thereby putting the blame solely on politics and policy makers (Shilliam 2014). It will not be sufficient to implement changes on an operational level, such as enrollment, curricula, and staff hires. This does not mean that it is not necessary to reflect on strategies of changing the curriculum to include more diverse writers and critics as well as attracting and retaining BPoC faculty. It means that these questions should not overshadow a much deeper issue, that is, the *systemic* reproduction of white middle-class normativity in the German academy. Postcolonial critics have long demonstrated that academic language and thought are located within this violent history of knowledge production. They have argued that how, why, and under which circumstances we speak about 'the Other' does *not* merely reflect the desire for more knowledge and insights. Rather, the act of speaking about 'the Other' in fact creates 'the Other' (Steyerl/Gutiérrez Rodríguez 2003: 10). Yet, the German academy has been too reluctant to disclose its involvement in nation-building processes, while its archives and genealogies have been produced vis-à-vis colonialism, anti-Semitism, anti-Romanyism, and heteropatriarchy.

In order to arrive at a comprehensive understanding of the problem, it is necessary to take into account the experiences of People of Color within the academy because BPoC students and academics find themselves, time and again, to be the sole Person of Color in seminars or entire departments. For them, academia translates into a feeling of constant uneasiness. In the short story "98 Percent Truth" ("*98 Prozent Wahrheit*"), Sarah Mouwani, a Black German writer, successfully puts into words how threateningly uncomfortable academia often is for BPoC students and academics. She depicts Humboldt University in Berlin as a space that, on the one hand, is haunted by its legacy of colonialism, racism, and anti-Semitism, and on the other hand, haunts Black and PoC students. "The larger than life-size Brothers Humboldt tower over Jasmin's weak knees. White marble statues, the gaze deep in a book. Brothers – a strange word. Equality, Liberty, and Brotherhood. White, blood-related statues. Blood" (Mouwani 2015: 78; my translation). While the protagonist Jasmin, a Black German student, is being constantly reminded that knowledge, epitomized by the Humboldt brothers, is white, male, cis-gendered, and able-bodied, she tries to make sense of her experiences as a Black woman at university. She reflects on how the racist-sexist assaults of her professor are trivialized by different institutional bodies, including the white gender-equality officer, the white secretary, and the white student representative. These institutional (white) bodies seem to apprehend "The Truth," which her Black female body cannot fathom, thus urging her, in a patronizing and colonial fashion, to "[…] please understand, he didn't mean it like that" (Mouwani 2015: 86; my translation). Mouwani paints a traumatizing image of Humboldt University, which is not only expressed in Jasmin's nervousness and alienation but also as physical pain, in the form of a constant headache. Mouwani's short story is a reflection on how normative and non-normative bodies are ideologically marked, but more importantly, observes that all bodies are cultural signifiers with particular historical, political, and social contexts.

Many BPoC students and academics can share stories of being dismissed by white professors and white students alike as 'too poor,' 'culturally under-developed, 'too African,' and/or 'too Muslim.' Thus, they apparently lack the essential qualities of 'culture,' 'knowledge,' and 'critical thinking.' To compensate, many social climbers often unconsciously adopt a white middle-class *habitus* to be able to integrate themselves into the new socioeconomic environment. However, at other times, imitating white middle-class culture is a deliberate coping strategy, which I will refer to as a 'utilization of mimicry.' In postcolonial theory, 'mimicry' is a desire to sever the ties with the 'self' in order to move towards the 'Other.' As Homi K. Bhabha points out, "mimicry is the desire for a reformed, recognizable 'Other,' as a subject of difference that is almost the same, but not quite" (Bhabha 1994: 122). In "Of Mimicry and Man," he argues:

> The effect of mimicry on authority of colonial discourse is profound and disturbing. [...] It is from this area between mimicry and mockery, where the reforming, civilizing mission is threatened by the displacing gaze of its disciplinary double [...]. What they all share is a discursive process by which the excess or slippage produced by the *ambivalence* of mimicry (almost the same, *but not quite*) does not merely "rupture" the discourse, but becomes transformed into an uncertainty which fixed the colonial subject as a "partial" presence [...] mimicry is at once resemblance and menace (1994: 123; original emphasis).

For me, one of the most absurd situations of mimicry of middle-class German whiteness I experienced as a student was when I applied for a scholarship for a Master's program in the United States. After setting up an appointment via e-mail, I dropped by the office, where I encountered a white German woman who was literally shocked to see me. She did not at all expect a young Black woman in *hijab*, and told me that she "had never seen someone like me [gesturing to my head] interested in the program." Part of the program entailed that the selected students would be "German ambassadors in the German American house," so the short interview revolved around me proving my 'Germanness.' The white interviewer asked if I would be willing to organize German cultural practices, such as "baking German Christmas cookies, coloring Easter eggs, organizing an Oktoberfest," and if I had "issues with working with men." By then it was obvious that she was not only surprised and confused by my presence, but clearly wanted to discourage me from applying to the program. My response to her neocolonial authoritative discourse around 'Germanness' and white feminism was to be as 'German' as possible, while also proving to be 'free' and 'liberated' enough. Adopting Bhabha's terms, I would now call this a strategy of mimicry and mockery: I pointed out to her that despite being Muslim, I had received a Catholic education in preschool and primary school and would have no issues whatsoever celebrating German Christian practices or working with men. Of course, I did not mention that among my white German friends, I did not know a single person who celebrated Easter, let alone still colored Easter eggs. I also did not mention that her question about "working with men" implied her belief in the Orientalist notion of Muslim women as being too oppressed by their Muslim fathers and brothers to freely mingle with other men. My mimicry of German Christian whiteness in a Black Muslim body caused a rupture in the colonial discourse by fixing the presence of the colonial Other. Furthermore, it threatened her notion, as it disclosed its arbitrariness. Yet, I was almost 'German,' but not quite, which meant I was not eligible to apply for a student exchange program. According to the white interviewer, they already had "much better qualified students," despite the fact that we had not discussed qualifications. When I now retell this racist incident, I confidently laugh about the situation and end with the sarcastic comment: "I bet I would have won a German cookie-baking competition." But my laughter cannot hide the lingering feeling of hurt and pain. After this experience, I felt humiliated and ashamed. We should never forget how racism degrades people by shattering a person's self-confidence and self-worth.

This conversation not only illustrates the absurdity and pain of an everyday racist experience on campus, but also indicates how much university and university-related institutions are involved in the process of nation-building, as they define who is 'German' and who is not. The delineating marker of 'Germanness,' as Fatima El-Tayeb points out, is whiteness and Christianity. She calls attention to how Europe, in particular the production of the imperial project 'Europe,' inevitably depends upon the existence of 'Others.' Furthermore, she convincingly shows that with the process of Europeanization, marked by the fall of border controls between European nation-states, the way of determining whether someone is European is increasingly recognized as their 'visible' appearance, i.e. whiteness (El-Tayeb 2003: 134). The narration of post-World War II Germany depends on the reproduction of perpetual 'Others,' such as so-called guest workers, migrants, Muslims, Jews, Roma and Sinti, refugees, and gender nonconforming people, who form a "constitutive margin" (Gutiérrez Rodríguez 2003: 24). The academy serves as a painful reminder of how racialized, non-normative bodies are stuck in marginal places. Nonetheless, poor/working-class, disabled, and racialized people continue to exist and produce knowledge from the margins, through and against white middle-class heteronormative German power/knowledge.

The figure of the *Kanak* can be fruitfully mobilized to articulate how non-white bodies, in particular those marked as both poor/working-class and Black, Muslim, of Color can subvert white middle-class German normativity. I am consciously deploying this term to draw on and underscore the strategies of resistance of second- and third-generation BPoCs in Germany. In his novel *Kanak Sprak*, which articulates the experiences of racialized communities, in particular the Turkish diasporic community, Feridun Zaimoglu draws an analogy to the Black consciousness movement. As a self-identifying expression, *Kanak* – similar to terms such as *Schwarzkopf* or *Ausländer* – refers to second- and third-generation children of so-called 'guest workers' and migrants who grew up in socioeconomically disadvantaged conditions. Frankfurt-based rapper H-Run argues: "For me every foreigner, may he be Turkish, Yugoslavian, Moroccan, or who knows what, is a *Kanak* because he is neither accepted in his own country nor here [in Germany]" (qtd. in Amipur 2005; my translation). Although H-Run claims every foreigner is a *Kanak*, there is the important qualifier that a *Kanak* neither feels at home in Germany nor in his home country. In other words, *Kanak* is a hybrid and racialized term situated at the intersection of race and class as well as at the intersection of racialization and migration. In *Undeutsch (Un-German)* Fatima El-Tayeb makes a similar case by showing how '*Kanak*' has turned from a self-designating term of Polynesian people, meaning simply "human," to a racial slur for those who are "not (really) 'German'," i.e. non-white and/or non-Christian. She argues that language as well as religion can also lead to a "kanakification of otherwise white racialized people" (El Tayeb 2016: 65; my translation). Feridun Zaimoglu and other *Kanaks* have stressed time and again that identifying as a *Kanak* is an honorific title, signifying: "I demonstrate and testify presence" (Zaimoglu 1995: 5; my translation). Kanak Attak, a diverse anti-racist and anti-nationalist activist group, redefined the term '*Kanak*' in the 1990s, writing in their manifesto:

> Kanak Attak is not interested in questions about your passport or heritage, in fact it challenges such questions in the first place. It challenges the conservative and liberal orthodoxy that good 'race relations' is [sic] simply a matter of tighter immigration control [...]. Kanak Attak is therefore anti-nationalist, anti-racist and rejects every single form of identity politics, as supported by ethnic absolutist thinking ("Manifest").

One of the few academics of Color identifying with the term is Reyhan Sahin, also known as Lady Bitch Ray, a Porn rapper, writer, and academic of Turkish decent. By repeatedly referring to herself as an "in Deutschland lebende Kanakenbraut" (*Kanak* bride living in Germany), she plays with the stereotype of the imported traditional and uneducated Turkish woman.

This image was most strongly promoted by Thilo Sarrazin in his controversially debated book *Germany Does away with Itself* (*Deutschland schafft sich ab*). In an interview he gave *Lettre International*, he states:

I do not have to accept anyone who lives off the welfare state, while rejecting it, who does not care enough for the education of his children, and who constantly produces more small veiled girls. This is true for 70 percent of the Turkish and 90 percent of the Arab population in Berlin (qtd. in Berberich 2009; my translation).

Sarrazin's comment illustrates the socially acceptable racist and sexist discourse around BPoC bodies, in particular Muslim women of Color, who are reduced to childbearing machines. Statements such as the one above rally both racial and sexual anxieties, urging us to critically reflect on how sexual and racial regulations are used to fend off perceived social threats. The interpellation of Muslim women as "*Kopftuchmädchen*" (veiled girls) and "*Importbräute*" (imported brides) illustrates that the "symbolic domain, the domain of socially instituted norms, is composed of *racializing norms,* and that they exist not merely alongside gender norms, but are articulated through one another" (Butler 2014: 182; original emphasis). Hence, intersectional approaches are not only fruitful but also necessary, since it is in many ways no longer possible to differentiate between axes of social regulation and power pertaining to race, gender, and sexuality. As Avtar Brah emphasizes, "racism is always a gendered and sexualized phenomenon," for the very idea of race is "an essentialist narrative of sexualized difference" (1996: 154).

By playing with these racist and sexist assumptions about women of Color, Lady Bitch Ray/Reyhan Sahin resists discourses of integration, assimilation, and dominant racialized gender politics. Her self-*Kanakification* and self-hypersexualization are a disruption of the "very possibility of considering her 'well integrated,' a status and social standing which appears in popular idiom often in the guise of the 'eine(r) von uns [Deutsche]' (one of us [Germans]) rhetoric" (Tuzcu 2017: 37). Moreover, it is crucial to be mindful of how the reproduction of heteropatriarchy and heteronormativity depends on how the reproduction of race is understood. The racial and sexual anxieties in the dominant discourse also carry the persuasive assumption that all BPoCs are necessarily heterosexual, thus neglecting the existence of queer BPoC. In her book *Bitchism,* Lady Bitch Ray/Rayhan Sahin coins the term "*Kanak_innen*" to turn away from male-centered hypermasculinity towards a feminization and queerness, indicated by the suffix '*-innen*' as well as the underscore (Tuzcu 2017: 44). Hence, she opens up a much-needed space for queer, transgender, and/or gender nonbinary *Kanaks*. Antke Engel's understanding of "queering" as a practice that does not only refer to gender and sexuality but more generally to all practices and processes which challenge norms and hierarchies, is conducive here. Considering this, it can be argued that the racially and sexually deviant body of *Kanaks* already has the potential to undertake a "queer-feminist and kanakist attack on the nation" (Engel 1999: 2; my translation).

One might be critical of the use of the term *Kanak*, yet it cannot be denied that this expression continues to be used widely within racialized poor and working-class communities in Germany. Having said that, I also acknowledge the concern of reestablishing pejorative terminology, which is why I propose that the term '*Kanak* academic' be understood as a concept *sous rature* (under erasure). In the preface to Jacques Derrida's *Of Grammatology* (Derrida 1976), Gayatri Spivak describes *sous rature* as to "write a word, cross it out, and then print both word and deletion"; in order to convey the meaning that "it is inaccurate yet necessary to say" (Spivak 1976: xiv). This is because the signifier in question is not wholly adequate for the signification, but it has to be used. The crossing out must be visible, and the signifier remains legible, as the constraints of our language offer no better replacement for it at the moment. Such partial erasure encourages one to re-problematize the meaning, applicability, and validity of the term, while at the same time opening up a space for creating new terminologies and categories (Slater 2004: 10). Thus, the ~~*Kanak*~~ academic inhabits a subversive space by performing a double move (St. Pierre 2000: 502). She is painstakingly conscious of how she is forced into subjectivity by sexist, racist, heteropatriarchal, and ableist discourses and practices. Despite the limitations, she continues to exhibit agency by constructing herself as a subject. The *Kanak*'s agency lies in what Walter Mignolo has termed "epistemic disobedience" (2013: 137), and in her potential for repeated renewal and reconfiguration, mutations and formations that are not determined in advance.

Considering that the academy is one of the most significant pillars of imperial Europe, it is high time to concede that structurally and discursively, there is no space for BPoC bodies and thought in a white, middle-class, heteronormative power/knowledge system. This does not mean that there are no BPoC students and academics, but that the higher-education system was never designed for non-normative bodies to begin with; it was intended first and foremost for white middle-class students and academics to thrive in.

~~*Kanaks*~~ in Academia: Racialized Bodies in White Academic Spaces

Some may view my very presence in academia as a ~~*Kanak*~~ academic, a working-class Black Muslima, as undermining my argument. In a way it may be used to negate the structures of exclusion in the educational system because it proves that some BPoCs – though few – have managed to be academically successful. Yet, Michelle Alexander points out in *The New Jim Crow* that "highly visible examples of black success are critical to the maintenance of a racial caste system in the era of colorblindness" (2010: 235). Similarly, the success of a handful of non-normative bodies is taken as evidence of the legitimacy of an educational system. These selected cases are used to strengthen the belief that the bodies trapped at the bottom of society still have the power to choose their fate. If a Black Muslim woman managed to reach this level of higher education, everyone else must be able to do so. If other BPoCs do not make it, it is because they are just lazy or intellectually incompetent.

A prominent example is the 2015 twitter campaign *#CadaanStudies*, in which Safia Aidid, a Somali-Canadian academic at Harvard University, drew attention to the fact that the international *Somaliland Journal of African Studies* had an editorial board that lacked a single Somali-identifying person. One of the board members, a white German anthropology scholar, defended the journal on social media by claiming that he had

> not come across many younger Somalis who would qualify as serious SCHOLARS – not because they lack access to sources but because they seem to not value scholarship as such. […] You have to work hard before you get out one piece of text and even then, you often get more criticism than praise. You certainly do not become rich quickly as a social scientist, at least if you have to pay your bills in Europe or North America. Now, where are all the 'marginalised' Somalis who do not get their share in academia? I guess you would have to first find all the young Somalis who are willing to sit on their butt for 8 hours a day and read and write for months to get one piece of text out (qtd. in Aidid 2015).

This white scholar's patronizing comment is not only racist, but also classist, as he mocks Somalis in the diaspora – a community with a predominantly refugee and working-class background – for focusing on instant riches and being too simple-minded for intellectually rigorous work. The white German scholar dismisses the fact that Somalis continue to be excluded from academic and policy-related discussions concerning them. Furthermore, he claims that the critique of Somali scholars is essentialist, and justifies their exclusion by valorizing scholarship as 'neutral' and 'objective.' He thereby reestablishes, in a colonial fashion, Western knowledge/power over research about Africa. The discussion of this incident has been largely ignored in German academia. The very few articles on it have, once again, downplayed the use of racist language and practices as a mere misunderstanding or an isolated incident by a 'black sheep' (pun intended). That this assumption is clearly not true can be seen in how the exclusion and dismissal of racialized bodies takes place in arguably more progressive academic spaces, such as those in which the discussion and analysis of racialized and gendered power structures are an integral part of the discipline, e.g. American, British, and Postcolonial Studies.

In 2015, scholars of the working group *Black Knowledges Research Group* (formerly *Bremen Black Studies*) at the University of Bremen decided to apply for a research grant to form a research group on the topic of *New Black Diaspora Studies*, which would be dedicated to the investigation of Black diasporic cultural productions. Despite previous criticism from the *Initiative of Black People in Germany* for consisting of an exclusively white research group (ISD 2014), the grant-application initiators once again ended up with an all-white research group. Black scholars and activists responded with a "Community Statement," emphasizing that the applied-for creative unit failed to conform to University of Bremen's diversity strategy, since the applicants seemed to mobilize 'diversity' and 'Blackness' as trendy buzzwords without questioning the positionality of white German scholars. They conclude that the way Black knowledge is employed at the University of Bremen "actively contributes to re-centering hegemonic white power structures as well as setting up a framework of research that is academically exclusionary, non-activist and potentially hetero-normative" ("Community Statement").

Irritated by the "Community Statement," a number of white academics in the fields of British, American, and Postcolonial Studies informed me that they found the "Community Statement" 'too radical' and its demands 'too exaggerated.' In response to my question whether they were familiar with the Black initiatives and movements in Germany, they disclosed that they had never heard of *Initiative Schwarzer Menschen in Deutschland e.V.* (Initiative of Black People in Germany), *Adefra e.V.*, or most of the BPoC activist-theorists such as Maisha Eggers, Fatima El-Tayeb, Kien Nghi Ha, Noa Ha, or Grada Kilomba, who have developed frameworks for thinking critically about the multifarious overlaps of race, colonialism, the Shoah, and the Roma genocide in Germany. When they are made aware of the work that has been done by BPoC movements in Germany, white Germans are seemingly shocked to see that BPoC people in Germany do think critically about whiteness. Yet, even this shock is an expression of racism, because "racist thinking perpetuates the fantasy that the Other, who is subjugated, who is subhuman, lacks the ability to comprehend, to understand, to see the working of the powerful" (hooks 1992: 339).

Shortly after the criticism, the *Black Knowledges Research Group* at Bremen University disbanded itself.[3] However, Black German scholars and activists also demanded that the research group engage in public and open discussions about the ethical questions and structural consequences of excluding marginalized communities from knowledge production. Despite the fact that the research group has also agreed to "a publicly held debate about these demands" (Black Knowledges), this has, as far as I know, still not occurred by late 2018. But the silence around systematic racist practices and exclusions speaks volumes: it is a strategy of epistemic violence, aiming at silencing Black resistance to co-option and appropriation. "Ultimately, this [strategy] is to preserve the racist status quo by all means […] so that colonial-racist normality can be maintained under the guise of white discretion" (Initiative Schwarzer Menschen in Deutschland 2014; my translation). The non-existing debate about the disbanded research network in Bremen is even more striking, given that 'diversity' and 'interculturality' have become buzzwords at German universities. However, the neoliberal academy steeped in white middle-class normativity only acknowledges and rewards a notion of diversity that remains exclusively in the realm of the symbolic. BPoC faculty are expected to be symbols of diversity, which means that they are supposed to be the face of diversity without promoting any actual political change in their departments or communities. Neoliberal multiculturalism creates, as Sofia Samatar poignantly shows, a "rupture between the diversity manifested in the body of the academic of Color" and the socioeconomic and political realities that affect their communities (Samatar 2015). Those academics that defy the commodification of their non-normative bodies as symbolic diversity and instead choose to continue working closely with their communities, are often rebuked and/or silenced for being too 'radical' or 'ideological.'

Apart from the many examples of the physical exclusion of BPoC academics from the German academy, it is important to highlight the many ways in which the discursive exclusion of BPoC voices is achieved. One of the methods which

---

3 The research group issued the following statement: "It has became [sic] clear to us that the *Black Knowledges Research Group* is thus more a part of the problem in combatting white enslavism and anti-Blackness than it is part of the solution. As such – this we have learned from our critics – our group is in effect blocking, rather than strengthening necessary structural change and a pro-active politics against the anti-blackness of German academia" (Black Knowledges).

has been used to subdue critical voices from the margins, mainly in the U.S. context but increasingly in Germany, as well, is to put the names of scholars and activists on online watchlists. *Professor Watchlist*, for instance, targets leftist and anti-hegemonic professors and, in particular, BPoC scholar-activists who are speaking out against, among other things, white supremacy, gender bias, Islamophobia, the police, and prisons ("Professor Watchlist"). African American philosopher George Yancy cautions that such lists have an impact similar to a Panopticon, as "a theoretical prison designed to create a form of self-censorship among those imprisoned" (Yancy 2016). The function of a watchlist is not to spy on or to out professors, but rather "to install forms of psychological self-policing to eliminate thoughts, pedagogical approaches and theoretical orientations" regarded as subversive (Yancy 2016). The acceptance of right-wing rhetoric on German campuses is worrisome and dangerous, in particular for BPoC students and academics. It is crucial to emphasize that the theoretical imprisonment of radical subversive BPoC thought takes place against the backdrop of globally increasing mass surveillance and mass incarceration of Black, Brown, and Muslim bodies.

So what does it mean to teach as a ~~*Kanak*~~ academic in Germany in an era of right-wing populism in Europe and the U.S., an era of the 'War on Terror,' which is used to justify mass surveillance and drone strikes in the Global South, an era of deadly border control both outside 'fortress Europe,' marked by drowned Black and Brown bodies in the Mediterranean sea, as well as inside through their subjection to racial profiling? As someone who ticks all the boxes of those Europe wants to get rid of (Black, Muslim, woman, working-class, and former refugee), teaching as a ~~*Kanak*~~ academic means teaching and researching in enemy territory. This brings me to Alexander Weheliye's uncomfortable question: "What does it mean to have to create oneself from scratch in environments where one is not supposed to exist?" (Weheliye 2016).

~~*Kanak*~~ Academics: Teaching in Enemy Territory

The curriculum, research methodology as well as classroom discussions in American and British Studies often tend to revolve around whiteness and the experience of white people. BPoC theorists, writers, and artists are rarely discussed, and if they are, they are reduced to being racialized minorities, ignoring the insights their texts give us into issues of gender, sexuality, bodies, or spirituality. Even departments with more progressive curricula and educators, including ~~*Kanak*~~ academics, tend to re-center whiteness by privileging questions and issues resulting from the lived experience of those positioned as white. This teaching practice assumes that whiteness and the workings of white normativity are invisible to all students. Hence, this approach ignores the experiences of BPoC students regarding how whiteness has inscribed itself into their bodies and how it re-inscribes itself again in the *very moment* a discussion on race is opened up in a predominantly white space. For instance, some educators tend to choose violently racist texts or graphic photos of racist assaults to make it as clear as possible to white students that racism, colonialism, and slavery have had real-life stakes for people. In doing so, they are oblivious to the devastating, traumatizing impact these words and images have on BPoC students, who are left alone in dealing with this triggering experience. This teaching practice does not take the diversity of the student body into account and is designed in such a way that the only students that gain from this discussion are white students. This often comes at the expense of BPoC students, who are expected to disclose their experiences of being subjected to racism, which in turn are analyzed, dissected, and assessed by the white gaze. In this way, their experiences are objectified. Feven Michaels, a Black German student of Ethnology, once told me that whenever there are discussions about Blackness or Africa in the classroom, she feels as if her body were turned into an 'object of study' for her fellow white students.

As a Black, *hijab*-wearing woman at the center of knowledge production, my mere presence puzzles students and colleagues alike. It requires justification of my existence in white, middle-class academic spaces, as it is viewed as a transgression. The reality of teaching while ~~*Kanak*~~ makes my experience qualitatively different from that of my white colleagues. Black British scholar Linda A. Palmer describes university campuses as "colonies where intellectual power and authority are [always] white" and in which BPoC academics are seen as "intruders" (qtd. in Grove 2014). Black British political scientist Robbie Shilliam illustrates how he never felt comfortable enough to walk slowly through the corridors of his institution and thus had to rush through and out of them. In the essay "Black Academia in Britain," he recalls how he "wanted to be at university, especially in the library, but all he could feel was that he was in enemy territory" (Shilliam 2014). Both statements resonate strongly with me, as I have also often experienced the academy as a hostile and violently white normative space, where I could never exist comfortably. That these are not singular isolated feelings is underscored by William A. Smith et al.'s groundbreaking psychological study on what he calls "racial battle fatigue." This term describes the impact of racial microaggressions, "defined as subtle, stunning, cumulative verbal and non-verbal insults layered with racism, sexism, elitism and other forms of subordination" on Black faculty at predominantly white colleges (Smith et al. 2015: 3). On the one hand, racial microaggressions maintain white normativity, and on the other hand, they put mental, physical, and emotional stress upon BPoC bodies in a manner similar to the stress soldiers experience on the battlefield (cf. Lili Rebstock's contribution in this volume for a more detailed discussion of the mechanisms of microaggressions). Smith et al. urge that more studies on racial battle fatigue be undertaken because otherwise, race-related stress responses go undiagnosed and untreated. It thereby adversely affects the health, academic success, professional productivity, and retention of faculty of Color (Smith et al. 2015: 15).

As racialized bodies disturb white spaces, academics of Color find themselves in a position in which they are at the same time the object and the subject of (racialized) classroom discussions. For instance, Fred Bonner, professor of education administration, noted that many Black professors experience "white classrooms filled with students who, on the one hand, question their academic credentials and, on the other hand, expect them to be funny and entertaining" (qtd. in Tuit et al. 2009: 72). Likewise, English professor Claire Garcia learned that her white students expected her to personally represent the fictional literary characters they were studying; she notes that she never felt as conscious of her race as when she stood before a class of 25 young men and women eager to learn about what it is like to be Black in America (1994: 2). While teaching a course on Muslim American fiction in the 2015 summer term, I had similar experiences. Although I have taught Black critical theory before, it seemed as if my students do not mark my Blackness as strongly as they do my Muslimness. Keen on learning about Muslims, my students subjected me to personal and intrusive questions while projecting their stereotypical notions of Muslim women on me. At the same time, they repeatedly questioned my credentials as an educator and academic. Black theorist and artist Grada Kilomba points out that speaking and listening emerge as an analogous project. Echoing Gayatri Spivak, she argues:

> The act of speaking is like a negotiation between those who speak and those who listen, that is, between the speaking subjects and their listeners. Listening is, in this sense, the act of authorization towards the speaker. One can (only) speak, when one's voice is listened. And those who are listened, are also those who belong, as well as those who are not listened, become those who 'do not belong' (2010: 86).

I have come to understand that as a ~~*Kanak*~~ academic, I do not have the privilege of walking into a classroom and having students assume that I am a capable and credible educator. I certainly have the privilege of speaking by setting my own thematic focal points in seminars, yet I do not have the privilege of being listened to by my students. The act of authorization towards me as a speaker in a white academic space is located in whiteness, and hence, remains with my white students. Thus, the question of who is heard and hurt in predominantly white-normative classroom discussions is a difficult and complicated one. As BPoC academics, we need spaces to discuss, exchange, and negotiate our multifarious racialized and gendered experiences in white academic spaces. In addition, educators in higher education have to familiarize themselves with inclusive teaching methods and tools which de-center whiteness and reflect various positions in a white German social context. It is necessary that we cultivate the classroom as a diverse, transgressive learning environment, where marginalized students can "name themselves, speak for themselves and participate in defining the terms of interaction, a situation in which we can construct an understanding of the world that is sensitive to difference" (Hartsock 2016: 210).

Conclusion

This article interrogated the idea of "state of embodiment" to reflect upon the figure of the *~~Kanak~~* academic in a predominantly white and middle-class space. While I certainly believe that it is pertinent to bring attention to some of the struggles faced by students and scholars of Color in German academia, the act of self-disclosure as a Black Muslim academic is filled with apprehension. For, as Audre Lorde aptly put it, "the transformation of silence into language and action is an act of self-revelation and that always seems fraught with danger" (1984: 42). Inhabiting a non-normative racialized body, the figure of the *~~Kanak~~* academic understands the relevance of creating counternarratives by drawing on racialized, gendered, and sexualized non-normative experiences to deconstruct and challenge the ways in which race and racism impact our pedagogical interactions. The challenge is to deconstruct the normative while considering the ways in which bodies are always already ideologically marked and the ways in which all bodies are cultural signifiers with deeply entrenched historical, social, and political contexts (Lacom/Hadley 2009: 55). We have to unlearn to admire and protect a neoliberal system that oppresses non-normative bodies, and demand transformations in order to create a more just society. As an academic who is aware that the way I research and teach is political, I believe that a continuous interrogation of the academy as a site is essential in order to encourage long overdue transformations of the status quo.

Works Cited

Aidid, Safia (2015): "Can the Somali Speak?: Open Letter to Dr. Markus Hoehne and the Somaliland Journal of African Studies." In: Warscapes April 1 (http://www.warscapes.com/blog/can-somali-speak-open-letter-dr-markus-hoehne-and-somaliland-journal-african-studies/).

Alexander, Michelle (2010): The New Jim Crow: Mass Incarceration in the Age of Colorblindness, New York: The New Press.

Amipur, Donja (2005): "Sprachvariationen in deutschen Ghettos." In: Heimatkunde Migrationspolitisches Portal September 18 (https://heimatkunde.boell.de/2005/11/18/sprachvariationen-deutschen-ghettos).

Amipur, Donja (2011): "Im Gespräch mit Prof. Merz-Atalik. Überrepräsentanz von Migrantenkindern auf Förderschulen." In: MIGAZIN: Migration in Germany December 12 (http://www.migazin.de/2011/12/12/uberreprasentanz-von-migrantenkindern-auf-forderschulen/).

Arndt, Susan (2005): "Mythen des weißen Subjekts: Verleugnung und Hierarchisierung von Rassismus." In: Susan Arndt/Maureen Maisha Eggers/Grada Kilomba/Peggy Piesche (eds.), Mythen, Masken und Subjekte: kritische Weißseinsforschung in Deutschland, Münster : Unrast , pp. 340-362.

Autorengruppe Bildungsberichterstattung (2012): Bildung in Deutschland 2012: Ein indikatorengestützter Bericht mit einer Analyse zur kulturellen Bildung im Lebenslauf, Bielefeld: W. Bertelsmann Verlag.

Berberich, Frank (2009): "Thilo Sarrazin im Gespräch: Klasse statt Masse. Von der Hauptstadt der Transferleistung zur Metropole der Eliten." In: Lettre International 86/1, pp. 197-201.

Berg, Christa (eds.) (1991): Handbuch der deutschen Bildungsgeschichte: 1870-1918. Von der Reichsgründung bis zum Ende des Ersten Weltkrieg, München: Verlag C.H. Beck.

Bhabha, Homi K (1994): The Location of Culture, London: Routledge.

Black Knowledges Research Group (2015): "Disbanding Black Knowledges Research Group." February 8 (http://www.fb10.uni-bremen.de/inputs/pdf/BKRG_Aufoesung-Disbanding_deu-engl.pdf/).

Bourdieu, Pierre (1984 [1979]) : Distinction. A Social Critique of Judgment of Taste. Richard Nice (trans.), Cambridge: Harvard University Press.

Brah, Avtar (1996): Categories of Diaspora: Contesting Identities, London: Routledge.

Butler, Judith (2014): Bodies that Matter: On the Discursive Limits of "Sex," London: Routledge.

"Community Statement: 'Black' Studies at the University of Bremen." Black Studies Germany, February 24, 2015 (https://blackstudiesgermany.wordpress.com/statementbremen/).

Cremer, Hendrik (2008): "'... und welcher Rasse gehören Sie an?' Zur Problematik des Begriffs 'Rasse' in der Gesetzgebung." In: Policy Paper Deutsches Institut für Menschenrechte 10/1, pp. 4-15.

Derrida, Jacques (1976): Of Grammatology. Gayatri C. Spivak (trans. and foreword). Baltimore: Johns Hopkins University Press.

El-Mafaalani, Aladin (2014): Vom Arbeiterkind zum Akademiker: Über die Mühen des Aufstiegs durch Bildung, Paderborn: Bonifatius GmbH.

El-Tayeb, Fatima (2003): "Begrenzte Horizonte. Queer Identity in der Festung Europa." In: Hito Steyerl/Encarnatión Gutiérrez Rodríguez (eds.), Spricht die Subalterne deutsch? Migration und postkoloniale Kritik, Münster: Unrast Verlag, pp. 129-145.

El-Tayeb, Fatima (2016): Undeutsch. Die Konstruktion des Anderen in der Postmigrantischen Gesellschaft, Bielefeld: transcript.

Engel, Antke (1999): "Queer-feministische und kanakische Angriffe auf die Nation Antirassistische Praktiken und das Konzept der Staatsbürger-Innenschaft." In: Antirassistische Öffentlichkeit-feministische Perspektiven 1/1, pp. 2-5.

Garcia, Claire (1994). "Emotional Baggage in a Course on Black Writers." In: Chronicle of Higher Education 40/47, pp. 1-3.

Geißler, Rainer/Weber-Menges, Sonja (2008): "Migrantenkinder im Bildungssystem: Doppelt benachteiligt." In: Aus Politik und Zeitgeschichte 49/1, pp. 14-22.

Grove, Jack (2014): "Black Scholars still Experience Racism on Campus." In: Times Higher Education March 20 (https://www.timeshighereducation.com/news/black-scholars-still-experience-racism-on-campus/2012154.article).

Gutiérrez Rodríguez, Encarnatión (2003): "Repräsentation, Subalternität und postkoloniale Kritik." In: Hito Steyerl/Encarnatión Gutiérrez Rodríguez (eds.), Spricht die Subalterne deutsch? Migration und postkoloniale Kritik, Münster: Unrast Verlag, pp. 17-37.

Gyamerah, Daniel (2015): "Der schulische Erfolg Schwarzer Schülerinnen und Schüler in Deutschland – Eine rassismuskritische Analyse des Mikrozensus." UN-Antirassismusausschuss – Parallelbericht Deutschland (https://rassismusbericht.de/wp-content/uploads/Hintergrundpapier-Daniel-Gyamerah).

Ha, Kien Nghi/Lauré al-Samarai, Nicola/Mysorekar, Sheila (eds.) (2016): Re/Visionen. Postkoloniale Perspektiven von People of Color auf Rassismus, Kulturpolitik und Widerstand in Deutschland, Münster: Unrast.

Hartsock, Nancy (2016): "Foucault on Power. A Theory on Women?" In: Susan A. Mann/Ashley S. Patterson (eds.), Reading Feminist Theory. From Modernity to Postmodernity, Oxford: Oxford University Press, pp. 210-212.

hooks, bell (1992): "Representing Whiteness in the Black Imagination" In: Lawrence Grossberg/ Cary Nelson/Paula A. Treichler (eds.), Culture Studies, New York: Routledge, pp. 338-346.

hooks, bell (1994): Teaching to Transgress: Education as the Practice of Freedom, New York: Routledge.

Hurston, Zora Neale (2016): "How It Feels to Be Colored Me." We Do American Studies. University of Virginia (http://xroads.virginia.edu/~ma01/grand-jean/hurston/chapters/how.html/).

Initiative Schwarzer Menschen in Deutschland (2014): "Gegen rassistische epistemische Gewalt an der Universität!" ISD Nov 18 (http://isdonline.de/gegen-rassistische-epistemische-gewalt-an-der-universitaet/).

Institut für Schulentwicklungsforschung/Technische Universität Dortmund/IfE der Friedrich-Schiller-Universität Jena (eds.) (2012): Chancenspiegel 2012 Regionale Disparitäten in der Chancengerechtigkeit und Leistungsfähigkeit der deutschen Schulsysteme, Gütersloh: Bertelsmann Verlag.

Kanak Attak (1998): "Manifest." November (http://www.kanak-attak.de/ka/about/manif_deu.html).

Kilomba, Grada (2010): Planation Memories: Episodes of Everyday Racism, Münster: Unrast Verlag.

Lacom, Cindy/Hadley, Susan (2009): "Teaching to Transgress: Deconstructing Normalcy and Resignifying the Marked Body." In: Maria del Guadalupe Davidson/George Yancy (eds.), Critical Perspectives on bell hooks, New York and London: Routledge, pp. 55-67.

Litter, Jo (1984): Against Meritocracy: Culture, Power and Myths of Mobility. London: Routledge.

Lorde, Audre (1984): Sister Outsider: Essays and Speeches, New York: Crossing Press.

Mignolo, Walter (2013): "Geopolitics of Sensing and Knowing: On (De)Coloniality, Border Thinking, and Epistemic Disobedience." In: Confero 1/1, pp. 129–150.

Mouwani, Sarah (2015): "98 Prozent Wahrheit." In: Emily N. Kuria (ed.), Eingeschrieben: Zeichen setzen gegen Rassismus an deutschen Hochschulen, Berlin: w_orten & meer, pp. 76-90.

OECD (2008): "Ten Steps to Equity in Education". OECD Observer Policy Brief, January (https://www.oecd.org/education/school/39989494.pdf).

OECD (2013): "PISA 2012 Results: Excellence through Equity (Volume II): Giving every Student the Chance to Succeed." OECD Publishing (https://doi.org/10.1787/19963777).

OECD (2016): "PISA 2015 Results. Volume I: Excellence and Equity in Education. Country Note Germany." OECD Publishing (https://doi.org/10.1787/19963777).

Phillips, David (1995): Education in Germany: Tradition and Reform in Historical Context. International Developments in School Reform, London: Routledge.

"Professor Watchlist". Turning Point USA, December 1, 2016 (https://www.professorwatchlist.org/).

Roig, Emilia (2017): "Uttering 'Race' in Colorblind France and Post-racial Germany." In: Karim Fereidooni/Meral El (eds.), Rassismuskritik und Widerstandsformen, Wiesbaden: Springer, pp. 613-627.

Samatar, Sofia (2015): "Skin Feeling." The New Inquiry, September 25 (https://thenewinquiry.com/skin-feeling/).

Schmoll, Heike (2012): "Doppelt so viele Schulabsteiger wie Schulaufsteiger." In: Frankfurter Allgemeine Zeitung October 30 (http://www.faz.net/aktuell/politik/inland/durchlaessigkeit-der-schulsysteme-doppelt-so-viele-schulabsteiger-wie-schulaufsteiger-11943746.html).

Shilliam, Robbie (2014): "Black Academia in Britain." The Disorder of Things, July 28 (https://thedisorderofthings.com/2014/07/28/black-academia-in-britain/).

Slater, David (2004): Geopolitics and the Post-Colonial. Rethinking North-South Relations, Oxford: Blackwell Publishing.

Smith, William/Yosso, Tara J./Solórzano, Daniel G. (2015): Challenging Racial Battle fatigue on historically white campuses: A Critical Race Examination of Race-related Stress, Bolton: Anker Publishing Company.

Steyerl, Hito/Gutiérrez Rodríguez, Encarnatión (eds.) (2003): Spricht die Subalterne deutsch? Migration und postkoloniale Kritik, Münster: Unrast Verlag.

St. Pierre, Elizabeth A. (2000). "Poststructural Feminism in Education: An Overview." In: International Journal of Qualitative Studies in Education 13/5, pp. 477-515.

Tuitt, Frank et al. (2009): "Teaching in the Line of Fire: Faculty of Color in the Academy." In: The NEA Higher Education Journal, pp. 65-74.

Tuzcu, Pinar (2017): Ich bin eine Kanackin – Decolonizing Popfeminism, Bielefeld: transcript.

Weheliye, Alexander G. (2016): "Black Life." In: ICI Berlin March 22 (https://legacy.ici-berlin.org/event/696).

Weishaupt, Horst/Kemper, Thomas (2009). "Zur nationalitätenspezifischen und regionalen Bildungsbenachteiligung ausländischer Schüler unter besonderer Berücksichtigung des Förderschulbesuchs." In: Ina Sylvester et al. (eds.), Bildung, Recht, Chancen. Rahmenbedingungen, empirische Analysen und internationale Perspektiven zum Recht auf chancengleiche Bildung, Münster: Waxmann, pp. 97-111.

Williams, Raymond (1958): Culture and Society, London: Hogarth Press.

Yancy, George (2016): "I Am a Dangerous Professor." In: The New York Times November 30 (https://www.nytimes.com/2016/11/30/opinion/i-am-a-dangerous-professor.html).

Zaimoglu, Feridun (1995): Kanak Sprak. 24 Mißtöne vom Rande der Gesellschaft, Hamburg: Rotbusch Verlag.

The Migrant Scholar of Color as Refugee in the Western Academy

Mahmoud Arghavan

Introduction

The modern refugee regime, which was established in the aftermath of the Holocaust based on the 1951 Convention on the Status of Refugees,[1] defines a refugee as "people who 'owing to a well-founded fear of persecution, on the grounds of race, religion, nationality or membership of a social group, find themselves outside their country of origin, and are unable or unwilling to avail themselves of the protection of that country' (Article 1a of the 1951 Convention on the Status of Refugees)" (Betts 2009: 5). Besides internally displaced people, border-crossing refugees constitute one type of forced migration who "vote with their feet." They have always been at the heart of world politics (Betts 2009: 6), because on the one hand, by crossing international borders the refugees symbolically discredit and delegitimize the sovereignty of their ruling states that either had persecuted or failed to protect them from persecution by others. On the other hand, beyond the national borders of their countries, these stateless border-crossers are subject to international law, and because of the 1951 convention, the international community is responsible for them.

1 The original from 1951 was geographically confined to Europe but its scope was made universal through the 1967 Protocol to the Convention. The core principle underpinning the regime is *non-refoulement*, which prohibits states from forcibly returning an individual to a country in which he or she faces a well-founded fear of persecution. Responsibility for monitoring and overseeing the implementation of the 1951 Convention lies with the Office of the United Nations High Commissioner for Refugees (UNHCR) (cf. Betts 5-6).

Alexander Betts in *Forced Migration and Global Politics* (2009) names the major contributing causes to displacement in the contemporary era as

> the two world wars, the colonial liberation wars, the proxy conflicts of the Cold War, a range of internal conflicts in the Balkans, Africa, and the Caucasus in the aftermath of the Cold War, occupation in Afghanistan and Iraq in the context of the "War on Terror," state partitions and nationalist claims to territory in South Asia and the Middle East, authoritarian regimes, human rights violations, large-scale development projects, and environmental disasters resulting from hurricanes, tsunamis, and climate change (Betts 2009: 2).

Perhaps inspired by W. E. B. Du Bois, who once correctly envisioned that "The problem of the twentieth century is the problem of the color line," (Du Bois 2007: 32) Thomas Nail in *The Figure of the Migrant* states that "The twenty-first century will be the century of the migrant" (2015: 1). The Office of the United Nations High Commissioner for Refugees (UNHCR) has estimated that "[b]y the end of 2017, 68.5 million individuals were forcibly displaced worldwide as a result of persecution, conflict, violence or human rights violations," of whom 25.4 million are refugees, the highest ever seen (UNHCR 2018).

Accordingly, Betts in an essay of 2015 states that because of two current global trends of fragility and mobility, "there are grounds to believe that refugees and displacement are likely to become a defining issue of the 21st century" (Betts 2015). The fragility of states which are unable or unwilling to ensure the human rights of their citizens, triggers the flow of survival refugees who flee wars, famine, religious harassment, ethnic cleansing, political persecution, etc.[2] Moreover, due to globalization the opportunity and inclination to move have generated greater human mobility than ever before. Migrant academics and skilled migrants who consciously, but not necessarily voluntarily, decide to depart their homelands in the Global South, hoping for a better life, academically and financially, in the West, will be placed in the latter category.

A note on this essay's perspective and structure: in the following I, who belong to the latter group and have been working with the former group of displaced people in refugee camps after I quit Western academia, will explore three issues regarding the experience of the migrant scholar of color who, for reasons such as political repression or lack of career opportunities in their home country, has taken refuge in Western academia and thus, in a white-dominated academic system. I draw on my 'me-search' on the subject position of the

2 The Fragile States Index places countries such as South Sudan, Somalia, Central African Republic, Syria, and Afghanistan towards the top of this list.

'migrant scholar of color,' a triangular figuration whose every angle indexes particular inclinations and raises critical questions regarding displacement, interruption, exclusion, utopian-dystopian struggles, racism, and shattered dreams in the West and its academic system. I perceive the mentioned experiences as similar to those of refugees in the West.

First, I will discuss the figure of the refugee as Agamben puts it, and its significance for developing more commitments to ethical standards in discussions about the current controversies in the world politics. Then, I will elaborate on the figure of the 'Migrant Scholar of Color' (MSC), and the important relationship between the subject position of the scholar of color and the subject matter of research when it concerns marginalized people of color across the globe. In the same vein, I will discuss the agency of the MSC in challenging the systemic constraints and the structural discrimination in his or her homeland and in the new country of residence. Finally, by comparing the situations of the MSC and the refugee, I will reflect on the role of both in diversifying the world. The MSC, I argue, contributes to the decolonization of the knowledge production systems through taking recourse to his/her insights into diverse parts of the world and also through his/her resilient battle against discriminatory structures both at home and in the host society; thus, s/he promotes the idea of global humanities. In a similar manner, the border-crossing refugees, I argue, diversify Europe through their 'unhomely' colorful presences.

The Figure of the Refugee

Giorgio Agamben, the Italian political philosopher, in "Beyond Human Rights" (2000) by adopting a biopolitical perspective and distinguishing between political life (*bios*) and biological life (*zoe*), discloses the pitfalls of the triangle Territory-State-Nation in the conception of human rights. From this view, the dependence of the rights of man on the nation-state appears problematic. The paradox is that the nation-state, which "makes nativity or birth [*nascita*] (that is, naked human life) the foundation of its own sovereignty"[3] (Agamben 2000:

3 "Human rights, in fact, represent first of all the originary figure for the inscription of natural naked life in the political-juridical order of the nation-state. Naked life (the human being), which in antiquity belonged to God and in the classical world was clearly distinct (as *zoe*) from political life (*bios*), comes to the forefront in the management of the state and becomes, so to speak, its earthly foundation. Nation-state means a state that makes nativity or birth [*nascita*] (that is, naked human life) the foundation of its own sovereignty" (Agamben 2000: 19-20).

19-20; original parentheses), has no "autonomous space in its political order" for the pure human in itself and will take no responsibility for its citizens beyond its borders, because crossing the geographical borders of the sovereign state, means transgressing the borders of its legal domain. This means once a citizen of a sovereign state takes off the garment of nationality and transforms into a refugee, his/her naked body loses "every quality and every specific relation except for the pure fact of human being" (Agamben 2000: 18) in need of survival.

Thus, as Agamben explicates, "precisely the figure that should have embodied human rights more than any other – namely, the refugee – marked instead the radical crisis of the concept" (Agamben 2000:18). He emphasizes that the un-representable 'figure of the refugee,' which threatens the very foundations of the nation-state and nationalism, must be the center for thinking the world anew and for the paradigm of a new historical consciousness (Agamben 2000: 14). He argues that

> given the by now unstoppable decline of the nation-state and the general corrosion of traditional political-juridical categories, the refugee is perhaps the only thinkable figure for the people of our time and the only category in which one may see today – at least until the process of dissolution of the nation-state and of its sovereignty has achieved full completion – the forms and limits of a coming political community. It is even possible that, if we want to be equal to the absolutely new tasks ahead, we will have to abandon decidedly, without reservation, the fundamental concepts through which we have so far represented the subjects of the political (Man, the Citizen and its rights, but also the sovereign people, the worker, and so forth) and build our political philosophy anew starting from the one and only figure of the refugee (Agamben 2000: 15).

Long before Agamben, during and in the aftermath of World War II, when Europe and the world were going through a humanitarian crisis accompanied by an unprecedented flow of displaced people, including Jews and other war refugees, Hannah Arendt in "We Refugees" contemplated that countryless/ stateless refugees "driven from country to country represent the vanguard of their peoples" because "[t]hey know that the outlawing of the Jewish people of Europe has been followed closely by the outlawing of most European nations" (Arendt 2007: 274). Having experienced the problematic dependence of human rights on the nation-state, she titled the chapter of her book *Imperialism* that concerns the refugee problem "The Decline of the Nation-State and the End of the Rights of Man" (Arendt 1951: 2).

Accordingly, in *Wasted Lives, Modernity and its Outcasts* (2004), Zygmunt Bauman conceives of the 'superfluous' population of migrants, refugees, and other outcasts as inevitable outcomes of modernity, capitalism, and economic progress. He looks at the situation of refugees through a 'biopolitical' lens, as well, conceptualizing a refugee's life as 'wasted life,' i.e. as the human residue of globalization and of what he calls 'liquid modernity' and 'liquid modern life,' which features permanent transience, absence of a fixed social role, and presence of a political vacuum (Bauman 2004). From Bauman's perspective, refugees as victims of negation, exclusion, and expulsion represent another form of 'wasted lives,' commodities which have lost their value in the political and economic market of conflicts and trade between nation-states. For Bauman, "Liquid modernity is a civilization of excess, exuberance, waste, and waste disposal" (Bauman 2005: 120), and the outcasts, refugees, migrants, and unemployed with their precarious lives are obvious cases of the 'waste of modernization.' To use Giorgio Agamben's terminology in *Homo Sacer, Sovereign Power and Bare Life* (1998) they struggle for survival of their biological lives (*zoe*) while being deprived of a political and socially qualified life (*bios*). Moreover, the sheltering of asylum seekers in a 'non-citizen space,' such as a refugee camp, and the mere fulfillment of their basic biological needs while depriving them of a political life and the human freedom to decide about their destiny, which Arendt in *The Human Condition* (1958) underscores to be the core of the human condition, makes refugee camps a new form of concentration camp. The stateless refugees, whose legal and political status has been suspended, are forced to live, to use Judith Butler's terms in *Precarious Life,* 'unlivable and ungrievable lives' (Butler 2006: xv). According to Butler, "the differential allocation of grievability that decides what kind of subject is and must be grieved, and which kind of subject must not, operates to produce and maintain certain exclusionary conceptions of who is normatively human: what counts as a livable life and a grievable death?" (Butler 2006: XIV-XV).

The status of refugee has always been a temporary condition and provisional arrangement that ought to lead to two alternative choices: an anthropophagic choice/naturalization or an anthropoemic choice/repatriation (cf. Agamben 2000: 19; Bauman 2008). The former choice is assimilation and absorption into the new body of the nation-state, whereby refugees shed their existence as refugee to become insider Others. The second solution is to expel and repatriate refugees. When a refugee resides in a new country and is 'semi-naturalized,' s/he is transformed into a noncitizen resident or 'denizen'[4] (Hammar 1990) of the land

4 Until the 19th century the term was used for a foreigner who was assigned the status of a subject by the king through 'letters of charter.' The longer these persons legally remained in the country, the more rights they obtained, becoming semi-citizens or

of residence. A denizen speaks the language, has a family and income there, but enjoys fewer rights than the citizens, and perhaps more than newly arrived asylum seekers. Denizens will be ghettoized in a vacuum sphere and live invisible, 'unsayable,' 'unimaginable,' and 'unthinkable' (Bauman 2008) lives. They become Others for the native people and assist the latter in consolidating their sense of national identity; the latter will often vote for nationalist parties which plot to silence these Others and ideally expel them. The refugee problem and the problem of denizens will be discussed in political campaigns and everyday discussions, but in real life they have neither a voice nor the right to vote.

Nevertheless, Bauman believes that there "will perhaps be a time when we will discover the avant-garde role of today's refugees – in which we will explore the taste of life in non-places and the pervatic permanence of transience that could become the common habitat of the citizens of this globalized and full planet" (Bauman 2004: 203). Agamben likewise emphasizes the presaging role of the refugee for the future of humankind by stating that "[o]nly in a world in which the spaces of states have been thus perforated and topologically deformed and in which the citizen has been able to recognize the refugee that he or she is – only in such a world is the political survival of humankind today thinkable" (Agamben 2000: 25).

As the stateless figure of the refugee directs us towards rethinking the world in order to make it more inclusive for everybody through curbing the power of the sovereign state, the figure of the migrant scholar of color (MSC), by possessing knowledge about corners of the world that are unknown to European scholars, will contribute to decolonizing knowledge production and destabilizing the white academy. In "The World and the Home" (1992), Homi Bhabha argues that "Where the transmission of 'national' traditions was once the major theme of a world literature, perhaps we can now suggest that transnational histories of migrants, the colonized, or political refugees – these border and frontier conditions – may be the terrains of World Literature" (Bhabha 1992: 146). According to him, we should neither focus on the 'sovereignty' of national cultures nor on the 'universalism' of human culture, but on those 'freak displacements' that have been caused within the cultural lives of postcolonial societies. Because displacement blurs the borders between home and world and confuses the divisions that require a noble way of thinking about current issues, something MSCs may be able to promote by drawing on their subject positions and lived experiences. The MSC can promote the discourse of global humanities through

denizens. Ultimately denizens had fewer rights than citizens but more rights than foreigners.

taking recourse to his/her insights into diverse parts of the world and through his/her resilient battle against discriminatory structures both at home and in the host society.

The Migrant Scholar of Color

Bearing in mind the fundamental differences between refugees and MSCs in terms of their living conditions and paths of arrival in the West, in this section I will reflect on the parallels between experiences and difficulties of MSCs, who struggle to make a home in the Western, white-dominated academy and experiences of refugees, who have to content with racism and discriminatory policies on an everyday basis in their struggles to make a home in the 'White West.' MSCs and refugees, both border-crossers and invaders of spaces which were not created for them, with their fragmented identities inhabit an interrupted temporality and spatiality at the intersection of leaving and not-arriving, the intersection of being rejected and refused in the lands of origin and the lands of residence.

Being aware of similar experiences of and alliances between scholars of color and migrant scholars of color in German academia, I should like to highlight some differences between the two and some extra obstacles that the MSC faces. The MSC's story differs from the story of the 'native scholar of color' along the same lines as refugees of color experience the world differently from both white refugees and 'native people of color.' Crossing geographical and epistemological borders entails dislocations in time and space which result in renegotiating the self-image by the refugee and MSC in accordance with shifting their borders and perspectives. Coming to terms with their new temporal and spatial environments takes time. Deracinated and not yet regrounded, but subjected to racialization in the host society, the MSC is constantly concerned with the identity question of "Who am I?" His/her identity process begins with being a 'migrant scholar,' but through the experience of "not passing through or passing by, of being stopped or being held up" (Ahmed 2012: 3), that is, through the experience of exclusion, ignorance, and colorblind racism, s/he becomes a 'migrant scholar of color.' The MSC dwells at the intersection of geographical, institutional, epistemological, and ontological displacements, the intersection of not-belonging and not-qualifying, and in the area of tension between the utopian imagination of a free academy and the dystopian reality of a white academy. The MSC lives at the margins of the Western academy like a 'denizen,' a 'margizen,' a temporary guest worker (*Gastarbeiter*) whose task is to fulfill the strategic diversity agenda of the institution and to consent to his/her repatriation afterwards. Just as in the case of the refugee, the dream of the MSC to make a home in the West is a mirage.

The MSC in a Western institution is a foreigner who does not speak the language, who is a stranger to the rules of the game of the institution and unaware of the ongoing discussions. The MSC does not belong and is never going to belong. The MSC dwells in between two spaces. Although s/he belongs neither to the land of origin nor to the land of residence, the MSC is burdened by challenges posed by systematic constraints and structural discrimination in his or her homeland *and* in the new country of residence.

Perhaps 'natives of color' experience ongoing microaggressions more than refugees or migrants since they speak the language of their society fluently, whereas refugees and migrants are made to feel the impact of racialization and discrimination more acutely. Statements such as "You speak English/German so well!" or "Where are you really from?" have the effect of making 'native' people of color feel alien in their own land. Because both 'native people of color' and migrant people of color are discriminated against in one way or another, I do not make this distinction in order to 'play discrimination Olympics'; nor is this statement meant to suggest that one group is less exposed to exclusion than the other.

It requires extra labor by the MSC to walk out of his vacuum space, adjust to a new institutional life and intellectual discourse, and establish a new social network. Of course, during this period the MSC also has to meet the expectations of the advisors, who usually take pride in mentoring an 'exotic' PhD candidate, but cannot be bothered to put extra effort in helping this stranger find his/her way in a totally new establishment.

The subject position of the MSC in the humanities differs in two respects from that of his/her counterpart in the natural/life sciences and in engineering. The latter MSCs are predominantly hired by the industry sector or research institutions financed and directed by corporations, which will profit from hiring competent MSCs. As a result, MSCs in the natural/life sciences and engineering have less difficulty in finding a job in white-dominated organizations. The myth of a postrace era might, in fact, materialize in the most important agents in the neoliberal order, i.e. capitalist giants, as long as a researcher of color contributes to the growth of the corporation's revenue. On the other hand, as soon as maximizing profit in capitalist/economic terms is not (proudly claimed) to be the main priority, however – as is the case in the humanities – the exact opposite effect sets in, and MSCs are not systematically invited to the marketplace, but structurally excluded from the academic field. Moreover, the object of study in the natural/life sciences and engineering differ minimally in the Global South and the West. Although in terms of infrastructure such as labs and research equipment in the natural/life sciences, the MSCs in the Global South in comparison to their Western counterparts may be in a disadvantaged position, the knowledge required for their education had been freely accessible in their lands of origin without being subject to state censorship, which controls the publications in the humanities and social sciences. Once these MSCs join a research institution in the West, they can cope more easily with their assigned tasks, whereas MSCs in the humanities and social sciences have to update their knowledge according to the standards and recent developments in their fields, to which they had only partial access in their countries of origin, however.

Contrary to the professional white Western scholars who, well equipped with the privileges of the First World's education system and steeped in its Eurocentric system of knowledge, rely on books to speak of the suffering of people of color, the MSC draws on his or her lived experience to speak of the unprivileged. In this sense, the self-reflective scholar of color does an intersectional me-search based on his/her lived experiences to critique the ongoing manifestations of racism, sexism, xenophobia, and heteronormativity.

In the following I will theorize from my lived experience of becoming a 'Migrant Scholar of Color' at an 'average institution' in German academia, which neither gave me residence, nor was it ever going to do so, to explain why I conceive of the MSC as a refugee in the Western academy.

In 2008, I completed my master's degree at a newly established American Studies program at the University of Tehran. Being labeled from day one as disloyal to the political agenda of the university, I struggled for two and half years with everyday challenges of studying in an ideologically charged atmosphere. Eventually I graduated with the conclusion that I could no longer live in my home country. Each systematic obstacle that I had to overcome to obtain my master's degree brought me one step closer to the borders of the Iranian academy, although I had had every intention to make a home there. Homeless and broken, in 2009, a three-year scholarship from the European Union facilitated my crossing the borders and taking refuge in an academic institution in the West. A white male professor, who happened to be one of the most influential figures in the field in Germany, agreed to supervise my PhD project. According to German academic culture, he adopted me as his '*Doktorsohn*,' since a supervisor is either a '*Doktorvater*' or '*Doktormutter*' to their PhD students, depending on the professor's gender. However, not all 'children' of these professors always receive equal treatment. I have come to learn that one can roughly differentiate between those PhD candidates who were socialized in a European or other Western educational system as biological children of these professors, and migrant students who grew up in the educational systems of the Global South as adoptive children. For instance, biological children are qualified to represent the institute in summer schools, conferences, and research stays, while the adoptive children need to stay home and perform in the institute's 'diversity shows' in order to fulfil the diversity measures of the program so that the latter can obtain further funding.

While I was probably not the most genius adoptive son of the white male professor, I asked him several times to provide more supervision in order to help me resettle in a German competitive PhD program. However, I hardly received any supervision at all. It took me the entire three and a half years of the PhD process to realize that the discrepancies I perceived between my utopian image of Western academia and my lived experiences were, in fact, real. For the white male professor, I was perhaps just an exotic name on the list of his PhD students who needed to finish his dissertation within the planned timeframe: somebody who will leave Germany after obtaining his degree and go back to where he belonged.

Initially, I used to perceive myself as a migrant scholar from a totalitarian country (Iran), who would make a home somewhere in Western academia. But through encountering 'institutional discrimination' and structural barriers as well as experiencing the invisibility of and invisible discrimination against scholars of color within German academia, I reached my subject position as a migrant scholar of color. As Mignolo would put it, I was "following the dreams and the life of the Spirit, only to realize, at some point, that the Spirit was not welcoming of Third World spirits" (Dabashi/Mignolo 2015: ix). Therefore building on Sara Ahmed's terminology in *On Being Included: Racism and Diversity in Institutional Life*, by 'white academics,' I refer here to those who inhabit a category of privilege, which protects them from all the obstacles that people of color have to encounter and overcome. I refer to those whose residence is assumed. In my view, what constitutes whiteness more than the color of the skin is epistemic racism,[5] which according to Walter Mignolo in *Can Non-Europeans Think*? "is built on classifications and hierarchies carried out by actors installed

5 Mignolo (2015) identifies epistemic racism as the logic "beneath the naturalization of certain ways of thinking and producing knowledge that are given the name Eurocentrism. Racism is not a question of one's blood type (the Christian criterion used in sixteenth-century Spain to distinguish Christians from Moors and Jews in Europe) or the color of one's skin (Africans and the New World civilizations). Racism consists in devaluing the humanity of certain people by dismissing it or playing it down (even when not intentional) at the same time as highlighting and playing up European philosophy, assuming it to be universal" (xi).

in institutions they have themselves created or inherited the right to classify and rank" (2015: xv) without ever being classified themselves in turn.

After receiving my PhD in June 2013, I insisted persistently 'on being included' in German American Studies mainly because I had underestimated the 'brick wall' of immobility in front of me and partly because I was optimistic to pass through it. For three semesters, I enjoyed teaching at that 'average institution' as an adjunct lecturer (*Lehrbeauftragter*), with little pay and great pleasure, hoping for a more rewarding position to come.

For a long time I was caught in a state of hesitation, of not knowing what was happening and how to react to frequent and deliberate exclusions. You need extra labor to comprehend a situation that others are structurally protected from, and respond to it (Ahmed 2012: 177). I experienced frequently what Eduardo Bonilla-Silva in *Racism without Racists* (2006) calls "smiling face" discrimination ("We don't have jobs now, but please check later"). It was only during this time while insisting on belonging to the institution that gives residence to others, that I discovered and developed my self-image as a MSC through the experience of being overlooked, excluded, and not being taken seriously by the white professors. Although I acknowledge the social and intellectual support that I received from a white female colleague at the same institute, this proves the fact that whiteness has more to do with inhabiting the norms and less with skin color.

Two applications cleared the air for me to reach the conclusion that diversity does not go beyond the form of official statements in German humanities. In theory, the university is required by law to take into account diversity measures, such as giving priority to women, ethnic minorities, and disabled people. However, in practice no one can question the institutions which do not implement these guidelines. The white professors can always justify hiring white Germans as their assistants or *Habilitanden*. I experienced minority studies, black studies, and postcolonial studies in German academia at large as a hypocritical business of capitalizing on people of color's suffering around the world without integrity in practice. There are plenty of cases that I perceive as a discrepancy between theory and practice on the part of white academics who engage in this progressive philanthropist business of studying disadvantaged sections of Western societies and disenfranchised people of the formerly colonized world.[6] Marginalized people of color from the Global South serve white academics as interesting objects of study to form a research group and get

6 White researchers' attempt to establish an all-white Black Studies program at the University of Bremen (cf. Kimberely Alecia Singletary's contribution in this volume) is only one known example.

it financed by the DFG (German Research Foundation) in order to sponsor more white junior scholars. But migrant scholars of color from the Global South are hardly qualified to be members of these research networks. As Kishore Mahbubani (2011) in *Can Asians Think?* and Hamid Dabashi (2015) in *Can non-Europeans Think?* explain, the Eurocentric system of knowledge production naturalizes certain ways of thinking.

My application for a postdoctoral position as coordinator of a research network sponsored by a grant from DAAD (German Academic Exchange Service), which was to establish the average institution as a hub for research on the literary self-expression of formerly colonized societies, was one of the two decisive and eye-opening occasions. According to the job ad, the director of the network (surprisingly a white male professor) was seeking a candidate who already held a PhD in literary studies and had a postdoc project thematically related to the Global South. Coming from the Global South, having immersed myself in postcolonial literary theory, and being able to present a well-developed postdoc project titled "American Transcendentalism and Persian Sufism in a Global Civil Society," I applied for this position fairly optimistic. Nonetheless, a male native speaker of German who was still working on his PhD project, which addresses a German literary figure of the enlightenment era, got the job. The legality of this selection process was questionable, but I could not make a case against the white professor because the words of an outsider MSC against an established white professor would not count. Looking at this from the metalevel, what disillusioned me most was the fact that the job ad had explicitly included *Das Allgemeine Gleichbehandlungsgesetz (AGG)* or Germany's "Anti-discrimination law," which "outlaws discrimination based on race or ethnic background, gender, religion or worldview, disability, age or sexual identity" (my translation).[7] But preaching is one thing, practicing is quite another.

On another occasion, in 2015 I applied for one of two 50 percent lecturer positions at that average institution. I was encouraged to apply for this position because I was already teaching two seminars there as an adjunct lecturer (*Lehrbeauftragter*) and the position required nothing more. For me it was more a matter of recognition and less about full membership in a white academic club. Nonetheless, I was not astonished at all to be overlooked again. The department decided to staff one of the positions with a woman PhD student of color, and leave the other one vacant – which suggests that despite excellent qualification

7 *Das Allgemeine Gleichbehandlungsgesetz (AGG)* is a federal law that outlaws "Benachteiligungen aus Gründen der Rasse oder wegen der ethnischen Herkunft, des Geschlechts, der Religion oder Weltanschauung, einer Behinderung, des Alters oder der sexuellen Identität."

on my end, the department's 'commitment' to diversity seemed to be exhausted once the spot of a token scholar of color, who served to 'prove' this commitment, was filled.

In the last week of my stay at the average department, I had a very upfront conversation with one of the white professors of the that 'average institute,' pointing out my points of criticism. They had no idea about my grievances because they had never had to view the world from the unprivileged position of the MSC.

According to Sara Ahmed, one tactic of diversity workers is to not 'quite' inhabit the norms of an institution and the norms of a society. When you do not conform to institutional norms, or you aim to transform them, "you notice them as you come up against them. The wall is what we come up against: the sedimentation of history into barrier that is solid and tangible in the present, a barrier to change as well as to the mobility of some" (175). Therefore, white academics as well as German citizens constantly need to question themselves whether they belong to the class of those ignorant of the wall. Because "when a category allows us to pass into the world, we might not notice that category. When we are stopped or held up by how we inhabit what we inhabit, then the terms of habitation are revealed to us. We need to rewrite the world from the experience of not being able to pass into the world" (176).

Ever since I started embracing the emancipatory potentials of my failure[8] as a MSC and took refuge in refugee camps, to work with and for fugitives like myself (being aware of the differences between my privileged situation compared to those trying to enter Europe on overcrowded boats while facing the risk of drowning), I have appreciated being in communion with refugees in our homelessness, in our brokenness. During my first days at work, before I had the chance to introduce myself, the white colleagues always mistook me as another refugee or in other cases as an interpreter of Arabic or Farsi/Dari. They were in a way right. I was and I am a refugee. Just like refugees from Syria, Afghanistan, Nigeria, Somalia, and many other war zones, who had fled life threatening-situations in their homelands to reach a safe place to live, I had escaped the totalitarian educational system of Iran, which had negated me and many other subversive intellectuals and denied us career opportunities. The refugees' human rights "to life, liberty, security, employment, and education" had been denied in their lands of origin; my home country had denied my right to academic freedom and the free flow of knowledge. We both are looking for a home in societies and institutions which reserve residency predominantly for whites. We are Others,

8 For approaches that theorize the liberating and resisting potentials of failure, see Moten/Harvey (2013); Halberstam (2011).

"space invaders," to borrow Nirmal Puwar's (2004) wonderfully evocative expression. We have invaded "the spaces reserved for others. We sometimes even experience ourselves as space invaders, a way of experiencing spaces as if they are not reserved for us (and, indeed, they are not)" (Ahmed 2012: 13). We both have been stopped from fulfilling our dreams in our lands of origin. Now we are struggling to make a living in the West, longing for belonging.

Nonetheless, our struggles have bright sides, too. We, the migrant scholars of color and refugees, are agents of diversity. The mere presence of us, the 'Others,' in Europeans' institutional and everyday lives brings the 'brick wall' between them and us into view. We remind the European institutions and established citizens of the thriving walls of racism, xenophobia, Eurocentrism, and homophobia. We attract the public's attention to these walls. We insist that looking away from the walls does not mean that we are over race, nationality, religion, gender, and class. We remind white Europeans that a postracial society and the discourse of colorblindness is nothing but a myth. As Sara Ahmed phrases it, "The very idea that we are beyond race, that we can see beyond race, or that we are 'over race' is how racism is reproduced; it is how racism is looked over" (Ahmed 2012: 183). Diversity work is a refusal to look away from what has already been looked over (Ahmed 2012: 183). We remind white Europeans of their 'buts' as they say: "I am not a racist, but …" "I am not a nationalist, but …" "I am a feminist, but …" "I am not a homophobe, but …" We challenge them by responding "but what?" There is no 'but' in living beyond race, gender, religion, ethnicity, and nationality. The veiled Muslim woman who is struggling to find residence in my comfort zone brings my wall into my view. So do I bring the wall into the white professors' view as I persist on finding a residence in their institutions.

Sara Ahmed states in *On Being Included* that diversity workers are not just philosophers[9] who reflect on the institutions, become conscious, and sometimes criticize the institutions.

> Rather, diversity workers acquire a critical orientation to institutions in the process of coming up against them. They become conscious of 'the brick wall,' as that which keeps its place even when an official commitment to diversity has been given. Only the practical labor of 'coming up against' the institution allows this wall to become apparent. To those who do not come up against it, the wall does not appear – the institution is lived and experienced as being open, committed, and diverse (174).

---

9 Referring to Marx's oft-quoted statement in *Theses on Feuerbach* that "The philosophers have only interpreted the world in various ways. The point, however, is to change it."

Migrant scholars of color, refugees, and their allies in doing diversity work try new strategies and tactics for getting through the wall. The aim is to transform the institution, society and the world through transforming the wall into a table, as Sara Ahmed suggests. The diversity workers engage a "critique of the way whites explain, defend, and ultimately justify the contemporary racial order" (Bonilla-Silva 2006: xv) through exposing the "color-blind racism" in institutional and everyday lives.

Conclusion

In recent elections in Europe and the U.S., nationalist and right-wing extremist parties have celebrated their victories either by taking office or by forming a powerful opposition voice against the ruling party/coalition. In Europe, parties such as *Alternative für Deutschland* (AFD) in Germany, The League in Italy, Swiss People's Party in Switzerland, and Freedom Party in Austria, among others, have gained significant electoral votes. In the U.S. since 2016, a populist holds the presidential office who propagates xenophobia, racism, white supremacy, nationalism, and sexism at the same time. In all these right-wing extremist campaigns, the anti-immigration rhetoric of the politicians who capitalize on people's xenophobic fears of the current so-called 'refugee crisis' in Europe and unprecedented waves of migration from the Global South into Europe and North America more generally, played a major role in winning the votes of those silent racist factions of the society who shy away from speaking their xenophobic minds openly but silently vote for the populist politicians who will propose plans for stricter border controls and deportation of 'illegal' people, namely refugees. It is perhaps ironic that public racism is illegal in the West and somehow not socially accepted, but voting for racist parties is legal, which allows people to democratically express their 'fear' of strangers.

In conclusion, in light of the resurgence of right-wing extremism in Europe and the U.S. on the one hand, and the re-emergence of Islamic extremism in the Middle East on the other, I want to return to the critical role of the stateless figure of the refugee who, in search of survival, has either fled the disastrous consequences of neoliberal economies, which can be found in each and every corner of the world today, or the war zones resulting from confrontations between various types of extremism, be it religious or ethnonationalist, in inviting us to rethink the world from the point of view of the illegal and unwanted people who have, once or several times, lost their place in their 'life worlds' (*Lebenswelt*) (Edmund Husserl) and are trying to reground themselves somewhere else. Rethinking the world means freeing ourselves from unipolar ways of viewing the world and taking a critical position towards local and international law, towards neoliberal local and global economies, towards Eurocentric as well as religio-centric (Corsini 1999: 827) and ethnocentric epistemologies. This practice of an epistemic disobedience, as Mignolo suggests, aims at "epistemic pluriversality; or if one wishes to maintain some kind of universality, one might refer to 'pluriversality as a universal project,' which today is one of the ultimate decolonial horizons" (Mignolo xlii), and will diversify our world and our world views. If the mere presence of the figure of the refugee challenges white domination in Western societies on an everyday basis, the endeavors of the migrant scholar of color in claiming an academic space in the West can assist our goal of decolonizing the knowledge production system.

Works Cited

Agamben, Giorgio (1993): The Coming Community. Michael Hardt (trans.), Minneapolis: The University of Minnesota Press.

Agamben, Giorgio (2000): Means Without End: Notes on Politics. Cesare Casarino/Vincenzo Binetti (trans.), Minneapolis and London: University of Minnesota Press.

Agamben, Giorgio (2005): State of Exception. Kevin Attell (trans.), Chicago: The University of Chicago Press.

Ahmed, Sara (2012): On Being Included: Racism and Diversity in Institutional Life, Durham: Duke University Press.

Arendt, Hannah (1943): "We Refugees." In: Menorah Journal 1/1, p. 77.

Arendt, Hannah (1985[1951]): Imperialism, Part II of The Origins of Totalitarianism, San Diego and New York and London: Harcourt Brace & Company.

Arendt, Hannah (1998): The Human Condition, Chicago: The University of Chicago Press.

Arendt, Hannah (2007): We refugees. In: Arendt, Hannah (ed.), The Jewish Writings, New York: Schocken Books, pp. 264-274.

Bauman, Zygmunt (2002): "Los nuevos intocables." In: EL País February 10 (https://elpais.com/diario/2002/02/10/opinion/1013295609_850215.html).

Bauman, Zygmunt (2004): Wasted lives: modernity and its outcasts, Cambridge: Polity.

Bauman, Zygmunt (2008): La societá sotto assedio, Roma: Laterza & Figli.

Benhabib, Seyla (2004): "The rights to have rights: Hannah Arendt on the contradictions of the nation-state." In: Seyla Benhabib (ed.), The rights of others: aliens, residents and citizens, Cambridge: Cambridge University Press, pp. 49-69.

Betts, Alexander (2009): Forced Migration and Global Politics, Chichester: Wiley-Blackwell.

Betts, Alexander (2015) "Human migration will be a defining issue of this century. How best to cope?" In: The Guardian September 20 (https://www.theguardian.com/commentisfree/2015/sep/20/migrants-refugees-asylum-seekers-21st-century-trend).

Bhabha, Homi (1992): "The World and the Home." In: Social Text: Third World and Post-Colonial Issues 31/32, pp. 141-153

Bonilla-Silva, Eduardo (2006): Racism Without Racists: Color-blind Racism and the Persistence of Racial Inequality in the United States, Lanham: Rowman & Littlefield.

Butler, Judith (2006): Precarious life: the powers of mourning and violence, London: Verso.

Corsini, Raymond J. (1999): The Dictionary of Psychology, London: Psychology Press.

Dabashi, Hamid,/Mignolo, Walter (2015): Can Non-Europeans Think?, London: Zed Books.

Du Bois, W. E. B (2007): The Souls of Black Folk. Brent Hayes Edwards (eds.), Oxford: Oxford University Press.

Halberstam, Jack (2011): The Queer Art of Failure, Durham: Duke University Press.

Hammar, Tomas (1990): Democracy and the Nation State: Aliens, Denizens, and Citizens in a World of Intenational Migration, Brookfield.: Gower.

Harney, Stefano/Moten, Fred (2013): The Undercommons: Fugitive Planning & Black Study, Wivenhow and New York and Port Watson: Minor Compositions.

Mahbubani, Kishore (2011): Can Asians Think?, Singapore: Marshall Cavendish Editions.

"Refugee Statistics" In: UNHCR The UN Refugee Agency, 2017 (https://www.unrefugees.org/refugee-facts/statistics/).

Schuilenburg, Marc (2008): "Citizenship Revisited: Denizens and Margizens." In: Peace Review – A Journal of Social Justice 20/3, pp. 358-365.

"United Nation's International Migration Report" In: highlights, 2017 (http://www.un.org/en/development/desa/population/migration/publications/migrationreport/docs/MigrationReport2017_Highlights.pdf).

Keeping Academia White: A Case Study

Kai Linke

Introduction

There are many institutionalized ways in which German academia stays white. Students of Color[1] are subjected to institutionalized racism throughout their educational careers. Low expectations from teachers, biased grading, tracking into remedial and occupational tracks, and non-recognition of education completed and degrees earned outside of Germany all conspire to prevent many Students of Color from obtaining the necessary qualifications that would allow them to even enter the university. International students additionally face bureaucratic and financial hurdles[2] as well as a restrictive immigration system on their path to enrollment. Once enrolled, Students of Color typically have less access to elite scholarships, study-abroad opportunities, prestigious internships, and high-quality academic mentoring from an overwhelmingly white professoriate than their white peers. People of Color who seek to climb the academic ladder also

---

1 In accordance with the common usage by many writers who belong to the respective groups, I capitalize terms such as "Students of Color," "Black," "Indigenous," etc. to highlight that these terms refer to self-definitions of groups of people that have been developed in resistance against white supremacy. Like the term "white," they do not refer to any type of biological understanding of 'different races' that could be differentiated on the basis of skin color, physiognomy, hair texture, etc. I do not capitalize the term "white" because it indicates the social location of people who benefit from and perpetuate white supremacy and thus lacks any kind of resistant or liberatory potential.

2 In addition to having to prove that they possess at least 8700 € to cover the cost of living in Germany for one year, beginning in 2017, non-EU citizens also have to pay an additional tuition fee of 1500 € per semester in Baden-Württemberg and North Rhine-Westphalia.

have to contend with racist hiring practices all the way from student assistant positions to professorships.[3] Apart from these institutionalized forms of racism that cumulatively work to keep People of Color out of the academy, white professors, administrators, and students also engage in subtle and not-so-subtle forms of everyday racism that serve to make academia comfortable and welcoming for white people, while making it a hostile and exclusionary space for People of Color.

In this article, I will examine one specific instance of everyday academic racism and its aftermath in order to analyze the many strategies that even well-meaning, anti-racist white people engage in on a daily basis not only to keep academia white, but also, simultaneously, to uphold the illusion that our whiteness is innocent and non-racist. The incident I will analyze occurred at a public conference at a German university and led to a series of public statements. I chose this particular instance of academic racism for my analysis precisely because the comparatively public nature of both the incident itself and its aftermath also means that it is fairly well-documented and therefore more accessible to analysis than other such instances.

However, at the same time, this is but one example of common strategies of white dominance that are used in academic settings. In this article, I want to focus on these strategies rather than on the specific people or institutions involved. Sara Ahmed offers an important caveat when dealing with racism in institutional settings:

> Racism should not be seen as about individuals with bad attitudes (the 'bad apple model'), not because such individuals do not exist (they do) but because such a way of thinking underestimates the scope and scale of racism, thus leaving us without an account of how racism *gets reproduced.* […] [E]liminating the racist individual would preserve the racism of the institution in part by creating an illusion that we are eliminating racism (2012: 44; original emphasis).

This article is decidedly not about 'a few bad white people,' but about how all of us who are white help to reproduce institutional racism at the university through "an ongoing series of actions that shape institutions" (Ahmed 2012: 45). For this reason, I anonymized the setting as far as possible and chose not to provide links to any public statements.

3 For more information on structural and institutionalized racism in the German education system, see Auernheimer (2013), Diefenbach (2010), Fereidooni (2011), Gomolla and Radtke (2009), Gyamerah (2015), and Kuria (2015).

I am offering my analysis from the specific viewpoint of a white junior scholar who has worked as a lecturer (*Wissenschaftlicher Mitarbeiter*) for two years and is currently working on his dissertation to obtain a PhD. Whereas my current academic rank positions me as marginal within academic hierarchies, I nevertheless receive the benefits of whiteness within German academic contexts and (unwittingly) participate in keeping academia white. Because I belong to the very group I write about (white academics), I sometimes use first-person plural pronouns (i.e. "we," "our") to denote that the white ignorance and defense strategies I describe are often enough my own. With regard to the specific case I discuss in this article, it is also relevant to point out that when the incident occurred, I was among the audience members who left the room in protest, and when the first statement about this incident was published, I was among the student activists who tried to hold the responsible department accountable for what had transpired at the conference. I am therefore neither impartial nor without blame, since, as my analysis will show, I was also criticized for the role I took on as the events unfolded. I am offering this analysis to try to understand what this incident and its aftermath can reveal about the many ways in which white academics such as myself keep academia white, while trying to keep ourselves blameless. My analysis is obviously only one possible way among many others to read these events and to try to learn from them. Given that I am offering this analysis from a white perspective, i.e. from the perspective of a signatory of what Charles M. Mills termed "*the Racial Contract* [which] *prescribes for its signatories* […] *an epistemology of ignorance* […] *producing the ironic outcome that whites will in general be unable to understand the world they themselves have made*" (1997: 18; original emphasis), it is, in fact, quite likely that I am missing or misreading important lessons that could be learned from what transpired. I decided to write this article despite the limitations of my perspective because I believe that the lessons I learned through my involvement in these events might be helpful for other people trying to dismantle racism in academia.

White Spaces

I want to begin my analysis by looking at how the conference was already marked as a white space, even before anything 'out of the ordinary' had happened. In this section, I will describe the often unremarked-upon ways in which racism *ordinarily* structures academic spaces as white. These quotidian practices of racism rarely elicit critique or interventions. They are simply 'business as usual,' the habitual, often unnoticed "sea of whiteness" (2012: 35) that Ahmed describes in *On Being Included: Racism and Diversity in Institutional Life*. Even though the conference was about epistemic violence against People of Color, women, and trans people, the conference organizers had invited a majority of white speakers, commentators, and facilitators to discuss white epistemic violence against People of Color as well as the resistance of People of Color against these racist and colonialist regimes. During the conference, these white scholars were invited to perform the role of "the Expert," which Malia Villegas and Adreanne Ormond define as follows in their analysis of relationships between white and Indigenous women:

> The Expert claims personal or professional relationships with Indigenous people and tends to speak on behalf of Indigenous peoples, particularly in forums where there is little to no Indigenous representation. The Expert guards her authority to speak for Indigenous peoples and may do so in order to 'help,' 'advocate for,' or 'inform' others. At the extreme, the Expert may even position herself as telling Indigenous peoples about themselves (2012: 31).

As my analysis will illustrate in a moment, this is exactly what happened at this conference with white people speaking as experts about Black people and People of Color. Also, as Aretha Schwarzbach-Apithy reminds us, the performance of white expertise is not only already a form of epistemic violence in and of itself:

> Doch nicht zuletzt geht es dabei auch um das Monopol, das Wissen und Leben der bereits Vernichteten bzw. noch Diskriminierten in verschriftlichter oder Vortragsform zum Besten geben zu können. Muss es erwähnt werden, dass dies bezahlt und gewürdigt wird? Wer erhält hier auf Kosten Anderer den Arbeits-/Lehrplatz, die Tantieme usw.? Die Komplizenschaft mit diskriminierenden und neokolonialen Verhältnissen ist für *weiße* Menschen immer noch lukrativ, selbst wenn scheinbar dagegen 'angeschrieben' wird (2009: 252; original emphasis).[4]

4 "Not least of all, this is also about the monopoly to be able to use the knowledge and lives of those already annihilated and those still discriminated against for one's own texts and talks. Does it have to be mentioned that this practice earns the authors remuneration and validation? Who receives jobs, the authority to teach, payment at the expense of others? Complicity with discriminatory and neocolonial conditions is still lucrative for *white* people, even if they are seemingly 'writing against' these conditions" (my translation).

It is precisely forums such as this conference that ultimately allow white people to build up and 'prove' our credibility and formal 'expertise' on matters of race, racism, and postcolonialism, which then leads to us being hired for the very few positions with a focus on postcolonial studies or critical race studies. As hiring decisions in this field demonstrate time and again, we are hired even when we compete against People of Color who not only possess the same formal qualifications but also bring a lifetime's worth of finely honed skills in identifying and effectively addressing racism to the table, or, in Mills's words, "a perspectival cognitive advantage that is grounded in the phenomenological experience of the disjuncture between official (white) reality and actual (nonwhite) experience" (1997: 109). These skills are vital to producing relevant research and being a successful teacher in the fields of postcolonial and critical race studies. White people, as signatories of the "Racial Contract [which] prescribes, as a condition for membership in the polity, an epistemology of ignorance" (Mills 1997: 93), do not develop these skills in the same way. Thus, the white people comprising hiring committees in Germany routinely dismiss the importance of these skills and hire more white people who are qualified on paper only (see Kimberly Alicia Singletary's and Priscilla Layne's contributions in this volume). This common practice further excludes People of Color even from the very fields that should be by, about, and for them, and thus cements institutionalized racism for years, if not decades to come.

In the aftermath of the conference, when the problematic whiteness of the conference was pointed out to the department hosting it – which I will call 'the average department' for the purpose of this article – the department responded by listing every single participant of the conference whom they perceived as a Person of Color by name, even though three of the five people they named had previously cancelled their participation in the conference. As this response shows, the average department viewed these participants largely as tokens,[5] whose very invitation was enough to automatically absolve the department of all subsequent charges of racism. Inviting the contributions of People of Color as tokens instead of handing over the authority to conceptualize and organize conferences and edit anthologies is another common strategy by which white people preserve the whiteness of academia while trying to appear non-racist. Frances E. Kendall refers to this white tendency to invite People of Color to join

5 Tokenism "describes an intergroup context in which very few members of a disadvantaged group are accepted into positions usually reserved for members of the advantaged group, while access is systematically denied for the vast majority of qualified disadvantaged group members" (Wright/Taylor 1998: 648).

in already existing projects, groups, and institutions as "come-be-part-of-what-we're doing syndrome" (2013: 66).

In addition to inviting a majority of white scholars as 'experts' on racism and treating Scholars of Color as tokens, the average department also scheduled the conference at exactly the same time as a yearly conference on racism in the German education system that was organized at the same university by a group of Students, Scholars, and Activists of Color. This scheduling conflict made it impossible for the large number of attendees at the latter conference, among whom were many People of Color, to attend the conference hosted by the average department, and it conversely made it impossible for people affiliated with the average department to attend the conference on racism in the German education system – even though both conferences addressed very similar topics. Through the scheduling of its conference, the average department communicated that neither making the knowledge shared and generated at the conference on racism in the German education system accessible to members of the department nor inviting the participation of people involved with that conference were important enough concerns to warrant a rescheduling of its conference. This decision all but ensured that the conference hosted by the average department would be a white-dominated space, where white people talked amongst themselves while many of the People of Color who are experts on the topics the conference purported to discuss congregated in another room on the very same campus.

Lastly, the conference organizers chose to host a conference which proclaimed in its title to address "decolonial challenges" in the most 'representative' room of the university, right underneath a larger-than-life-size oil portrait of the German colonizer after whom the university is named. This unremarked-upon portrait visually communicated that the participants were invited to enter a space in which colonizers were celebrated and the colonized were made invisible.

Racist Incidents

It was in this environment, which was already, through its choice of speakers, its choice of date, and its choice of a room, marked as a space of white dominance, that the following incident occurred. A white professor gave a talk about anti-Black racism in the German academy, to which a white assistant professor (*Juniorprofessor_in*) had prepared a response. The discussion was supposed to be presided over by a white moderator. It is already indicative of white people's insensitivity to our own practices of racist epistemic violence that apparently none of the conference organizers or even the invited white 'experts' on epistemic violence perceived this set-up as problematic. As white German academics, we are so desensitized to the routine exercise of white people analyzing, discussing, evaluating, and judging the lives and words of People of Color that we cannot even perceive the violence inherent in this practice. Schwarzbach-Apithy comments that "Konstellationen, in denen uns *weiße*, stets gutmeinende WisserInnen über uns und unser Leben aufklären [...] sind schlicht unerträglich" (2009: 252; original emphasis).[6]

The professor began their[7] talk by theorizing anti-Black racism, using theories developed by Black theorists such as Frank Wilderson, Saidiya Hartman, and Rinaldo Walcott. In doing so, the professor used phrases taken from their work that described the reality of anti-Black racism in extremely graphic, stark, and violent terms. I attended this talk with a friend, who identifies as a Person of Color, and they immediately perceived the violence inherent in a white person using these violent phrases over and over again to give an 'objective' description of anti-Black racism.[8] The white professor apparently felt that it was necessary to describe anti-Black racism with a veritable deluge of extremely violent phrases that made it seem as if their talk was primarily geared towards shocking white people. The talk did not give the impression that they had in any way anticipated that there would be Black people and People of Color in the audience who were intimately familiar in their very bodies with the violence the professor described. In their own white body, the professor felt

6 "[C]onstellations, in which *white*, always well-meaning knowers educate us about ourselves and our lives [...] are simply unbearable" (my translation).

7 I use singular "they," "them," and "their" as gender-neutral pronouns.

8 The problem with much scholarship on racism done by white people in Germany is that we have a tendency to see ourselves as completely separate from the reality we study, as if racism had no bearing on us, our research, and the material effects our research can have in the world. We tend to presume that the lived experiences of Black people and People of Color are 'research objects' like any other that we can access at will and describe, measure, and analyze however we see fit, completely disregarding the subjectivity, agency, and expertise of the people we treat as convenient sources of quotes and data. While the myth of academic objectivity compromises the viability and usefulness of all academic research, it is particularly damaging when people who benefit from and uphold a particular system of oppression assume that they can 'objectively' study the lives of the very people they help to oppress instead of making sure that they have the opportunity to speak for themselves on their own terms and have their voices be heard and respected. For an extended discussion of the problems with objectivity and critical alternatives, see Harding 2004.

entitled to reenact this violence through their words, apparently without stopping to consider how the meaning and impact of these words and phrases change when they are taken out of the context of Black people analyzing and protesting the racism they face, but are instead incorporated into a white professor's speech, 'objectively' and 'academically' describing Black lives. In this move from Black protest to white description, the professor's words performatively interpellated Black people once again as the 'abject other' – professedly in the service of critiquing this very interpellation. Listening to the professor, it seemed to me that they relished the chance to find the most horrific description possible of how Black people had been brutalized, and that as a white audience member I was in a certain sense invited to share this lust in wallowing in the horrendousness of Black people's suffering.

A similar dynamic is currently being discussed with regard to the incessant broadcasting of videos capturing police murders of Black people in the U.S. April Reign, the Managing Editor of *Broadway Black*, the Editor at Large for *Nu Tribe Magazine,* and the creator of #OscarsSoWhite, for example, sees a double standard at play. She asks,

> when reporter Alison Parker and cameraman Adam Ward of WDBJ-TV were gunned down on live television, the consensus by the news media was that the video was too graphic to be shown. [...] Why is the video of white people being killed considered too graphic for replay, but videos of black women, men and children are replayed on a seemingly endless loop to the point of numbness? (Reign 2016)

She states that there is no benefit for her as a Black woman in watching these videos: "For me, these videos are debilitating. [...] To watch videos of people who look like me being killed only increases my fear that someone I know may be next" (ibid.). She also charges that, for the general viewership, "the dehumanization of black bodies becomes some sort of perverse entertainment as images of our pain are broadcast for the world to see" (ibid.). This very dynamic of placing People of Color in the position of the 'helpless victim' and white people in the position of the shameless voyeur of their suffering is one of the many common ways in which academia asserts itself as a space for the gratification of white people at the expense of People of Color.

On yet another level, the professor's detailed descriptions of racism attempted to fulfill an additional white desire, which Audrey Thompson identifies as "the desire to be and to be known as a good white person" (2003: 9). The professor established themselves as somebody who was 'in the know,' somebody who had read a lot of Black theorists and who could competently describe and identify racist structures, thus ostensibly proving themselves to be, in Thompson's words, "demonstrably different from other, racist whites" (2003: 9).

As all of this was going on, my friend was not the only person who viscerally responded to the talk. Alerted by their reaction, I noticed that all around me, people were whispering and shifting uncomfortably in their seats. After the professor had already used the n-word in English in a quote, the tipping point came when they used it in German in their own words to critique other white people's racism. My friend immediately got up to leave the room and after a moment's hesitation, I followed them and noticed that six other people were exiting the room with us. To be honest, I know for a fact that my reaction would not have been as immediate if my friend had not gotten up first. I even doubt whether my discomfort would have materialized as a noticeable reaction at all, had I attended the panel on my own. Even though I care a great deal about dismantling racism, as a white person, neither my sensitivity to racist violence nor my willingness to act is as acute as it could be. The analysis I just offered of what exactly made the professor's talk feel so violent and racist is the product of months of reading and conversations with Friends of Color (who chose to remain anonymous for the purpose of this article). It is not 'my own' analysis by a long stretch of the imagination. My delayed, muted, and confused immediate reaction to the talk illustrates that, "as we are materially privileged in particular ways, our epistemic abilities are suspect" (Hoagland 2007: 112). Sarah Hoagland explicates that we who lack "epistemic privilege lack critical abilities" (ibid). It is precisely this white lack of critical epistemic abilities that should give hiring committees pause when they are about to hire yet another white person as an 'expert' on colonialism and racism.

White Defenses

Our protest marked the point where the racist practices of white 'business as usual' shifted gears and immediately turned into 'defense of white innocence.' This defense of white innocence mostly took the form of denial, which Grada Kilomba defines as "an ego defense mechanism that operates unconsciously to resolve emotional conflict, by refusing to admit the more unpleasant aspects of external reality and internal thoughts or feelings. It is the refusal to acknowledge the truth" (2010: 22). At the conference itself, this denial first took the form of the professor attempting to continue their talk as if nothing had happened. The professor and the other panelists only reluctantly agreed to end the talk and take a break after Nadezda Krasniqi, a Romni well-known for her anti-racism work and activism, who was part of the audience, insistently asked them to do so. Denial then took the form of the white respondent reading their prepared response – again, as if nothing had happened – and actually opening their response by thanking "their esteemed colleague for their very interesting talk." In a remarkable display of white solidarity, which Robin DiAngelo defines as, "the unspoken agreement among whites to not talk openly and honestly about race and to avoid causing another white to feel racial discomfort by confronting them when they say or do something racially problematic" (2012: 154), the professor could not only count on the outspoken solidarity of the other white panelists. They could also count on the solidarity of white audience members who began to discuss the talk and the response without the slightest reference to the fact that the talk had been disrupted because of the racist violence it had enacted.

After the conference, denial took the form of a dead silence on the part of the average department and all the individual people involved in this panel. None of them had anything to say about the fact that they had hosted and/or participated in a conference on epistemic violence that had turned out to be so unbearably violent that audience members left the room in protest. This silence was only broken when student activists put pressure on the average department to commit to a concrete course of action that would change the fact that 100 percent of the senior professors in the department are white. In response to student demands, the average department put together an internal document detailing all their anti-racist 'achievements' in order to argue that no further action was necessary. This document rewrote institutional history to a remarkable degree. While there was no mention of roughly a decade of anti-racist student activism pushing against institutionalized racism at the department, the department actually listed the conference on epistemic violence as proof positive of its continuous, successful engagement against racism.

When an anonymous group finally released a public statement concerning the events at the conference, demanding a public apology and a change in the racist practices that had been manifest at the conference, the defense of white innocence became much more aggressive and deployed a host of silencing strategies. The statement's critique of the extremely violent language used to theorize anti-Black racism was largely ignored and the discussion focused instead on the much narrower issue of the professor's use of the n-word. Denial turned into outright negation when the professor claimed in a public e-mail that they had, in fact, never used the n-word outside the context of a quotation. In an instance of what Karen L. Dace calls "the whiteness of truth," which leads to white people always being "'found innocent' of improper or racist motives, judgments, intentions and actions" (2012: 42), the leadership of the average department believed the professor rather than the authors of the statement and the student activists who backed the account offered in the statement. Pointing to the professor's denial of their use of the n-word, the average department discredited the statement by claiming that it "spread falsehoods" and did absolutely everything in their power to silence its critique. First, the department leadership outright refused to forward the statement even to the members of the department, let alone to the much larger listserv of the department, where it could have been offered for discussion to all people interested in the work of the department. After facing sustained pressure from students, the average department finally put the statement on the department website – *after* the department itself, the professor who gave the talk, as well as the respondent had all written their own counterstatements, which in effect denied all responsibility for what had transpired at the conference. The group who had written the original statement responded to all three counterstatements, but while the counterstatements could be accessed in full on the department website, none of the responses were posted. As is so often the case, the last word belonged to white people in powerful positions. Even in sharing the statement, the department still tried to effectively silence it by drowning it in their denial and not allowing the critics to address this denial.

Not only did the defense of white innocence take the form of trying to ignore and silence the protest against and critique of academic racism. It also took the form of tone policing, where white people use what they perceive as "das Auftreten, die Sprechweise, die Arroganz, die Unverschämtheit, die Aggressivität, die schlechte Erklärweise, das Polarisieren oder de[n] Ton Schwarzer Studierender" (Schwarzbach-Apithy 2009: 254)[9] as an excuse for not engaging with any argument or critique People of Color offer. The tone policing already began at the conference itself when, after a short break after the talk had been brought to a halt, the group of people who had initially left the room in protest returned and asked the panelists to skip the rest of the professor's talk for fear that the latter would perpetuate further racist violence and to move on directly to the response from the assistant professor instead. While the panelists themselves readily agreed, a white person from the audience loudly protested and shouted, "Wir lassen uns doch von Ihnen hier keinen Maulkorb verpassen!"[10] By framing the request to discontinue a racist talk as an illegitimate and extreme infringement on the right of white people to persist in racist

9 "[T]he attitude, the way of talking, the arrogance, the brazenness, the aggression, the poor explaining, the polarizing, the tone of Black students" (my translation).

10 "We will not allow you to put a gag order on us!" (my translation).

speech, they brought up "the figure of the censoring student" (Ahmed 2015). Ahmed analyzes the function of this figure: "When you hear a challenge as an attempt at censorship you do not have to engage with the challenge. You do not even have to say anything of substance because you assume the challenge is without substance" (ibid). In this case, the audience member proceeded from the assumption that it was the professor's right to inflict the very epistemic violence the conference was supposed to address and their right as an audience member to listen to and enjoy this violence. As their remark shows, they did not think that People of Color (and their white allies) had the right to name racist violence as such and to demand that it be stopped. Apparently, the audience member felt that the People of Color in the audience were under the obligation to endure this violence so as not to infringe upon the professor's right to free speech. This is a grave misunderstanding of the right to freedom of speech. The right to freedom of speech protects people against state censorship, not against contention. It neither obliges people to give a platform to speech they deem violent and hateful nor does it protect people from resistance to their speech. At this particular conference, one would assume that the conference organizers, speakers, and audience members share a commitment not to perpetuate the very same racist violence that the conference seeks to critique. It is well within the rights of audience members to demand that racist speech not be tolerated at this event and that a speaker who has already engaged in racist speech acts not be given further opportunities to persist in the same vein. By framing this act of resistance as censorship, the audience member who protested against ending the professor's talk located the problem not in the professor's racist speech, but in the challenge issued to this speech, thus attempting to reverse the relation of victim and perpetrator of violence.

The tone policing continued at the first department meeting after the initial statement was released. A group of student activists, to which I belonged, came to the department meeting to discuss why the department leadership refused to make the statement public. Instead of becoming angry at the department leadership for trying to keep the statement secret even from the members of the department, many members of the department became angry at *us* for 'disrupting' their meeting. As had often been the case in previous conflicts with the department, the department then used formal rules that they seemed to invent on the spot and that could not be found in writing anywhere to decide that they would not include a discussion of the department leadership's silencing strategies in the meeting's agenda. The white professors thus used their institutionalized, procedural power to silence critiques of racism by keeping them off their agenda. When we collectively left the meeting after this decision, members of the department loudly and angrily accused us of being aggressive and of making all dialogue impossible – the very same dialogue that they had just decided not to have.

The psychological mechanism operative in this reaction has been described as "splitting and projection" (Kilomba 2010: 22): "[T]he subject denies that she/he has such-and-such feelings, thoughts or experiences, but goes on to assert that someone else does. The original information [...] is denied and projected onto the 'Others'" (ibid). The department had, in fact, already inflicted violence on People of Color through and during the conference. It exacerbated the violence by attempting to portray the very site of this violence as a badge of honor proving their *anti-racist* commitment. When they were criticized, they continued their violent behavior by aggressively refusing to engage with the student activists who tried to hold them accountable. Instead of recognizing and addressing the violence they were perpetuating, however, they projected their own violence onto the student activists and claimed that we were too violent to be heard.

In yet another instance that I would see as tone policing, the department wrote in their counterstatement: "Die Diffamierung von namentlich genannten Wissenschaftlerinnen im Netz durch anonyme Autor_innen finden wir grundsätzlich problematisch; wir halten dieses Vorgehen für wissenschaftlich und menschlich unredlich und dem gemeinsamen Anliegen keinesfalls dienlich."[11] In my experience, this outraged response to the anonymity of critique is not at all uncommon in powerful white people. It seems that to some, the mere fact that a critique was voiced anonymously is enough to completely discredit its content. In the response that the department refused to publish, the authors of the statement explained why they chose to remain anonymous:

> Dazu ist zu sagen, dass wir im Gegensatz zu vielen von Ihnen keine auf Lebenszeit garantierten festen Stellen haben. Darüber hinaus haben Sie als Lehrende ganz konkret Macht über einige von uns. Wir haben oft genug erlebt, dass Studierende, die Kritik an Lehrenden üben, daraufhin fortgesetzter Gewaltausübung ausgesetzt sind. Dem können

11 "We find the online defamation of academics, who are identified by name, by anonymous authors generally problematic; to us, this is academically and morally reprehensible and not at all helpful in furthering our common goals" (my translation). I do not provide links to the original statements so as to preserve the anonymity of all involved.

und wollen wir uns nicht aussetzen. Diese Auseinandersetzung findet nicht auf Augenhöhe statt. Sie haben in vielerlei Hinsicht deutlich mehr Macht als wir.[12]

In the anthology *Presumed Incompetent: The Intersections of Race and Class for Women in Academia,* a very similar explanation is offered for why many women did not feel that it was safe for them to contribute to the volume:

> [A] significant number of women decided not to contribute to the anthology for fear of retaliation. They believed that they would be penalized for airing their home institution's dirty laundry in public, and they were not prepared to become pariahs. One woman felt that writing about her experiences would not only burn bridges at her current institution but also undermine her future career prospects (Harris/Gonzáles 2012: 11).

These fears are clearly not unfounded. In another highly publicized conflict, which took place at the education department at Humboldt University in Berlin, the *entire* professoriate signed a statement denouncing an anonymous group of first-semester BA students who had dared to publicly criticize one of their professors for his racism, his cis_hetero_sexism, and his ableism (Jerusalem et al. 2014). In a personal conversation with one of the signatories of this statement, the professor once again cited the anonymity of the group as one of the (if not *the*) main reason that caused them to see the critique as illegitimate and to sign the statement against the group of students. Ahmed reminds us of "the immense power that academics have *over* students: They grade student essays and exams; they have discussions about students in meetings that are closed; they sit on committees that decide funding; they have access to confidential files that hold personal information" (2015; original emphasis).

DiAngelo also asserts that "whites are almost always racially comfortable and thus have developed an unchallenged expectation to remain so" (2012: 179). When white people are criticized for our racism and thus become uncomfortable, we often respond by subjecting the critics to "a socially sanctioned array of moves against the perceived source of the discomfort, including penalization, retaliation, isolation, ostracization, and disengagement" (ibid). Since, as white

12 "Contrary to many of you, we do not have secure jobs, guaranteed for life. Furthermore, as administrators and teachers at this department, you have very concrete power over some of us. We experienced often enough that students who criticize teachers will then be subjected to continuing violence. We cannot and do not want to risk this. This debate does not happen on a level playing field. In many respects, you have much more power than we do" (my translation).

academics, we obviously have a lot of power and this power has been used numerous times against people further down in the institutional hierarchy who dared to criticize us – why do so many powerful white academics find it so incomprehensible that their critics would seek the protection of anonymity? Why are so many of them seemingly incapable of dealing with criticism simply on the merit of its content, without having to know who exactly formulated it? If the criticism is justified, why does it even matter to know the names of the authors? Why are we complicit in "creat[ing] a picture of professors or academics as the ones who are 'really' oppressed by students" (Ahmed 2015)?

Trying to Become Less of a White Problem

After having analyzed at length how whiteness asserted and defended itself both at the conference and in its aftermath, I now want to attempt to turn my critical gaze on my own role in the aftermath of the event. As I mentioned before, I was part of the group of student activists who attempted to hold the average department accountable after the first statement had been released. One of the biggest problems of this group was that it was majority-white. While there are certainly several reasons for why this was the case, I want to focus here on the fact that we, the white members of the group, also reproduced racism as we were attempting to fight it, which would obviously be a deterrent to People of Color who might have wanted to be part of the protest.

As I pointed out above, as white people we are signatories to "an epistemology of ignorance" (Mills 1997: 18), i.e. we are often unable to recognize racism as racism even when it manifests itself right in front of us. This inability to identify racism is compounded when we ourselves are the ones behaving in racist ways. Because the "*racist = bad / not racist = good* binary" (DiAngelo 2012: 167; original emphasis) teaches us that only 'bad' people are racist and we typically see ourselves as 'good' people, white people usually have an extremely hard time even entertaining the possibility that we *could* do something racist, let alone identify the specific instances in which we *did*, in fact, do something racist. Since I am no exception to this rule, it is likely that there were many instances in which I did something racist, which I was unable to identify as such and which was not pointed out to me, either. For this reason, I can only discuss two specific examples where Nadezda Krasniqi invested the time and energy to help me see the racism in what I had done.

At one point, she attempted to hold some white members of the group accountable for their racist exclusionary practices and instead of simply backing her account of what had happened, I exhibited a common pattern that DiAngelo identifies as "Explaining Away/Justifying/Minimizing" (2012: 209) by stating that I had been similarly excluded in the past. While this was not my intention, the impact of my statement was clearly to deny the racist dimensions of what had happened. As DiAngelo reminds us, it is the impact that counts, not the intention (2012: 180-181). Just because I did not think well enough before speaking and did not anticipate that my statement would have the effect of minimizing the critique that had just been voiced, my statement did not cease to have this very effect. My statement allowed other white members of the group to claim that, yes, regrettable mistakes had been made, but racism had not played a role in why and how a Romni had been excluded.

Krasniqi also pointed out to me that I had sometimes assumed a leadership position in our discussions with the department that was not mine to take. In their piece "Ten Things to Remember: Anti-Racist Strategies for White Student Radicals," Chris Dixon describes a tendency in white people to

> aggressively direct[ ] everything with an overwhelming sense of entitlement. Too often white students covet and grasp leadership positions in large campus activist groups and coalitions […] myths of 'merit' cloak these racial dynamics, but in reality existing student leaders aren't necessarily the 'best' leaders; rather, they're frequently people who have enjoyed lifelong access to leadership skills and positions – largely white, middle-class men (Dixon).

Our previous dealings with the average department had already revealed that white, middle-class people like me not only have more access "to leadership skills and positions," but are also listened to more and are taken more seriously in a racist system, even if we do not have any particular leadership skills at all. In an earlier altercation with the average department, students put forward a motion demanding that the department change the fact that 100 percent of the senior professors are white. The exact same motion was read out loud twice in two different department meetings. At the first meeting, when Krasniqi read the motion, the department rejected the motion as "completely unreasonable," "too aggressive," "not realistic." At the next meeting a few months later, a white student read the same motion, which had not been changed at all between the first and second reading, and the department suddenly thought the motion was "much more reasonable" and "a good basis for discussion." This situation exposed the blatant racism in the department where People of Color are delegitimized from the start as "too angry" and "not academic enough" to be listened to and white people are welcomed as "more reasonable" discussion partners about *matters of racism*, even though People of Color are clearly in a better position to understand the workings of the very system of oppression that targets them. Even though I did not 'covet' a leadership position, I did buy into these racist logics when I felt that I 'had to' speak up in discussions with the average department because I was 'best suited' to do so. I did not pause to think that I was 'best suited' only in a racist system that valued the academic training and the comparative emotional detachment from matters of racism that my whiteness affords me. Instead, I could have insisted that the department face their own racism and begin to listen to People of Color regardless of how they perceive their tone of voice or choice of words. The fact that I sometimes contributed to these discussions in a way that I can only describe as self-righteous in hindsight additionally served as an attempt to position myself as a 'good white person' in opposition to the 'bad white people' in the average department who were not as 'knowledgeable' about racism as I was (cp. Thompson 2003; DiAngelo 2012).

Clearly, these types of behaviors and dynamics are a problem in a group that attempts to dismantle racism in the academy. I offer these examples to emphasize the fact that even though some of us white academics might be actively involved in anti-racist or decolonial initiatives, we are still also part of the problem. As Barbara Applebaum writes, "I can reproduce and maintain racism even when, and *especially when*, I believe myself to be morally good" (2015: 6; original emphasis). Under the current conditions, which continue to constitute those of us who are white as racist, it is impossible for white people not to be "a white problem" (Yancy 2015). Being a white problem does not excuse us from taking action against racism, but it means that we have to proceed with particular care, humility, and a willingness to be criticized and work hard to become less of a problem in the process.

Conclusion

As my discussion of the aftermath of the conference has shown, there is a strong tendency among white academics to "suture" ourselves, i.e. to "install forms of closure, forms of protection, from counter-white axiological and embodied iterations, epistemic fissure, and white normative disruption" (Yancy 2015: xv). George Yancy asks white people to instead engage in active processes of "un-suturing," of keeping ourselves vulnerable to the fact that we are the problem, so that we can "change the conditions, and the repetitions that call/hail a different kind of subject – a different me" (2015: xxiv).

What would happen, I wonder, if white people learned to understand criticism as a gift that allows us to learn and to change, instead of as a personal attack that we have to deflect at any cost? In my experience, it is usually not the initial racist act that causes the most harm, but white people's inability to respond to criticism constructively. In our misguided attempts to absolve ourselves of the charge of racism, we heap layer upon layer of racist attack onto the people who dared to criticize us, thus sending a clear message to them and everybody who is watching that the racist status quo will be defended at all costs and that anti-racist perspectives are decidedly unwelcome and will be punished. It is my firm belief that a lot would be gained if we could only learn to listen hard for the truth in any criticism gifted to us, take responsibility for what we did, apologize, try our best to repair the harm done, and do whatever we can to ensure that we will act differently in the future. What would happen if those of us who are white were open to interrogating and having interrogated the way we do 'business as usual' – the way we do research, what we do research on, how we present our research, what we teach, how we teach, who and how we mentor, how we organize talks and conferences, how we publish books and journals, how we fill job openings, how we treat our colleagues and students – in order to identify all the many ways in which those of us who are white contribute to keeping academia white? I am sure that we could learn to identify many more than just the few ways sketched out in this article and I am equally sure that it is entirely possible for us to step by step transform our practice and the institutions of which we are a part so that doing things differently would become a pleasant habit, not an annoying chore or a seemingly insurmountable challenge. Could we collectively, white people and People of Color together, change the conditions so that academia could become a space that is co-owned and co-created by and for People of Color?

Works Cited

Ahmed, Sara (2012): On Being Included: Racism and Diversity in Institutional Life, Durham: Duke University Press.

Ahmed, Sara (2015): "Against Students." In: The New Inquiry June 29 (https://thenewinquiry.com/against-students/).

Applebaum, Barbara (2015): "Flipping the Script … and Still a Problem: Staying in the Anxiety of Being a Problem." In: George Yancy (ed.), White Self-Criticality beyond Anti-Racism: How Does It Feel to Be a White Problem?, Lanham: Lexington Books, pp. 1-9.

Auernheimer, Georg (eds.) (2013): Schieflagen im Bildungssystem: Die Benachteiligung der Migrantenkinder, Wiesbaden: Springer VS.

Dace, Karen L. (2012): "The Whiteness of Truth and the Presumption of Innocence." In: Karen L. Dace (ed.), Unlikely Allies in the Academy: Women of Color and White Women in Conversation, New York: Routledge, pp. 42-53.

DiAngelo, Robin (2012): What Does It Mean to Be White?: Developing White Racial Literacy, New York: Peter Lang.

Diefenbach, Heike (2010): Kinder und Jugendliche aus Migrantenfamilien im deutschen Bildungssystem: Erklärungen und empirische Befunde, Wiesbaden: Springer VS.

Dixon, Chris: "Ten Things to Remember: Anti-Racist Strategies for White Student Radicals." Colours of Resistance Archive n.d. (http://www.coloursofresistance.org/749/ten-things-to-remember-anti-racist-strategies-for-white-student-radicals/).

Fereidooni, Karim (2011): Schule – Migration – Diskriminierung: Ursachen der Benachteiligung von Kindern mit Migrationshintergrund im deutschen Schulwesen, Wiesbaden: Springer VS.

Gomolla, Mechthild/Radtke, Frank-Olaf (2009): Institutionelle Diskriminierung: Die Herstellung ethnischer Differenz in der Schule, Wiesbaden: Springer VS.

Gyamerah, Daniel (2015): "Der schulische Erfolg Schwarzer Schülerinnen und Schüler in Deutschland – Eine rassismuskritische Analyse des Mikrozensus." UN-Antirassismusausschuss – Parallelbericht Deutschland (https://rassismusbericht.de/wp-content/uploads/Hintergrundpapier-Daniel-Gyamerah).

Harding, Sandra (2004): The Feminist Standpoint Theory Reader: Intellectual and Political Controversies, New York: Routledge.

Harris, Angela P./González, Carmen G. (2012): "Introduction." In: Gabriella Gutiérrez y Muhs/Yolanda Flores Niemann/Harris González (eds.), Presumed Incompetent: The Intersections of Race and Class for Women in Academia, Boulder: University Press of Colorado, pp. 1-14.

Hoagland, Sarah Lucia (2007): "Denying Relationality: Epistemology and Ethics and Ignorance." In: Shannon Sullivan/Nancy Tuana (eds.), Race and Epistemologies of Ignorance, Albany: New York Press, pp. 95-118.

Jerusalem, Matthias (2014): "Rassismus-Vorwurf in der Vorlesung 'Grundbegriffe und Theorien pädagogischen Denkens und Handelns' (WS 2013/14) oder Über einen Versuch, wissenschaftlichen Diskurs in politische Agitation zu transformieren." Stellungnahme, Humboldt-Universität zu Berlin.

Kendall, Frances E. (2013): Understanding White Privilege: Creating Pathways to Authentic Relationships Across Race, New York: Routledge.

Kilomba, Grada (2010): Plantation Memories: Episodes of Everyday Racism, Münster: Unrast.

Kuria, Emily Ngubia (2015): Eingeschrieben: Zeichen setzen gegen Rassismus an deutschen Hochschulen, Berlin: w_orten & meer.

Mills, Charles M. (1997): The Racial Contract, Ithaca: Cornell University Press.

Reign, April (2016): "Why I Will Not Share the Video of Alton Sterling's Death." In: The Washington Post July 06 (https://www.washingtonpost.com/posteverything/wp/2016/07/06/why-i-will-not-share-the-video-of-alton-sterlings-death/?utm_term=.7e555027fbd6).

Schwarzbach-Apithy, Aretha (2009): "Interkulturalität und Anti-Rassistische Weis(s)heiten an Berliner Universitäten." In: Maureen Maisha Eggers/Grada Kilomba/Peggy Piesche/Susan Arndt (eds.), Mythen, Masken und Subjekte: Kritische Weißseinsforschung in Deutschland, Münster: Unrast, pp. 247-261.

Thompson, Audrey (2003): "Tiffany, Friend of People of Color: White Investments in Antiracism," In: International Journal of Qualitative Studies in Education 16/1, pp. 7-29.

Villegas, Malia/Ormond, Adreanne (2012): "On Friendship, Kinship, and Skinship: Healing Relationships between Indigenous and White Women Scholars." In: Karen L. Dace (ed.), Unlikely Allies in the Academy: Women of Color and White Women in Conversation, New York: Routledge, pp. 27-41.

Wright, Stephen C./Taylor, Donald M. (1998): "Responding to Tokenism: Individual Action in the Face of Collective Injustice." In: European Journal of Social Psychology 28/4, pp. 647-667.

Yancy, George (2015): "Introduction: Un-Sutured." In: George Yancy (ed.), White Self-Criticality beyond Anti-racism: How Does It Feel to Be a White Problem?, Lanham: Lexington Books, pp. xi-xxvii.

III. Diversity-Conscious Approaches to Academic and Pedagogical Practice

On Racism without Race: The Need to Diversify *Germanistik* and the German Academy

Priscilla Layne

While the majority of this volume poses a critique of the lack of diversity in American Studies in Germany, this essay is concerned with what the field of *German Studies* in Germany would gain from more diversity. As an African American working in the field of German Studies, who has spent several years living in Germany and has experienced German academia as both a student and a professor, I have seen first-hand the lack of diversity in German higher education compared to the U.S. While the U.S. is arguably a more racially diverse country, it also boasts a handful of academic disciplines which in part exist for the study of minority groups' experience in the U.S. and contribution to American history and American culture.[1] The liberation struggles fought by minority groups in the U.S. since the 1960s permanently changed the landscape of higher education in the U.S., leading to the establishment of fields such as African American Studies, Latinx Studies, and Ethnic Studies. Furthermore, the existence of such disciplines can largely be attributed to the important role of race in American society; a society that was arguably built on the antagonism between whites and non-whites.[2]

In contrast, in Germany the term 'race' (*Rasse*) was dismissed following the end of World War II, because it was associated with Nazi ideology, anti-

---

1 When I use the terms 'race' or 'racial diversity,' I do so in full acknowledgement that 'race' is not a biological essence. Rather, I follow the social constructivist theory that 'race' is a social construction, which though not real, has real effects for racialized people.

2 The notion that the U.S. was founded on the antagonism between blacks and whites is central to Afro-pessimism and articulated by Frank Wilderson (2010).

Semitism, and an obsession with 'racial purity' (Chin/Fehrenbach 2009: 6). As Anke Wischmann writes, "The claim that using 'race' as a critical concept reproduces racial stereotypes is combined in Germany with a taboo of 'race.' The widespread belief is that racism has been overcome with the end of National Socialism in 1945 and the Re-education program of the late 1940s" (2008: 476). And yet, many People of Color (POC) in Germany would argue that despite the absence of the term 'race' from everyday conversation, racism very well exists, and not just on the margins of society, but in its very center, and not just on an interpersonal level, but as a structural problem.[3] In this essay, I argue that a failure to acknowledge privilege, racializing practices, and racist structures in Germany continues to contribute to the lack of diversity within the academy. And if this is a problem in American Studies, it is likely an even larger problem in *Germanistik*, where scholars often still work within a paradigm of German literature and culture that reflects the homogenous ideal of Herder's *Volk*, and marginalizes the important contributions of minorities, including Black German and Turkish German authors and other people with a "migrant background."[4]

In the American academy, scholars have started addressing issues of privilege and structural inequalities by beginning presentations and scholarship with their own positionalities. Positionality refers to the idea, introduced by Linda Alcoff in 1988, "[...] that certain important aspects of our identity – for example, gender, race, social class, age, and national origin – are markers of relational positions rather than essential qualities [...]" (qtd. in Thompson

3 For more extensive discussions on structural racism in Germany, see El-Tayeb (2005a); Sow (2008); Kuria (2015); Kilomba (2016); Jones (2017); "Germany: UN Rights Panel."

4 "Persons with migration background (in the broad sense) are defined as in the micro-census 'all persons who immigrated after 1949 to the present territory of the Federal Republic of Germany, as well as all foreigners born in Germany and all those born in Germany of at least one immigrant or a foreign parent' (Statistisches Bundesamt 2012, 6)" (Wischmann 2008: 473). Despite my use of this term, I agree with Wischmann that the description "*Menschen mit Migrationshintergrund*" is highly problematic because it is so unspecific and does not acknowledge the structural inequalities between different people in Germany who either migrated themselves or are the children or grandchildren of migrants. "This definition is problematic because of its broadness and due to its exclusion of many people who do not belong in this particular group, but might have similar experiences, e.g. people of colour who are neither 'foreign' nor have a 'foreign' parent, but are nevertheless discriminated against and hence disadvantaged" (Wischmann 2008: 473). For a discussion of the nationalist origins and goals of *Germanistik*, see Norberg 2018.

Tetreault 2012: 1676). I position myself as an African American, ethnically Caribbean, professor of German Studies who studies and teaches Black German history and culture and who was trained in and currently works in the U.S. As such, I have a unique position in this discussion about race in the German academy because despite having spent three years in Germany as a foreign exchange student, I did not receive any degrees from a German university, and I have only taught in Germany as a guest assistant professor and also given a few academic lectures. Thus, I am aware that I cannot speak on behalf of Black Germans or even Black people living in Germany, nor would I want to.

When I come to Germany, I am a visitor. As an American citizen, I have the privilege of American currency and an American passport, which translates into the privilege of effortlessly moving across European borders without interrogation, though I have sometimes been asked to produce my passport to prove I am not a refugee or undocumented. And I do not *have* to deal with the daily incidents of racism and microagression in Germany with which people of color who live here are faced. Furthermore, because I am only ever in Germany as a guest for a brief period of time, I am aware that white Germans may interact with me differently. It is often easier for white Germans to accept a foreign *guest*,[5] rather than a German citizen who identifies as a Person of Color and who

5 The term 'guest' of course invokes the history of *Gastarbeiter* (guest workers) who came to West Germany from the 1950s through the 1970s from Southern European countries (Greece, Italy, and Portugal), as well as Turkey and Yugoslavia. Initially, West German politicians were enthusiastic about the arrival of these workers. Following the economic wonder of the 1950s that occurred thanks to the aid from the U.S. via the Marshall Plan (1947), West Germany needed manual laborers. The construction of the Berlin Wall in 1961, which cut off workers who commuted from East Germany, exacerbated this need. However, West Germany arguably did not do enough to integrate these foreign workers in order to make their transition to a life in Germany easier. They did not receive language instruction and they were often segregated from Germans and designated to live in dormitories in close proximity to the factories where they worked. West German politicians had hoped that these workers, the majority of whom were men, would come and work for a limited amount of time and then return to their home countries. However, as more and more guest workers participated in family reunification programs to bring their spouses and children to West Germany and seemed inclined to settle there for a longer period of time, xenophobia and racism towards guest workers increased, especially in the shadow of economically difficult times like the oil crisis of 1973. Max Frisch famously stated about this moment in history, "We called for workers, but people are coming" ("*Man hat Arbeitskräfte gerufen, und es kommen Menschen*") (Frisch 1967:

demands to be recognized as part of the national community. This we can see as far back as W.E.B. Du Bois's positive experiences in Germany both in the 1890s, while simultaneously black men were being lynched in the U.S., and in the 1930s, while German Jews were being systematically oppressed, dehumanized, and murdered (Du Bois 2007).

Finally, I know that as an African American *academic* with a Ph.D. and a tenured position at an American university, white German academics likely treat me with more respect than they treat People of Color who are local students and researchers, who often have more precarious positions at universities. It always irks me that when I give a lecture in Germany based on my research about the literature of a Black German artist, white Germans are immediately eager to listen and are surprised to encounter this material. But I know that if a Black German student, professor, or non-academic wished to present the same material, white Germans would likely question their expertise and consider whether they are too *betroffen* (affected) to be able to speak 'objectively' about the topic. Several Black German scholars, artists, and activists have addressed the problem of how Black people are seen as only objects of knowledge, rather than subjects thereof. In *Afrokultur: »der raum zwischen gestern und morgen,«* Natasha Kelly writes: "European scholarship is based on a Eurocentric worldview, according to which Africa is treated as the counter-image (*Gegenkonstrukt*) to Europe and African scholarship is denied (*abgesprochen*) any academic validity (cp. Kilomba 2008: 26ff)" (Kelly 2016: 13).[6] And this counts not only for Africans, but all African-descended people, including Black Germans. In *Undeutsch*, Fatima El-Tayeb writes:

> The implicit assumption that racialized (*rassifizierte*) people can never produce analysis, only 'consternation' (*Betroffenheit*) is actually even less problematic than the corresponding conviction that white, heteronormative research is (*sei*) never political or subjective […] This attempt and rationalization for excluding racialized researchers

100); because Western European countries viewed guest workers purely as temporary labor power, this quote highlights, they were not equipped to deal with people with families and complicated lives who needed help integrating into this new society. For more information on this history, as well as the presence of foreign workers in the GDR, see Göktürk et al. (2011).

6 "[…] europäische Wissenschaft [beruht] auf einer eurozentrischen Weltanschauung, wonach Afrika als Gegenkonstrukt zu Europa verhandelt wird und afrikanische Wissensformationen eine wissenschaftliche Gültigkeit abgesprochen bekommen." All translations from German to English are those of the author unless otherwise noted.

(*Wissenschaftler_innen*) [...] is much more symptomatic of the academic dealing with (the history of) race and racism in Germany (El-Tayeb 2016: 22).[7]

One of my favorite challenges to this kind of racism in academia is Black German poet Philipp Khabo Köpsell's poem "The Brainage," which is an excellent example of how an aesthetic work can *also* do theoretical work. Köpsell linguistically plays with the German language with a first line that recalls Heidegger's convoluted writing style in *What is Called Thinking?* (*Was heißt Denken?* [1959]).[8] If according to E. Grunert, Heidegger's *What is Called Thinking?* argues that "We have to learn how to think again" (1959: 101), Köpsell's possible riff on the text implores us to consider the questions: think *how* and according to *whose* standards? At the start of his poem, Köpsell plays on the word *Wissenschaft* (knowledge) by separating it into a noun and verb: "Knowledge creates borders creates definition" ("Wissen schafft Grenzen schafft Definition") (2011: 51). This first line evokes the problem of knowledge and power (Foucault 1980); that those who have the power to create knowledge often use it to define people "creates people creates thought creates people and nation" ("schafft Menschen schafft Denken schafft Mensch und Nation") and by doing

7 "Die implizite Annahme, dass rassifizierte Menschen nie Analyse, sondern nur 'Betroffenheit' produzieren können, ist hier sogar noch weniger problematisch als die dazugehörige Überzeugung, daß weiße, heteronormative Wissenschaft nicht politisch und subjektiv sei [...] Ansatz und Begründung des Ausschlusses rassifizierter Wissenschaftler_innen [...] sind vielmehr symptomatisch für den wissenschaftlichen Umgang mit (der Geschichte von) Rasse und Rassismus in Deutschland."

8 Martin Heidegger (1889-1976) is one of many white German thinkers whose problematic, racist past may be acknowledged, but it is not always used to contextualize his ideas, many of which continue to be popular with academics today. Heidegger, who began his career in philosophy as a lecturer in Freiburg in 1919, became rector in Freiburg in April 1933. A month later he joined the Nazi Party and "until he resigned as rector in April 1934, he helped to institute Nazi educational and cultural programs at Freiburg and vigorously promoted the domestic and foreign policies of the Nazi regime [...] Heidegger continued to support Hitler in the years after his rectorship, though with somewhat less enthusiasm than he had shown in 1933–34" (Duignan 2010: 278). There are numerous academic texts that attempt to reconcile his philosophy with his fascist past, including most recently Faye (2012), O'Brien (2015), and Feinmann (2016).

so, *Wissenschaft* also includes and excludes groups of people from the process of creating knowledge.[9]

And since Black German scholars, activists, and artists have already addressed this problem in their work – work readily available for white Germans to read – I simply would like to add to the conversation by reflecting on my own particular experience as an African American working in German Studies. I hope my experiences in Germany and teaching German in the U.S. can offer some useful points of reflection that can simply add another dimension to the very necessary discussion connecting the essays in this volume about *how* and *why* Humanities departments in Germany can and should be diversified, and structural inequalities should simultaneously be addressed.

I have often witnessed in Germany that white academics not only fail to position themselves when working on topics dealing with race, but they also assume they can occupy an objective position from the outside, which allows them to analyze an object without anything at stake besides intellectual inquiry. I witnessed this at a recent conference, when a white, female German scholar presented on the history of blackfacing in Germany and posed the question whether blackfacing by whites is necessarily *always* offensive or if there are sometimes instances where it could be productive. I am pretty sure all of the scholars in the room who were People of Color were appalled at such a suggestion. Blackfacing is an offensive act that is traumatic for People of Color to encounter. It is already problematic that a white scholar would conduct research on blackfacing without positioning herself within structures of racialization, because she assumes an objective viewpoint on the basis of a supposed objectivity. But it is also offensive that said white, female scholar would suggest blackface could be productive without acknowledging that she is not targeted by blackfacing's parody and therefore does not experience the trauma it invokes.[10]

As Arghavan, Hirschfelder, and Motyl state in the introduction to this volume, "[…] the presuppositions with which Humanities scholars approach their object of inquiry is influenced by the ways of knowing they have acquired by having occupied particular subject positions throughout their biographies" (15). Thus, when a white German professor works on a topic like blackfacing, it is important that they position themselves first and consider how their privilege and their own experiences may influence their perspective. This very problem is discussed in a conversation between Black German scholar Peggy Piesche and

9 Emily Ngubia Kuria also discusses this problem of white German academics assuming an objective point of view (2015: 20; 108).

10 For more information on the history of blackfacing in Germany and its continued practice, see Milagro (2012). Cf. Wipplinger (2011) and Sieg (2015).

white American scholar Sara Lennox in the volume *Remapping Black Germany: New Perspectives on Afro-German History, Politics and Culture*. Lennox is a feminist German Studies scholar who has been actively researching Black German Studies since the 1980s. She is also a friend and colleague to several prominent Black German scholars and activists. Nonetheless, Piesche is steadfast in her skepticism about whether white scholars should work in Black German Studies. Piesche maintains that if a white scholar is *not* going to treat the field as just a passing trend, they have a certain responsibility to acknowledge their white privilege. She states, "responsibility starts with undertaking a critical consideration of where they are situated and of the way they consider their objects and include that as part of their analyses. Otherwise […] they are just continuing to contribute to an epistemological canon that rests on our collective experiences […]" (2016: 279). In the following three sections, I wish to share some of the experiences I have had with the lack of diversity in German academic spaces, and use them to consider how power is deployed to keep People of Color out or keep them from speaking. I will also analyze the mechanisms behind and the implications of the lack of ethnic and racial diversity in *Germanistik* in Germany and consider possible next steps for improving the situation.

Race and German Studies/*Germanistik*

One way to fight racism, discrimination, implicit and explicit bias, and white fragility is to have more People of Color at the university, both in positions of power (in order to actually change something in the structures) and as students (in order to hold those in power accountable). Nevertheless, there are a lot more reasons why having more diversity at the university has positive effects. In this section, I will reflect on the benefits of diversity for my own discipline, German Studies.

From my experience, German Studies or *Germanistik*, depending on whether we are discussing the U.S. or Germany, tends to see itself as being free of racial problems, aside from recognizing the importance of German Jewish Studies, which includes teaching the work of German Jewish intellectuals and artists, addressing the problem of Anti-Semitism since the Middle Ages, teaching about the Shoah and Germans' attempts since World War II to come to terms with it, and discussing the experiences of German Jews in contemporary society. It should perhaps not be surprising that German Studies/*Germanistik* has largely been a discipline where whiteness goes unmarked while racialized authors are marked. As Jakob Norberg argues in his essay "German Literary Studies and the Nation" (2018), from its nascence *Germanistik* has been a nationalist enterprise;

an attempt to craft a distinct narrative of a German nation by way of a genealogy of German literature. Thus, if white Germans have long viewed the nation as a white one, then race would never figure an important role and non-white authors would be viewed as deviations from a white norm.

Therefore, if *Germanisten* (in Germany) have for some time viewed themselves as white scholars working on literature by white people, this could explain why for a long time, there was not much of an attempt to think about race in a more general way. I have also wondered to what extent the influence of postwar literary analysis has contributed to creating a feigned sense of objectivity. Following World War II, *Werkimmanente Literaturwissenschaft* (New Criticism) was the prominent form of literary analysis. Bettina Gruber defines *werkimmanent* as "[…] every attempt at literary criticism […] that limits itself to an occupation with the work, respectively the text, and dismisses everything else as 'peripheral' (*Umfeld*)" (2009: 763).[11] Thus, it is not far-fetched to imagine that race would be considered part of the "*Umfeld*" that was ignored, especially if the literary critics and authors at the time took their own whiteness for granted, viewing themselves as simply Germans and racially unmarked.

A more sustained attempt to address issues of ethnicity within *Germanistik* occurred with the creation of intercultural German Studies (*interkulturelle Germanistik*), the origins of which lie in the understanding that

> […] German *Germanistik* eo ipso is not in the position to sufficiently consider the interests (*Lerninteressen*) and learning conditions of foreign students […] from the very beginning, the conceptual goal was to supplement a study of culture based on language and texts with a 'culture-curriculum' […] In cooperation with Heinz Göhring (1975), plans began immediately to interrogate theoretical questions about processes of intercultural communication in favor of improving transcultural understandings (Wierlacher/Bogner 2003: 2).[12]

11 "[J]edes literaturwissenschaftliche Vorgehen […] das sich strikt auf die Beschäftigung mit dem Werk resp. Text selbst beschränkt und daher alles andere als 'Umfeld' ausblendet."

12 "[…] deutsche Germanistik, eo ipso gar nicht in der Lage ist, mit den Lerninteressen ausländischer Studierender auch ihre Lernbedingungen hinreichend zu berücksichtigen […] Konzeptionelles Ziel war von Anfang an ein sprach- und textbezogenes Kulturstudium mit einem 'Kultur-Curriculum' […] In Kooperation mit Heinz Göhring (1975) griffen die Planungen zugleich Fragen der Theorie von Prozessen zwischenkultureller Kommunikation zugunsten der Verbesserung transkultureller Verständigungen auf."

This idea of transcultural understandings reveals the discipline's weakness when it comes to addressing issues of race, ethnicity, and diversity *within* the nation. Unlike the category of "multicultural literature" in the U.S., which refers to people of diverse backgrounds writing about their American experience for American readers, intercultural literature (*interkulturelle Literatur*) was originally conceived as aiding efforts to spread German culture *abroad*, from a white-centered standpoint. Minorities writing in German could serve as messengers to their home communities about why Germany should be of interest to them. That is why *interkulturelle Literatur* has always had strong ties to German as a Foreign Language (*Deutsch als Fremdsprache*). But what does it mean to really acknowledge the "interests and learning conditions of foreign students"? Would that mean teaching African American students of German about Audre Lorde's time in Germany? Or would it mean presenting them with Black German texts about topics to which they could relate? And if the focus of *interkulturelle Germanistik* is on German Studies abroad, how does that help People of Color *in Germany*? How does one make *Germanistik* a more inviting, relevant discipline for them?

What I've learned from teaching German Studies in the U.S. is how important representation and diversity can be to students. In the U.S., German Studies is predominantly of interest to white American students who either have some German ancestry, are interested in a particular discipline for which German is useful (e.g. philosophy, music, literature), or have the desire to study abroad. However, we also have non-white students with varying backgrounds: African American, Latinx, and Asian American students, foreign exchange students, and immigrants who spent some time in Germany before moving to the U.S. And although the students who are People of Color make up a minority in my own department, it is nevertheless important that they see themselves reflected in the curriculum and that they are not pushed to the margins. At the end of each semester at my university, students are asked to fill out a survey (provided by the university) about their class. They are asked questions about their instructor's performance and one such question is whether or not the instructor valued diversity. On one such survey, a white American student responded that this question was useless in a German class, because there were no minorities in her class and, besides, German is a "white language." When I heard this statement, I was appalled, because such a statement negates the existence and experience of the hundreds of thousands of People of Color for whom German is their native or primary language.

In America, the danger of presenting a curriculum in German Studies that is predominantly white, is that we alienate People of Color and we do not teach our students to engage with a contemporary German society that very well includes many POC. But in Germany, the danger of presenting a curriculum in *Germanistik* that is predominantly white, is that you teach People of Color living in Germany that their experiences do not matter. You effectively erase their existence. That can also mean denying them a past, making it seem that every Person of Color is the very first (El-Tayeb 2005a). But it also means denying them a future, or at least limiting the possibilities they see for their future, because they do not see themselves reflected in the society, except when they are included as stereotypes. Thus, when American students are presented with a predominantly white curriculum in German Studies, they get a false view of Germany (its present and its past). But for students of Color in Germany, the effects are even more devastating.

When I raise the issue of diversifying a German Studies curriculum, many of the skeptics echo the same excuses. They say there are not enough texts by People of Color in German. Or they say they do not have the expertise to teach an entire class on "people with a migration background." Or perhaps they claim a unit on "diversity" would not fit into their syllabus. What they get wrong is the assumption that including texts in German by People of Color means you need an entire unit, lesson, or course about diversity or multiculturalism. While this has been the approach taken for decades, it also creates a literary 'ghetto' for authors of Color, keeping them at the margins. It is just as possible to incorporate more texts by People of Color in whatever lesson or syllabus you intend to teach, whether it is about literature and science, Romanticism, or poetry.

I once taught a class in the U.S., in German, on literature and film about Berlin. We began with poems from the 18th century about Berlin's emerging modernity and we ended with a novella by a Black German author, Michael Götting, called *Contrapunctus* (2015); a story about several Black Germans living in Berlin and grappling with personal problems like family and relationships, as well as everyday racism and the specter of German colonialism. On the first day of discussing *Contrapunctus*, one of my students, a woman of Color, stated that in eight years of learning German (in high school and college), this was the first time she saw herself reflected in a text. I then learned that the student was Senegalese American; her mother was from Senegal and her father was a white American. She said that *Contrapunctus* especially spoke to her, not just because the characters were Black, but also because one character suffered from an illness that was suspected to be sickle cell anemia and this student of mine also had sickle cell anemia. This was a powerful moment for me that nearly brought me to tears. Due to my own experience learning German, I could imagine how invisible this student must have felt, learning German language and culture for eight years and never seeing or hearing about someone who looked like her, even though there were plenty of people who *did* look like her and could speak to her experience as a woman of Color and a woman of African descent living in Germany. Those texts exist. But her instructors never thought to include them, because they are working with a canon of German literature that largely consists of dead white men. Although my anecdote comes from the U.S., it could be easily applied to Germany. In fact, displacing this scenario to Germany makes it even more urgent. What would it feel like to be Black German or Turkish German, to be born and raised in Germany, to spend years in public school and to never encounter a character who looks like you or an author who can speak from a similar position as you? Would you feel included in the nation? Would you feel like your opinion or your experience mattered? And how empowering could it be for a student who came to Germany as an immigrant or a refugee to be presented with texts that reflect their experience?

Thus, if we are going to combat racism in the German academy, we can start by better diversifying disciplines like *Germanistik*. And first, we have to start with *what* is taught and decolonize the canon. Decolonialization and decoloniality are terms that were introduced in the 21st century in an attempt to find new ways to combat the effects of Western imperialism.[13] If a postcolonial viewpoint assumed we had achieved an existence *after* colonialism, decoloniality acknowledges that even in the absence of colonial governments, many people across the globe continue to suffer from colonial relationships in and with the West. As Karishma Desai and Brenda Nyandiko Sanya write, in contrast to colonialism,

> [...] coloniality is an epistemological frame that binds these historical projects [colonialism and imperialism] to contemporary times in an integral way. Coloniality recognizes how colonial power relations left lasting marks in the areas of authority, sexuality, knowledge, economy, and in our naturalised understanding of what it means to be human. Decoloniality, then, requires delinking from coloniality and modernity [...] (2016: 712-713).

Decolonizing the German canon does not mean erasing all of the dead white men from the syllabus. I also teach Schiller, Goethe, Kleist, Brecht, and Kafka. But in every course I teach, when I sit down with my syllabus I ask myself "What texts by women and People of Color can I include?" I do not have to teach a class on

13 An important scholar who has contributed to the theoreticization of decoloniality is Walter Mignolo. See Mignolo, *The Darker Side of Western Modernity* (2011).

race in German Studies to include authors of Color. I do not have to teach a class solely on women and feminism to include female authors. And the same can go for texts with LGBTQ or differently abled perspectives. When I teach "Introduction to German Literature," we begin with Lessing and Goethe in the 18th century, but by the time we get to the 20th century, students have read Turkish German author Emine Sevgi Özdamar and Black German poet May Ayim. By integrating texts by authors of Color into a syllabus with a more general theme like "Introduction to German Literature" or "Literature of Berlin," students do not view these texts as 'extra' or as some kind of required diversity component. Rather, by presenting texts in this manner, I help students see female authors and authors of Color as *belonging* to the canon of German literature, as I see it. This is something that every professor of German Studies can do. It might be more difficult in earlier periods like the Middle Ages or Early Modern period. However, in those cases, when one teaches a text like *Parzival* or *Simplicissimus*, where Blackness is not the central issue but there are marginal Black characters, it is important for the professor to address and contextualize the issue of representing race head-on and not just gloss over these representations. Rather, German professors should dare to challenge their students and themselves to think more critically about race and think about how they can work to deconstruct racial stereotypes in every context.

In my experience of guest teaching and guest lecturing in Germany within the past four years, I have often found that while German scholars are curious about my work on critical race theory and postcolonialism, they act as if I was the first person to pursue these topics. However, there are established fields of German postcolonial studies and Black German Studies. The texts are available, but it is up to white professors to *want* to read them and to purposefully incorporate them in their teaching and research. Furthermore, it is just as important that white German professors not only look to People of Color for aesthetic works, but also engage with theoretical texts by People of Color. From African American scholar W.E.B. Du Bois, who actually studied at Humboldt University in the 1890s, to Frantz Fanon and Audre Lorde, and more recently Michelle Wright and Fatima El-Tayeb, there are plenty of Black scholars engaging in theoretical work. While some scholars, like Achille Mbembe, are thought of as addressing more universal, philosophical issues, a scholar like Lorde might be considered of no use to analyzing German literature because she is coming from another field, namely African American Studies. Nevertheless, I believe in Alexander Weheliye's assertion that Black Studies is not a niche discipline; rather, because of how Blackness has been defined vis-à-vis the Western category of the human, Black Studies is actually the study of the human condition (2014). Therefore, there is no reason why Lorde's reflections on intersectionality, race, and gender cannot be useful for discussing the same topics within a German context, while also acknowledging the differences. As Fatima El-Tayeb states in *European Others*, it can be both possible and productive to "creolize" theory from the American context so that it might help explain how race, both its presence and its absence, functions in a European context (El-Tayeb 2011: xviii).

Of course, the curriculum would also become more diverse if the people who are teaching were more diverse. Some Germans might argue that it is not possible to hire POC for academic positions if POC are not choosing to pursue PhDs. But the problem might in fact lie not in disinterest from POC, but in the *way* in which PhD candidates in Germany are chosen. In the U.S., students who would like to pursue a PhD in a particular field apply to a number of programs across the country. Then a committee of professors at each program evaluate the applicants yearly and choose whom they will admit. The number of students admitted to a PhD program depends on the size of the department, the size of the applicant pool, and how much funding the program has available. In Germany, although some *Graduiertenkollegs* (graduate schools) have been funded by the German Research Foundation at German universities, the majority of PhD candidates in the Humanities and Social Sciences are still identified by professors. Studies in the U.S. have shown that when it comes to hiring practices, American faculty choose the candidate who looks/thinks like them (Sensoy/DiAngelo 2017). This is one of the barriers to diversifying the faculty; faculty have a tendency to reproduce themselves when given the power to choose who joins the department.[14] Thus, it is likely this is also at work in Germany when PhD candidates are identified. Is a white German academic going to have any reservations about choosing a student of Color to pursue a PhD? Do white German professors see their POC students as less qualified? Do white German professors assume their POC students will want to work on a topic related to race/ethnicity? If white German professors themselves do not research race/ethnicity in their field, will they be wary of taking on students who do? I do not have the answer to these questions. And I cannot say for certain that a system of application according to the U.S. model would work in Germany, since the structures and hierarchies at German universities are so different. Nevertheless, these are the kinds of questions white Germans need to ask

---

14 Kuria also documents such practices happening on the level of white administrators' seemingly arbitrarily deciding whether or not to admit a POC to a Masters' program based on whether they expect a person of that race or ethnicity to have the right training (2015: 45).

themselves in order to work towards a future in which students of Color are encouraged to pursue a PhD. More POC PhD students will create a pipeline of possible candidates for future academic jobs.

There are some who might argue that including more faculty of Color will just *reshape* the problem rather than fix it. These critics are likely referring to "opportunity hoarding," which Charles Tilly defines as follows:

> If members of a network acquire access to a resource that is valuable, renewable, subject to monopoly, supportive of network activities, and enhanced by the network's modus operandi, network members regularly hoard access to the resource, creating beliefs and practices that sustain their control. If that network is categorically bounded, opportunity hoarding thereby contributes to the creation and maintenance of categorical inequality (1998: 99).

Opportunity hoarding is not strictly a phenomenon among whites. As Tilly continues, "Opportunity hoarding often rests on ethnic categories, members of which reinforce their control over hoarded resources by means of their power to include or exclude other members with respect to language, kinship, courtship, marriage, housing, sociability, religion, ceremonial life, credit, and political patronage" (1998: 99). Opportunity hoarding can also happen among professions, including in academia. Professors on exams committees, who are in charge of who does or does not successfully defend a Master's thesis or dissertation, can and often do function as 'gatekeepers,' because we set the norms that determine who can become a part of the profession. But to those who suggest that a faculty of several POC will necessarily serve the same gatekeeping function as white academics, choosing only those who look or think like them, I offer a few reasons why that is unlikely: 1) Having one or even a few POC faculty will not give them the ability to overturn the power of their white counterparts. 2) There has never been a case in Germany where POC were the majority of a faculty. 3) Even in the U.S., in a field where one might expect to find a majority POC faculty, like in African, African American, or African Diaspora Studies, this is often *not* the case. And even in the rare chance of there being a majority POC faculty, they are often still beholden to white superiors, whether that be a Chair of the department, Dean, Provost, or President of a university. Thus, POC faculty have never been in the position to make decisions that would solely benefit themselves. POC faculty have never wielded that kind of power, not in the U.S. and especially not in Germany.

The fact of the matter is studies show that just because you get more POC students and faculty into the academic pipeline, this neither ensures they will remain nor does it ensure they will thrive as much as their white colleagues. And as Minefee et al. argue, this is *not* because POC simply choose to exit the pipeline voluntarily or because they do not have the skills to succeed. Minefee et al. assert that people do not just "'passively flow' through the pipeline. Instead, individuals are agentic and traverse the institutional, social, and discursive practices that help sustain the leaky pipeline" (2018: 80). Thus, a "leaky" pipeline often results from the structural inequalities and racism students and faculty *continue* to face, even once they have gotten their foot in the door of an institution. As Tomaskovic-Devey et al. demonstrate, even after a student completes their PhD, just being a POC with a higher degree does not mean the playing field has been leveled for them:

> Perhaps the most disturbing finding of the analysis is that not only is the race/ethnic gap greater at higher levels of education, but the white advantage and black and Hispanic disadvantage in early career earnings trajectories is severest for those with the highest levels of education. Thomas (1993), focusing only on the black-white wage gap, suggests that this particular race-class interaction "may be at least partially explained by the inability of blacks with high levels of education and occupational attainment, to translate these 'assets' into income due to discrimination in the labor force. Whites, who are untouched by these discriminatory barriers, can more easily convert these sources of 'human capital' into income" (p. 341) (Tomaskovic-Devey et al. 2005: 82).

In light of this research, if white people say they cannot give POC more power in academia, because *we* POC faculty may use it to discriminate, they are either perpetuating the myth that white researchers are objective and POC researchers are automatically biased *or* they are admitting to having discriminated against POC all of this time and are fearful of what could happen if the power imbalance shifts. To make the argument that POC should be blocked from power because they *may* misuse it, while simultaneously ignoring the fact that white faculty *have* been misusing it, constitutes both racism and gaslighting.[15]

15 Gaslighting refers to "[…] a specific type of psychological and emotional abuse […] a varying cluster of manipulations used to undermine an intended target person's mental stability (Welch 2008). *The Gaslight Effect* (Stern 2007) is an interaction between a *gaslighter*, who desires to maintain control, power, and a sense of being right in the relationship; and the *gaslightee* 'who allows the gaslighter to define her sense of reality because she idealizes him and seeks his approval' (p. 3)" (Hightower 2017: 2).

Conclusion: Staying Optimistic, But Also Realistic

As a professor at an American university, I am not demystified by the apparent cult of diversity at work within American universities. In *On Being Included: Racism and Diversity in Institutional Life*, Sara Ahmed discusses how diversity can become "[…] a convention, or a conventional way of speaking about the university. Diversity becomes a ritualized or polite speech" (2012: 58). In that way, diversity can be "a sign of the lack of commitment to change and might even allow organizations such as universities to conceal the operation of systemic inequalities" (2012: 53). While I agree with Ahmed's intersectional feminist critiques of the term 'diversity,' I disagree with the argument Walter Benn Michaels makes in *The Trouble with Diversity: How We Learned to Love Identity and Ignore Inequality*, namely that this obsession with racial diversity is an attempt to deflect discussions about class (2006: 6). First of all, several of the arguments Michaels makes about race seem naïve and misguided at best and at worst disingenuous, such as: (1) alleging that racial diversity is as arbitrary as hair color diversity, which does not acknowledge that People of Color are in fact *also* policed about their hair; (2) insisting that the concept of different races exists because we ALL want to hang on to them, rather than acknowledging that white supremacy is invested in them in order to maintain white privilege; (3) saying that anti-racism is about insisting equality is already there, instead of having to create it through economic policies.

In America, part of anti-racist work is removing the purposefully racist barriers (red lining, banks not lending to Blacks, hiring discrimination, disparity in pay) that have led to the extreme gap in wealth between Blacks and whites. As I will elaborate below, this is an intersectional problem. And Michaels' claims like "[…] what American liberals want is our conservatives to be racist" or "[…] racism has been pushed to the fringes of public life" (2006: 82) are dated, especially twelve later, in the era of Trump's presidency. Finally, his focus on what allegedly keeps blacks out of college – wealth – ignores the problem of the leaky pipeline; that even if the number of Black students in universities increases, this does not guarantee they will stay or graduate, if the environment of the university is plagued by Eurocentric thinking, structural inequalities, and racism.

If we target the core of Michaels' argument, while it is true that Americans are more likely to identify with a racial group than with the lower or upper class – as he says "The class we like is the middle class" (6) – there is a difference between being *willing* to identify with a group and *having* to identify with a group. Class is something that we can hide and that we can lie about. Racial categories, on the other hand, are harder to ignore, not only because American society insists on categorizing its citizens this way, but also because race is so often tied to physical appearance. Americans' obsession with race is evident from the "one-drop rule," which when adopted as the Racial Integrity Act in 1924 argued that having just "one drop" of Black blood – that is, having any African ancestors – meant you were Black. If you are Black in America, you can lie about what class you belong to. You can even buy things on credit in order to feign belonging to a wealthier class. However, you cannot lie about being Black. In fact, Black celebrities, whether or not they voluntarily identified as Black, have learned that even having wealth and belonging to the upper class did not save them from racial discrimination and racist violence. Which brings me to what is also misleading about Michaels' statement; it ignores intersectionality. Not only does being middle or upper middle class *not* exclude you from racist treatment, but it is often the students of Color who are *also* lower class, and therefore, their experience of the academy is a result of both their class *and* race.

Many universities seek to promote diversity both among their student body and among their faculty. In the U.S., demographic trends predict that by 2045, the U.S. will become "minority white," because People of Color will make up the majority (Frey 2018). Already, states like California have learned how increasingly diverse the student body can become, and universities there have taken steps to try to better accommodate students of Color and better meet their needs (Tatlock 2010). Studies also show that being among a diverse cohort positively benefits students, and it is therefore no wonder that many American students, regardless of their background, would prefer to go to an institution that boasts some form of diversity (Gurin et al. 2002). Previously, universities may have concentrated much of their efforts on recruiting and admitting students of Color and centering them in their advertisements and public platforms. I personally have participated in such campaigns more than once. First, when I was a senior in college at the University of Chicago and had just returned from studying abroad for a year in Germany. I was asked to contribute a blurb about my background and my study-abroad experience for an ad campaign seeking to increase donations for student scholarships. The second time was more recently, when the study-abroad magazine at my institution wished to publish a similar, but more elaborate story about how I came to my field and what my experience abroad has meant for my life and career.

But American universities are starting to realize that this is no longer enough. Because once they have lured more students of Color to campus, what those students often realize is that 1) they are in the minority 2) they do not have very many professors who look like them 3) their disciplines do not account for or appreciate diversity and instead stick to the same Eurocentric curriculum that privileges the experiences of white people. Thus, what is becoming increasingly centered in discussions around diversity in the U.S. is not just how to diversify the student body *and* the faculty, but how to address the structural racism and dominance of whiteness in the institution. Özlem Sensoy and Robin Diangelo write:

> As Eduardo Bonilla-Silva (2015) explains, White-oriented and -led institutions reproduce Whiteness through their curriculum, culture, demography, symbols, and traditions, while they simultaneously pass as neutral spaces free of race and racialized perspectives. Only people of color are racialized and seen as 'bringing' race into race-neutral (White) spaces. If there are no peoples of color present, race remains unnamed and is not presumed to be an organizing institutional factor (2017: 560).

But as Sensoy and Diangelo demonstrate in their essay, "'We are all for Diversity, but…'," simply hiring more diverse faculty or admitting more students of Color will not solve structural problems if the racism and whiteness responsible for them are not addressed. Sensoy and Diangelo's highly informative essay, which includes extensive suggestions for how to hire more People of Color, shows the kinds of pitfalls and obstacles that are often invisible to the majority-white faculty, staff, and administration. Job candidates of Color must often prove that they are not only experts in their field, but that they have not simply benefitted from some kind of "minority bonus." They must bear much of the burden of counseling students of Color or mentoring minority junior faculty; work that is often uncompensated and *does not* count towards promotion. And faculty of Color may sometimes be denied entry into a department, simply because, as I mentioned above, hiring committees tend to reproduce themselves; to choose the candidate who seems like the "best fit," which translates into denying faculty of Color a job opportunity, for fear that their minority status might translate into conflicts in the group. And when a person of Color is hired, if they are the 'only one' in their department, they can suffer from isolation and a lack of the support and community needed to thrive.

These are all reasons why institutions who seek to diversify their faculty and student body must simultaneously take a look at how racism and whiteness manifest themselves within the structures of the institutions. Are only *certain* kinds of knowledge valued, i.e. Western, positivist epistemologies? Is experiential knowledge undervalued or not valued at all? Are People of Color viewed as 'tokens' rather than as equally competent as their white peers? These are the kinds of questions institutions must ask themselves at every level. As

Wischmann argues about the engrained white supremacy in the concept of *Bildung*, when one looks at the origins of this philosophical concept in Kant, Hegel, and Humboldt, one sees that "Everybody can become *gebildet*, but with different efforts and necessarily the same goal, namely to become like a 'white' European" (2008: 479). And as Emily Ngubia Kuria asserts, "meanwhile Black people and POC have knowledge at their disposal which *white* colleagues (*Kolleg_innen*) have not tapped into" (2015: 100). Instead of maintaining an educational structure with the goal of making POC appease a white hegemony, white Germans have to work towards dismantling white supremacy in the academy and restructuring it to create a space of learning that truly values diverse backgrounds, diverse opinions, diverse experiences and skills. Only when the universities, departments, and disciplines can commit to deconstructing white privilege, will any real progress be made towards building an academy that better reflects our globalized world and multicultural societies.

Works Cited

Ahmed, Sara (2012): On Being Included: Racism and Diversity in Institutional Life, Durham and London: Duke University Press.

Chin, Rita/Fehrenbach, Heide (2009): "Introduction: What's Race Got to Do with it? Postwar History in Context." In: Rita Chin/Heide Fehrenbach, After the Nazi Racial State: Difference and Democracy in Germany and Europe, Ann Arbor: University of Michigan Press, pp. 1-29.

Crenshaw, Kimberlé (1989): "Demargilizing the Intersection of Race and Sex: A Black Feminist Critique of Antidiscrimination Doctrine, Feminist Theory and Antiracist Politics." In: University of Chicago Legal Forum 1/8, pp. 139-167.

Desai, Karishma/Sanya, Brenda Nyandiko (2016): "Towards Decolonial Praxis: Reconfiguring the Human and the Curriculum." In: Gender and Education, 28/6, pp. 710-724.

Du Bois, W.E.B. (2007): The Autobiography of W.E.B. DuBois, New York: Oxford University Press.

Duignan, Brian (2010) (ed.). The 100 Most Influential Philosophers of All Time, New York: Britannica Educational Publishing.

El-Tayeb, Fatima (2005a): "Dangerous Liaisons: Race, Nation and German Identity." In: Patricia Mazón/Reinhild Steingröver (eds.), Not so Plain as Black and White: Afro-German Culture and History, 1890-2000, Rochester: University of Rochester Press, pp. 27-60.

El-Tayeb, Fatima (2005b): "Vorwort." In: Maureen Maisha Eggers/Grada Kilomba/Peggy Piesche/Susan Arndt (eds.), Mythen, Masken und Subjekte: Kritische Weißseinforschung in Deutschland, Münster: Unrast Verlag, pp. 7-9.

El-Tayeb, Fatima (2011): European Others: Queering Ethnicity in Postnational Europe, Minneapolis: University of Minnesota Press.

El-Tayeb, Fatima (2016): Undeutsch: Die Konstruktion des Anderen in der postmigrantischen Gesellschaft, Bielefeld: transcript.

Faye, Emannuel (2012): "Being, History, Technology, and Extermination in the Work of Heidegger." In: Journal of the History of Philosophy 50/1, pp. 111-130.

Feinmann, José Pablo (2016): Heidegger's Shadow. Joshua Price and Maria Constanza Guzmán (trans.), Lubbock: Texas Tech University Press.

Foucault, Michel (1980). Power/Knowledge – Selected Interviews and Other Writings 1972-1977. Colin Gordon (ed.), New York: Pantheon Books.

Frey, William H. (2018): "The US will Become Minority White." In: Avenue, March 14 (https://www.brookings.edu/blog/the-avenue/2018/03/14/the-us-will-become-minority-white-in-2045-census-projects/).

Frisch, Max (1967): Max Frisch: Öffentlichkeit als Partner, Frankfurt am Main: Suhrkamp.

"Germany: UN Rights Panel Highlights Racial Profiling Against People of African Descent," February 27, 2017 (https://news.un.org/en/story/2017/02/552282-germany-un-rights-panel-highlights-racial-profiling-against-people-african).

Göktürk, Deniz/Kaes, Anton/Gramling, David/Langenohl, Andreas (eds.) (2011): Transit Deutschland: Debatten zu Nation und Migration: eine Dokumentation, Konstanz: University Press.

Gruber, Bettina (2009): "Werkimmanente Literaturwissenschaft/New Criticism." In: Jost Schneider (ed.), Methodengeschichte der Germanistik, Berlin and New York: Walter de Gruyter, pp. 763-776.

Grunert, E. (1959): "M. Heidegger: Was heisst denken?" In: Freiburger Zeitschrift für Philosophie und Theologie 6/1, p. 101.

Gurin, Patricia/Dey, Eric L./Hurtdado, Sylvia/Gurin, Gerald (2002): "Diversity and Higher Education: Theory and Impact on Educational Outcomes." In: Harvard Educational Review 72/3, pp. 330-366.

Hightower, Erin (2017): "An Exploratory Study of Personality Factors Related to Psychological Abuse and Gaslighting." PhD dissertation, William James College.

Jones, Bruce (2017): "A Body Does Not Just Combust: Racism and the Law in Germany," In: World Policy Journal 34/2, pp. 31-35.

Kelly, Natasha (2016): Afrokultur: »der raum zwischen gestern und morgen,« Münster: Unrast.

Kelly, Natasha (2017): "Rassismus betrifft alle Fächer." In: Der Tagesspiel, August 30 (https://www.tagesspiegel.de/wissen/uni-dozentin-natasha-a-kelly-rassismus-betrifft-alle-faecher/20255456.html).

Kilomba, Grada (2016): Plantation Memories: Episodes of Everyday Racism, Münster: Unrast.

Köpsell, Philipp (2011): "The Brainage." In: Susan Arndt/Nadja Ofuatey-Alazard (eds.), Wie Rassismus aus Wörtern spricht: (K)Erben des Kolonialismus im Wissensarchiv deutsche Sprache. Ein kritisches Nachschlagewerk, Münster: Unrast, pp. 51-53.

Kuria, Emily Ngubia (2015): eingeschrieben: Zeichen setzen gegen Rassismus an deutschen Hochschulen, Berlin: w_orten & meer.

Michaels, Walter Benn (2006): The Problem with Diversity: How We Learned to Love Identity and Ignore Inequality, New York: Metropolitan Books.

Mignolo, Walter (2011): The Darker Side of Western Modernity: Global Futures, Decolonial Options, Durham, North Carolina: Duke University Press.

Milagro, Lara-Sophie (2012): "Die Blackfacing-Debatte III – Man muss kein Neonazi sein, um rassistisch zu handeln. Die Bequemlichkeit der Definitionshoheit." In: nachtkritik March 28 (https://www.nachtkritik.de/index.php?option=com_content&view=article&id=6740:die-blackfacing-debatte-iii-man-muss-kein-neonazi-sein-um-rassistisch-zu-handeln&catid=101:debatte&Itemid=84/).

Minefee, Ishva/Caridad Rabelo,Verónica/Stewart, Oscar Jerome C. IV/Jones Young Nicole C. (2018): "Repairing Leaks in the Pipeline: A Social Closure Perspective On Underrepresented Racial/Ethnic Minority Recruitment and Retention in Business Schools." In: Academy of Management Learning and Education 17/1, pp. 79-95.

Norberg, Jakob (2018): "German Literary Studies and the Nation." In: The German Quarterly 91/1, pp. 1-17.

O'Brien, Mahon (2015): Heidegger, History and the Holocaust, London and New York: Bloomsbury Academic.

Ortega, Mariana (2006): "Being Lovingly, Knowingly Ignorant: White Feminism and Women of Color." In: Hypatia 21/3, pp. 56-74.

Piesche, Peggy/Lennox, Sara (2016): "Of Epistemologies and Positionalities: A Conversation, Berlin, October, 21, 2014." In: Sara Lennox (ed.), Remapping

Black Germany: New Perspectives on Afro-German History, Politics and Culture, Amherst: University of Massachusetts Press, pp. 274-281.

Sensoy, Özlem/Diangelo, Robin (2017): "'We Are All for Diversity, but…': How Faculty Hiring Committees Reproduce Whiteness and Practical Suggestions for How they Can Change." In: Harvard Educational Review 87/4, pp. 557-580.

Sieg, Katrin (2015): "Race, Guilt and Innocence: Facing Blackfacing in Contemporary German Theater." In: German Studies Review 38/1, pp. 117-134.

Sow, Noah (2008): Deutschland Schwarz Weiss: der alltägliche Rassismus, München: Bertelsmann.

Srivastava, Sarita (2005): "'You're Calling Me a Racist?' The Moral and Emotional Regulation of Antiracism and Feminism." In: Journal of Women in Culture and Society 31/1, pp. 29-62.

Tatlock, Lynne (2010): "USA: German in the Changing Landscape of Postsecondary Education." In: Die Unterrichtspraxis/Teaching German 43/1, pp. 11-21.

Thompson Tetreault, Mary K. (2012): "Positionality and Knowledge Construction." In: James A. Banks (ed.), Encyclopedia of Diversity in Education, Thousand Oaks: Sage Publications.

Tilly, Charles (1998): Durable Inequality, Berkeley: University of California Press.

Tomaskovic-Devey, Donald/Melvin, Thomas/Johnson, Kecia (2005): "Race and the Accumulation of Human Capital across the Career: A Theoretical Model and Fixed-Effects Application." In: American Journal of Sociology 111/1, pp. 58-89.

Weheliye, Alexander G. (2014). Habeas Viscus: Racializing Assemblages, Biopolitics, and Black Feminist Theories of the Human, Durham, NC: Duke University Press.

Wierlacher, Alois and Andrea Bogner (eds.) (2003): Handbuch für interkulturelle Germanistik, Heidelberg: JB Metzler.

Wilderson, Frank (2010): Red, White and Black: Cinema and the Structure of U.S. Antagonisms, Durham: Duke University Press.

Wipplinger, Jonathan (2011): "The Racial Ruse: On Blackness and Blackface Comedy in fin-de-siècle Germany." In: The German Quarterly 84/4, pp. 457-476.

Wischmann, Anke (2008): "The Absence of 'Race' in German Discourses on Bildung. Rethinking Bildung with Critical Race Theory." In: Race, Ethnicity and Education 21/4, pp. 471-485.

"So You Want to Write about American Indians?" Ethical Reflections on Euro-Academia's Research on Indigenous Cultural Narratives

Amina Grunewald

Abstract

Taking up the provocative-ironical question of Choctaw scholar Devon Abbott Mihesuah, it is paramount to focus on non-Indigenous Euro-academia[1] and its research and teaching on Indigenous peoples and their narratives, because these practices are articulated from a privileged and authoritative academic position, and therefore from a platform of power of an assumed Eurocentric scholarly 'superiority' based on supremacist thinking in almost all Western academic disciplines. Are there opportunities and practices for a critical reassessment of non-Indigenous scholars and their work when dealing with representations of Indigeneities in research and education? Linked to this is the question of Indigenous knowledge transferability, translation, and the potential for dialogue between Indigenous and Western epistemologies and research due to epistemological and ontological differences, for example, a relational approach to all inhabitants of Native land confronting a Western anthropocentrism.[2] The aim of

1 Terms like 'Western' or 'Euro-academia' are umbrella terms that are, in this paper, understood as labels for uncritical Eurocentric, supremacist research mentalities.

2 There are Indigenous cultural knowledge and belief systems as well as related practices that challenge Eurocentric logics and global capitalist resource management. Diverse Indigenous cosmologies challenge non-Native exploitative profit thinking, in which land is understood as an economic resource that can be controlled, mapped, distributed, sold, and agricultured. In short, land is used as a colonized commodity and property without considering traditional Indigenous concepts of land and its rooted

this article is to discuss how Indigeneities can be negotiated by Western agents in particular places of education and therefore in spaces of power. Which cultural, sociopolitical, and ethical contexts need to be taken into account within a decolonizing discourse when dealing with Indigenous peoples, their knowledge and belief systems, and their self-representations? I, for example, am aware that my privileged position is that of a white European academic-learner when dealing with Indigenous people(s) and their narrated lived experiences. Therefore, Indigenous voices (of scholars, of artists, of activists) are given an authoritative position to represent their agendas in my paper in order to let Indigenous cultural insiders have their say about problematic Western cultural outsiders' involvements – Western outsiders that often enough have insensitively trespassed invisible borders regarding Indigenous spaces of protection or protocols, and disregarded issues that arise for Indigenous people(s) who are forced to live within Westernized settler nation-states. This paper not only draws on Indigenous authors'/scholars' works and linked methodologies and philosophies (Fixico, Mihesuah, Champagne), but also on my experiences during research stays in Vancouver and Montréal – all dealing with Indigenous issues regarding the power of (self-)representations and therefore alterNative political and cultural agendas and visibilities in North American public spaces.

Once Siri Hustvedt, as a conference keynote speaker, told a story about a Western scientist and his research focus: an Indigenous tribe somewhere in a rural area of the world. The researcher was settling down near the tribal village and started observing the tribe, taking notes for his work. After a couple of days, the tribe silently packed up its belongings and moved on. The researcher cleverly assumed that this must be a traditional nomadic tribe. This

reciprocal relationalities: land as an animated other-than-human entity, containing its own ontologies and aesthetics that find expression in Indigenous storytelling, without falling back into naïve eco-romanticisms of the 'Noble Savage'/'Greenpeace mascot' type. In the North American colonial context(s) of land thefts, disputes, and First Nations' sovereignties, these stories are not some folklore from a mythical past but contemporary and real for First Nations to recover, and to draw resilience, resistance, and legal bases for sovereignties from: these stories are important place-narratives that create Native spaces when inserting them prominently, for example, as visualized artworks, into Western gallery and museum spaces.

> mobility pattern was repeated a couple of times: the tribe moved on, the researcher followed. Until … it finally dawned on the researcher that the tribe was fleeing *him*, silently criticizing his insistent presence.[3]
>
> *(German Association for American Studies conference 2013, Mainz, Germany)*

Introduction

From a European academic-learner's perspective, I have found it challenging to write this article, trying to offer a satisfying or even stable conclusion on Eurocentric/Western academic research practices regarding Indigeneities. For discussions on critical academic self-positioning are ongoing, especially in settler nation-states like the U.S. or Canada, but also beyond the Atlantic – as the 2016 workshop "Who Can Speak and Who Is Heard/Hurt? Ethnic Diversity, 'Race' and Racism in American Studies in Germany" at University of Tübingen demonstrated. There are key issues that every non-Native academic (teacher, scholar, student) needs to be aware of: stereotypes, the politics of representation in a settler nation-state and beyond, ethics and accountability in research and language use, debates about authoritative voice, capitalization, and appropriations of diverse Native modes of knowledge. Accordingly, this paper addresses ethical questions and strategies that challenge Euro-academic privilege and philosophies as platforms of power in research and education on Indigeneities in North American colonial contexts. In this context, Indigeneities/Indigenous cultures also need to be recognized as fluid and ever-evolving collective and individual concepts marked by particular locally diverse inflections and cross-cultural contacts.

3 Especially life scientists and, formerly, anthropologists have been criticized for collecting knowledge on Indigenous people(s) for publication, commodification, and their own economic gain without respecting the concept of reciprocity that might include financial compensation for less privileged Indigenous informants. Life scientists still seem to struggle with the demand to respect Indigenous rules of knowledge transfer, which run counter to practices of commodification, publication, copyrighting, patenting, and selling their knowledge within global market places. This appropriation of and capitalizing on Indigenous knowledge for the benefit of Western life sciences without respecting reciprocity has been identified as biopiracy and (neo-) colonialism.

I can't offer any final remedies to questions of who might investigate or teach Aboriginal narratives in the ethically most correct way. However, I am aware of my responsibilities: to resist Eurocentric terminologies, methodologies, imageries, and appropriations when writing about and teaching Native cultural self-representations. I can certainly propose ideas on how to problematize ethical shortcomings, and I'm convinced that this text would be a more comprehensive one if it had been the result of a collaborative Euro-Indigenous co-creation. Ideally, this paper should be a continuously evolving text, inviting updated findings and further enhancements, thus functioning like an interculturally knitted network.

My own interest in dealing with Indigenous community-based self-representations was triggered by a conference at Humboldt University of Berlin featuring Indigenous speakers from North America. From an otherwise fantastic conference, one awkward moment is still present with me: during the concluding session, the Native representatives were asking us students what we would do with the knowledge they had shared with us. Only one student understood the question in its ethical dimension: she thanked the Native participants for sharing, and respectfully asked what we could do for them in exchange for offering us so much insider knowledge. The common ground between them was the overall concept of reciprocity in contrast to centuries of ongoing colonial exploitation, occupation, genocide, ecocide, land theft, displacement, forced assimilation, and stereotyped misrepresentations. An ethical research methodology will always include reciprocity, responsibility, and respect for First Nations, Métis, and Inuit societies[4] and their cultures, as well as a critical awareness of settler nation-state policies and the interruption of Native sovereignties on Turtle Island (North America).

Consequently, researching and teaching alterNative[5] narratives calls for ethical frameworks that include reciprocity and sensitivity regarding colonial sociopolitical and cultural contexts. This research ethics foremost implies awareness of Native alterities and mindfulness of often asymmetric transcultural contacts. Reciprocity might include financial remuneration, material gifts, and

4 I hereby employ the official terminology of Canada's policy of multiculturalism. Note that these terms constitute umbrella terms that do not mirror the vast diversity of Native Nations on Turtle Island. For further terminological choices, see the part "Methodology and Self-Positioning."

5 I use the term "alterNative" in the sense of "Native alternatives" of equal value to Western dominant narratives, paradigms, and constructions. My reference here is Katja Sarkowsky's monograph *AlterNative Spaces. Constructions of Space in Native American and First Nations' Literatures* (2007).

other offers. This is, of course, a sensitive issue: in certain Native communities, these offers might be understood as condescending gestures of pity that perpetuate a victim status. Note also the use of the word 'gift' instead of 'donation,' suggesting a more respectful relation on equal footing. In order to carry out embedded research within First Nations, one needs to inform oneself thoroughly about their political, social, cultural, and economic situation and about gift offers that might be deemed decent in the respective First Nation. Alcohol and chocolate confection as guest presents might, for example, be regarded as highly insensitive (due to disproportionate rates of alcoholism and diabetes on some reservations). Reciprocity might include volunteer work within Native communities. It might comprise sharing one's own knowledge (contacts, scholarship, and project-application support). Furthermore, a work in exchange might also be to research a topic which would be inquired by a First Nation itself in a collaborative effort; its outcome would exclusively serve that First Nation, such as researching its local history, language revitalization projects, delving into contract law studies that positively affect its land claims, or museum repatriation initiatives (especially when the Western cultural outsider acquired the project funding and an institutional research base). The personal career, prestige, and the European paradigm of the freedom to research (*Freiheit der Forschung*) should not be major stimuli for such collaborative research efforts. The center of attention will lie on the First Nations' interests rather than on Western authoritative knowledge accumulation in the form of scholarly text corpora. However, the employment of a non-Native cultural outsider could clash with the question of why an outsider of all people should do the research while Indigenous students, scholars, curators, and artists are rarely promoted to build professional careers.

Canadian-based scholars Aman Sium, Chandni Desai, and Eric Ritskes acknowledge and focus on the same issues as follows:

> Alongside the question of how we engage colonial geographies is the question of how we engage colonial institutions. We are cognizant that we write from a (relatively) privileged position of the Western academy, an institution born from – and premised on – knowledge theft, muzzling, and selective storytelling. It is an institution that helped draw up the first blueprints of colonization, both here and abroad. […] Can we in fact decolonize the Western Academy and its global appendages? Further, what are the costs of our membership in these institutions? What are we forced to give up and what is taken from us? Lastly, what are the possibilities and limitations of using the "master's tools" to destroy and rebuild this house? (Sium/Desai/Ritskes 2012: IV).

First, non-Native academics need to accept Native academics as professional equals when it comes to authority and professional positions. Second, non-Native academics need to come to terms with their fears that their own career might be jeopardized by Native 'competitors,' who are still underrepresented in leading positions in the academy in North America and especially in Europe. Hence, the introduction of transnational collaborative models, as in North American museum practices since the Civil Rights Movement era, and more recently, in European museum exhibition policies, has been a first step to include Aboriginal academics, artists, and other First Nations members in processes of representational decision-making and curatorial implementations. Countering European white supremacy and discursive authority entails that non-Native academics step back as they might want to acknowledge decolonizing work as priority. For example, they should refrain from applying for job positions requiring specialization in Indigenous subject matters in the disciplines of American Studies, Cultural Studies, Ethnic Studies, and especially Native Studies in North America as well as in Europe, should any calls be posted in the future.

Methodology and Self-Positioning

This paper draws not only on the textual analysis of Indigenous scholars' works (Fixico, Mihesuah, Corntassel, Alfred, and others), but also on short-term research stays in British Columbia and Montréal, Canada. They resulted in various informal interviews with Indigenous and non-Indigenous agents from First Nations cultural centers and educational institutions as well as from Euro-Canadian galleries, museums, and universities. These personal conversations/open interviews all dealt with alterNative political and cultural agendas and with the appropriateness of applying categories of analysis from North American Cultural Studies to Indigenous contexts (dimensions such as race, class, and gender, and 'master' narratives vs. alterNative narratives, as well as work opportunities for Native personnel, for example).

Based on a self-reflexive approach in dealing with interpreting cultural productions by Indigenous agents, I chose the research position for this text to be a dialogical one, which means that it considers ideas from Indigenous scholars a cross-cultural effort, a sound basis, and key for analyzing Native narratives. This self-reflexive approach and self-critical thinking also challenges Western academic centralities of whiteness as hegemonic power base in order to counter the often unconscious self-conception of European academics as the norm and center, which (still) attributes racist markings to the 'Native Other.'

Consequently, Indigenous epistemologies, ontologies, and positions form the theoretical base and center for my research ethics and self-positioning process. Many Native authors have been particularly noteworthy offering methodological and ethical frameworks. While immersing myself as a German PhD student in a North American decolonizing discourse, I found Gerald Vizenor (Anishinaabe heritage) and his cultural concept of "postindian survivance" particularly helpful. This concept emphasizes diverse Indigenous epistemologies and related ontologies as paramount for decolonizing self-conceptualizations, while aiming for ultimate sovereignties through tribally diverse storytelling. Oral (hi-)stories of Indigenous communities are stories that artists continuously and publicly retell to Native and non-Native audiences alike, who witness these visual or textual narratives as tribal evidence production, Native survivance (a combination of 'survival' and 'resistance'), and continuing presence in settler societies.

Endorsing this agenda, Maori scholar Linda Tuhiwai Smith (Ngāti Awa, Ngāti Porou) in her influential study *Decolonizing Methodologies: Research and Indigenous Peoples* (1999) offers a radical criticism of Western academia as unrightfully claiming a center position in knowledge production. She consequently propagates Indigenous knowledge systems as primary sources to investigate Indigenous narratives. Western thought systems are perceived as valuable but mostly of secondary importance. Furthermore, scholars like Devon A. Mihesuah (Choctaw) and Donald L. Fixico (Shawnee, Sac and Fox, Muscogee Creek, Seminole) are regarded as major voices in this paper. In their article "Being Indigenous – Resurgences against Contemporary Colonialism" (2005), Kahnawake Mohawk scholar Taiaiake Alfred and Cherokee writer Jeff Corntassel propose a decolonial imperative: they demand a recuperation of Indigenous paradigms that do not reflect or critique colonialism, but that instead emphasize Native distinctiveness such as unique communal kinship relations, that is relations with clan and nation members, but also with other-than-human entities (land and its social relations). Corntassel and Alfred furthermore highlight Aboriginal languages, storytelling, and spiritualities (Alfred/Corntassel 2005: 608). Native unique and diverse worldviews contrast with or even go beyond European human exceptionalism, empiricism, and rational logics. Native epistemologies and ontologies are especially vital for interpreting works of art, such as novels by Laguna Pueblo writer Leslie Marmon Silko (*Ceremony; Almanac of the Dead),* Haisla/Heiltsuk author Eden Robinson's *Monkey Beach,* and Cree writer Tomson Highway's novel *Kiss of the Fur Queen.* These Native contextual approaches help to acknowledge a wide range of Aboriginal concepts, values, standards, and concerns that can be utilized to decolonize Western practices of research and education. What needs to be addressed here, though, is that the non-Native scholar/instructor can by no means be the teacher of a Native epistemology if that scholar/instructor adheres to ethical research guidelines. This should definitely be left to representatives of that very First Nation due to secret/sacred knowledge issues and protocols/permits to tell or circulate stories.

In addition, when centering Aboriginal peoples' viewpoints, my privileged position as a Western European academic-learner in American Studies is suddenly to be found at the margins or even outside of a North American Indigenous decolonizing discourse. This perspective can be a very troubling experience for European academics: one may be excluded from academic field trips in a Native Studies course for being a non-Native cultural outsider, for example. These protective boundaries around safe spaces need to be respected and must not be overstepped by non-Native agents. Consequently, limits of knowledge sharing due to cultural-religious spaces of protection need to be accepted. Furthermore, the self-positioning and the interpretation of works by Indigenous artists sometimes say more about non-Native academics than about the works themselves: one's own prejudices, deficits of knowledge, lacking research ethics, and cultural and academic modes of perception are all put to the test.

Against this background, this paper seeks to offer and invite an individual self-reflection on what it means to discuss Indigenous works when oneself is not of North American Aboriginal descent, and instead comes from a Western, European urban academic background. Not being a member of a First Nation, I can only rely on the narratives of Indigenous artists, academics, community workers/educators, Elders, etc. Although I am able to draw on short-term visits, I only obtained a small impression of actual living conditions and lived experiences of First Nations members, not even coming close to what it means and feels like to be Indigenous in a settler nation-state and having to deal with ongoing experiences of structural and operational settler racism.

Only if one acknowledges one's dominant Eurocentric position as an ideological construction against the Indigenous objectified 'Other,' one can recognize its artificiality. Linda Alcoff's concept of the white double consciousness is useful here:

> Perhaps white identity needs to develop its own version of "double consciousness" [...]. [...] for whites, double consciousness requires an ever present acknowledgement of the historical legacy of white identity constructions in the persistent structures of inequality and exploitation, as well as a newly awakened memory of the many white traitors to white

privilege who have struggled to contribute to the building of an inclusive human community (Alcoff 2000: 280-281).

This certainly concerns European academics, as well. Thus, the question is what scholars can do in Euro-academia, for instance at their departments and universities? There are already leading examples in Euro-academia who are willing to serve as trailblazers (by inviting Aboriginal voices and perspectives to academic platforms and collaborative projects).

Terminological Reflections

Prior to the gradual colonial invasion and occupation of Turtle Island until today (for the settlers and their descendants have never left the continent), no 'Indians' or 'Americans' existed. Indigenous tribal[6] societies of diverse cultures and cross-cultural contacts have been living on Native lands. Only after Columbus's landing were hundreds of Aboriginal societies/tribes uniformly turned into 'Indians.' Aboriginals should primarily be named after their tribal affiliation. For example, the author and filmmaker Sherman Alexie is not simply Native American, but primarily Spokane/Coeur d'Alene. 'Native American,' 'American Indian,' 'Indigenous,' 'Native,' and 'Aboriginal' are general, official terms for Indigenous populations – they have been assigned to them by Europe, Euro-America, and Euro-Canada, however. The use of umbrella terms is obviously due to practical reasons. Yet, this employment of language is also problematic, for these umbrella terms belie the diversity of over six hundred Aboriginal tribes (numbers depend on the source), and thus also impose rather simplistic categories on highly complex colonial, hence asymmetric power relations based on white supremacist thinking. In Canada, the official differentiation of Aboriginals into First Nations, Métis, and Inuit was introduced by the Canadian state following the official concept of multiculturalism. North American Aboriginals also call themselves 'American Indians,' 'Indians,' or 'Skins' (!), if they do not refer to their proper tribal/national affiliation (Lakota Sioux, Haida, Lenape, etc.).

---

6 The term 'tribal' as used here denotes a distinct First Nation local political, social, and cultural center in a multi-centered world in contrast to policies of assimilation and marginalization. It should be noted, though, that the term 'tribe' implies an incorporation into a settler nation-state, hence the advantage of the term 'First Nation' over 'tribe,' with the first emphasizing and aspiring to a sovereign nation-to-nation relation.

The term 'Skin' can be thought of as working analogously to the n-word: Only cultural insiders may use it as a reappropriated term. The term is considered to be colonialist and derogatory. A verbal re-conquest has been taking place since and has turned the word into a cool slang word. For non-Natives, the s-word should be taboo, as its acceptable use is turned into an offensive speech act when uttered by a non-Native outsider. There is also another s-word: the word 'squaw' is a derogatory term for an Indigenous woman. Joy Harjo (Muscogee) claimed on TV in the 1990s that it means 'vagina' in Algonquian languages, and consequently reduces women to their genitalia (cf. Carstarphen/Sanchez 2012: 128-149). Other sources claim that the term 'squaw' means 'woman' and was appropriated by Western settlers for derogatory purposes (cf. Schilling 2017). Racist settler fantasies of exoticized Native women and offensive labeling of Indigenous women since contact times have played a role in cases of missing and murdered Indigenous women in North America.[7]

Tribal affiliations always have priority over Western designations. Identifying colonialist-racist hate terminology as in offensive words like 'squaw' or 'chief' and in monolithic conceptions of the 'Indian' is crucial to decolonizing language and abolishing anti-Native hate speech. Recognizing settler nation-states' reductive terminology ("terminal creeds" in the words of Gerald Vizenor, cf. Blaeser 1996: 52-53; Vizenor/Lee 1999: 19) constitutes an important first step of identifying supremacist settler propaganda against and related political and socioeconomic inequalities for Native Americans (and First Nations, Métis, and Inuit communities in Canada). For settler ideology has aimed at justifying land occupation, exploitative land resource management, and economic growth at the expense of an Indigenous population by dehumanizing/animalizing the 'subaltern Other' ever since the European invasion and settlement of North America. In my PhD thesis written in German, I use the tribal-national affiliations, the English terms 'Native American' and 'American Indian,' and the German terms '*Indigen*,' '*Ureinwohner_innen,*' and '*Urbevölkerung*' (the latter two terms meaning original inhabitants). Note also Adibeli Nduka-Agwu's term '*Erstbewohner_innen*' (first inhabitants) of the Americas (Nduka-Agwu/Hornscheidt 2013: 40-45). The stereotyped labels of '*Indianer*' and the English variants of 'Indians' and 'Injuns' are only employed (in quotation marks) when stereotypes of Indigenous people are discussed. Furthermore, English and

7 Consider the 2007 Pickton Trial in Vancouver: if there had not been racist bias among local police officers, numerous subsequent kidnappings, rapes, and murders of mostly Aboriginal women could have been prevented – thus runs the accusation of activist groups against Vancouver's police authorities.

German terminology applied to Indigenous peoples and concepts can prove to be quite problematic when labels such as 'object,' 'artwork,' 'thing,' and 'artefact' are used for entities that, from a Native perspective, may have qualities and agencies of personhood and narratively address other-than-human-related cosmologies. Art historian Ruth B. Phillips observes in her study *Museum Pieces. Toward the Indigenization of Canadian Museums* "that the terms [in Native languages] used for the same entities are often animate, referring to other-than-human persons, to use anthropologist Irving Hallowell's term" (2011: xv).

A final remark regarding designations: Indigenous people(s) have been living in tribally diverse societies based on cultural markers and common histories since before contact times. The ethnic categorization of Aboriginals into First Nations and especially by blood-quantum laws reflects a Western political, social, and biological classification practice that reintroduces Western settler logics and norms, thus endorsing settler nation politics. Tragically, Indigenous societies themselves have assimilated the blood-quantum rule in order to distinguish between full members and part-members of a First Nation. However, strategically, this last type of categorizing is important when it comes to land claims and population counts within a Western legal and geopolitical framework of a settler nation-state.

Indigenous Academic Place-Making[8]

In the following, I will address ethical imperatives for non-Native academics that are problematized by Native scholars. Sensitive issues on representation that are teaching- and research-related can be boiled down to the (ab-)use of representational powers over 'the Other.' Mihesuah accordingly states:

> The problem with many books and articles about Indians is not with what is included but with what is omitted. There are many works on tribal histories and cultures that are fine examples of library and archival research, but the research usually ends there (Mihesuah 1998: 4).

Research and education should thus focus on Aboriginal people *behind* the texts addressing Native issues. What is more, Western scholars must not accept

8 By employing the term 'place-making,' I refer to Mishuana Goeman's use of the word in the sense of storied remapping and redefining land as distinctly Native identity marker and thus as part of tribally diverse Native identity formation (Goeman 2008: 29).

mostly incomplete written accounts on 'Indians,' which means that when dealing with Indigenous issues and related narratives, European academics need to seek out direct encounters with Indigenous representatives and environments as part of their research. Many authors, Mihesuah cautions, believe that because they have talked with *a* Native person, this one voice is all they need to legitimize their work (2005: 75). European academics contacting Native nations also need to be aware of sensitive Indigenous context knowledge in stories, ceremonies, rituals, heterogeneous materials, and traumatic histories.

Consequently, another ethical dilemma when teaching and researching Native narratives at Western academic institutions surfaces: how is difficult, painful knowledge arising from the colonial context (such as the 'Missing and Murdered Women' issue or residential school traumata) adequately communicated to diverse non-Native audiences? With regard to the ongoing issue of missing Native women in North America, Anishinaabe artist Rebecca Belmore, whose performance *Vigil* (2002) also deals with this difficult knowledge, used the color white in her photographs to accuse white perpetrators of crimes against Native women, showing the female Native half-naked bodies as sites of violence strangled and covered by white sheets and surrounded by the color white (see, for example, Belmore's photographic works *Untitled* (2004) and *Fringe* (2007)). Showing Disney's *Pocahontas I* and *II* as films and pictures is also challenging, since the youngest viewers thus already come into contact with sexualized images of Native women in these mainstream media productions. These Disney films only reproduce stereotypes that implicitly invite and promote violent abuse of Aboriginal women as sex objects and continuously affect their self-esteem in a negative way. The Halloween costume 'Pocahotty' is further proof of severe misrepresentations of Native women in mainstream culture that need to be rectified. Many Hollywood films, in fact, depict 'Indian' women as sex objects (e.g., the cult movie *Apocalypse Now* and its 'Pocahotty' stage performance). A more analytical approach to these films featuring representations of 'Indian women' (i.e. the sexualized attractive Indian Princess stereotype) and their impact on non-Native diverse audiences would raise the following questions: Who should have the right to consume this cinematic imagery – a selected audience who could learn to revise corrupted representations? Is there a limit to an uncritical, uncommented presenting of stereotyped 'Indians' as film characters in public media? Deconstructing and critiquing stereotypical images still perpetuate these representations (even in academic spaces) by habituation and cumulative repetition. Presenting them puts viewers into the passive position of a consumer and even voyeur, which elicits expectations of Indigenous femininities as exotic, sensual, and sexual, subjecting 'the Indian woman' to the Western (mostly male white) gaze of power.

Moreover, Native scholars working in Western academic spaces are aware that Indigenous methodologies might not be accepted by non-Native scholars who prefer and apply Western empirical, 'rational' methodologies and philosophies, and who ignore or even ridicule Indigenous knowledge systems as unacademic lore and beliefs. Secondly, there is the problem that non-Native researchers reproduce 'master' narratives, reconfirming and teaching book knowledge and assumptions about Indigenous peoples that defy any efforts of broadening one's horizon on alterNative approaches. In addition, these 'master' narratives have for too long been internalized by Indigenous people. And, furthermore, is tribally diverse knowledge translatable when introduced into a transcultural contact space? For example, the contrast between a European human exceptionalism and other-than-human belief systems comes to the fore in the graffiti of the sea-wolf *Wasco* (2012) by Larissa Healey (Ojibway) and Corey Bulpitt (Haida), which shows a Northwest Coast figure design that is considered tangible or intangible, depending on the audience. The message is obvious, however: within the sprayed painting, a distinct political agenda can be detected; Native Northwest Coast relationalities and sovereignties over sea/land are expressed and claimed in this image. The question remains to be asked: can non-Native audiences translate Indigenous, tribally diverse visual and spiritual-religious concepts that derive from Native epistemologies and ontologies without any knowledge of the respective Native language(s) and culture(s)? Wahpatonwan Dakota scholar Angela Cavender Wilson's paragraph on the importance of Indigenous oral storytelling traditions may clarify problems and questions of limits of transferability:

> Stories in the oral tradition have served some important functions for Native people [...]. These stories, much more than written documents by non-Indians, provide detailed descriptions about our historical players. They give us information about our motivations, our decision-making processes, and about how nonmaterial, nonphysical circumstances (those things generally defined as supernatural, [...] and spiritual by Western thinkers) have shaped our past and our understanding of the present. [...] So while archival materials can offer a glimpse into the world-view of Native people, the degree to which they can provide information on the American Indian [...] is quite small relative to what can be gained through an understanding of oral tradition (1998a: 24-25).

Of late, Indigenous scholars have increasingly been invited to German American Studies conferences. Again, it is important not only to invite Indigenous agents but also to engage with them in collaborative projects or to offer employment opportunities at one's home institution. One example to be followed is the Indigeneities project based at Royal Holloway College of the University of London (2009-2014). Within this framework, various Indigenous scholars, activists, and/or artists from the Americas, Australia, and Aotearoa (New Zealand) were invited to have their say in colonial/decolonial matters. The Indigeneities network and conference series furthermore offered Aboriginal agents staff membership, a platform for networking activities, especially among each other (which I identify as special purpose here), and for collaborating with non-Native scholars and other stakeholders. Moreover, European scholars need to be open for exchange visits to Native lands and institutions, for example the Nicola Valley Institute of Technology (NVIT), the First Nations University of Canada (FNUniv), and the Institute of American Indian Arts (IAIA), which foremost follow and implement Indigenous educational agendas.

In his work *The American Indian Mind In A Linear World* (2003), Donald L. Fixico (Shawnee, Sac and Fox, Muscogee Creek, Seminole) strongly recommends collaboration and consultation with tribal communities when he formulates ethical tribal frameworks for Western academics. He suggests a protocol that "involves instruction for three phases of studying American Indians. The first phase involves contacting the tribe, the second phase is conduct of research and behavior in Indian Country, and the third phase is the postresearch accountability or the follow-up" (2003: 133-134). Fixico subsequently describes in detail and recommends all three phases of such a research project for non-Native scholars:

> During the first phase of research, the visiting scholar should act with respect in contacting the tribe. The executive branch of a tribal community should be contacted with the scholar presenting his or her request to research, showing of credentials, demonstration of purpose or objective of research and publication. The scholar should agree to a statement of sharing the published literature or reporting the results to the tribe. After contacting the tribe's executive branch or leadership, the scholar should patiently wait for a reply. As permission must be received for researching at private libraries and archives, the same patience should be exercised by the scholar researching an Indian community. This permission becomes more imperative due to the fact that various tribes and their communities will likely have different guidelines or rules or regulations for researching, or they may not have any guidelines at all. If a response from the tribe is not received after a month, then another request should be made by the visiting scholar. In preparation for contacting the tribal communities, the scholar may access the available web pages of the tribes in order to seek appropriate information such as the contact person and how to

contact such persons of the tribes. In addition, the web site may also have a set of guidelines or protocol in doing research on the reservation or in the Indian community. [...] For the tribe, an official designated "contact person(s)" should be assigned to the researching scholar. This person would be the liaison to ensure that the researching scholar does not violate a cultural code or tribal law. [...]
The second phase of actual research should involve how the scholar needs to act while researching. [...] Whether the scholar is researching in tribal archives or at a tribal library, interviewing tribal members or taking notes on cultural and political activities [...]. *A code of professionalism needs to be established so that scholars' actions while visiting Indian Country, and researching and writing about American Indians will respect both sides. Such codes and responsibilities need to be taught to undergraduate students and graduate students who want to research and write about Native Americans. Simultaneously, the history of exploitation and its blind nature should be explained to them. Furthermore, information on the ill relationship that has been created between academics and the Indian community at large needs to be shared with researchers. Unfortunately, academic freedom has been exploited in the case of American Indians, and it will continue to be this way until cultural guidelines are established. This same academic freedom has licensed the scholar to do as he or she pleases with little or no sensitivity to the impact of the research results or publication on the tribe or Indian community. For the present, such sensitivity is not of high priority among mainstream researchers.*
The third phase of postresearch involves accountability and final responsibility for the researching scholar. Accountability is a critical issue in studying American Indians. The first question is "what is the researcher accountable for?" The second question is "to whom is the researcher held accountable?" Properly crediting the tribe and giving credit by acknowledging the contact person and the individuals in the tribal community or Indian community who assisted the researching is important. In this manner, at least the tribal community and its members are respected for their information that they shared (Fixico 2003: 133-135; my emphasis).

Most of the time, the knowledge gained by Western academics first again serves... the Western academics' career and not Native agents such as scholars, social workers, educators, activists, translators, and especially younger members and Elders, who need protection from inquisitive outsiders. Connected to this are some key questions concerning academic projects, raised by Choctaw scholar Devon A. Mihesuah in the introductory chapter of her work, *So You Want to Write about American Indians? A Guide for Writers, Students, and Scholars*:

Before you get started on your project, ask yourself why you want to write about American Indians. Think carefully about these questions:

- Are you a student who must complete your thesis or dissertation in order to graduate? […]
- Are you a professor who wants to make a name for yourself in your field of study?
- Are you a professor who must write in order to keep your job? […] (Mihesuah 2005: 1).

Affirmative answers to these questions indicate self-interest as motivation for writing about Native Americans. The question might, however, also be if one would like to offer solutions through one's writing (Mihesuah 2005: 2), even if one positions oneself as cultural outsider.

This paper is first of all a call for individual attentiveness, for self-positioning, and for personal critical guidelines. Studying and writing on Indigeneities and, consequently, on Indigenous peoples requires cultural sensitivity, as a variety of cultural contexts must be taken into account. It should also be kept in mind that Native mediators have the right to determine who might have access and what sensitive, difficult contents might be publishable or not – even to oppose the freedom to research for Western academics/cultural outsiders.[9] Thus, James Clifford can be quoted as a non-Native U.S. historian when he recommends the following to non-Native scholars and students in *Returns*:

> To engage with these [Indigenous] histories requires representational tact, a patient, self-reflexive openness […]. […] an alert receptivity and willingness not to press for conclusions. A constant awareness of our own partial access to other experiences is required […] (2013: 134-135).

Clifford here highlights the potential for Western researchers to explore several Indigenous ways of thinking and to contemplate sites of translation where Western discourses are challenged by a decolonizing discourse that promotes Native knowledge and belief systems.

Angela Cavender Wilson states that research work on Indigenous concerns and narratives needs to be a long-term project that entails years of involvement and building trusting relationships with Native people. The rewards of this scholarship might not come from Western academic peers (1998b: 26). She emphasizes that those researchers who don't choose this alterNative route must be aware of and come to terms with the fact that the results of their work may be regarded as only limited knowledge issued from a white perspective, and thus

9 This opposition is known as "refusal": Native mediators simply refuse to cooperate as an act of protest and resistance (cf. Simpson 2014; Tuck/Yang 2014).

cannot be fully credited by American Indians, who might apply their own paradigms (such as reciprocity, relational community thinking, protocol rules).[10]

Alternative models of research and education need to be introduced to Euro-academic spaces and symmetric relationships need to be fostered between a diverse academic staff. Intersubjective cross-cultural practice takes time, but working proactively against Indigenous outsider status means e.g. to re-imagine relationships by promoting and especially funding cross-cultural cooperations on an equal level. Funding institutions and universities should not solely base their hiring and funding decisions on the research subject matter; they should also take into consideration what ethics a research project practices and whether an Indigenous agent can be employed.

Yet, cultural outsider perspectives can, from a critical distance, offer valuable insights to a decolonizing discourse, which would allow for a combination of multi-voiced perspectives, as Chippewa sociologist Duane Champagne suggests. Champagne favors an open, interdisciplinary platform for academic exchange between Indigenous and non-Indigenous scholars in the field of American Indian Studies. According to Champagne, one does not have to be a member of a specific culture in order to conduct research on it, thus offering the opportunity for complementary insights (1998: 181-182).

Final Thoughts

This paper aimed to highlight various ethical demands addressed to and responsibilities of Western academics when dealing with research, publishing, and teaching on Indigenous peoples and their narratives. Transforming Western

10 However, Indigenous scholars' opportunities should not be limited to instructing students solely on subject matters related to ethnicity/race, which would render them involuntarily ostracized and reduced to a tokenized representative of diversity (their ethnic/cultural/national origins) despite well-meaning intentions of the welcoming Western academic institutions. The question is also if decolonizing the university should and can be left solely to Eurocentric, middle-class ideas of what this will entail. Instead, one should take into consideration what ideas Indigenous agents have in mind about diversity and decolonial activities, and, for a change, listen to their propositions. How might Native agents influence and transform a Western academic space? Do they actually want to be included? Do First Nations agents wish to be categorized as 'Aboriginal' scholars/educators/instructors or as mere individuals who not only happen to teach Native Studies and Law from an Indigenous perspective but also work as tenured professors of Economics, Mathematics, or Physics?

academics into self-reflective critical learners is a step-by-step process. This paper is intended as a call for a more sensitive and combined approach that perceives Indigenous voices on equal footing with Euro-Western thought systems. Especially for a European research context, this means fostering transatlantic collaborations and exchange visits, particularly with Indigenous Studies Departments. Native educators (like artists, Elders, scholars, activists, graduate students etc.) need not only be invited to conferences, but should be employed at Western academic institutions, too. They should receive positions as lecturers in residence, on a long-term or even permanent basis – despite maybe lacking classical Western academic credentials, but instead offering valuable non-Western intellectual and educational methodologies and traditions existing beyond Western disciplines. This would mean a great opportunity for Western academics to broaden their horizons, step back, and listen carefully in order to let alterNative authorities have their say.[111] Again the question remains if Indigenous agents actually wish to travel to Europe and share their knowledge in transnational, transatlantic collaborations. They should of course decide for themselves on these opportunities, and whether they are willing to spend time abroad, separated from their families and communities.[112]

This paper's main conclusions favor ethical partnership collaborations to promote transnational collaborative exchange projects, to be aware of the authoritative, hierarchical, and privileged position from which non-Indigenous academics write/teach/perform, to pay attention to transcultural power imbalances in representations, and to attempt to interpret cross-culturally but at the same time recognize limitations and respect spaces of protection. More works and agendas by Indigenous artists-activists need to be investigated to learn (more) about diverse Native contexts and colonialist policies of settler nation-states. Finally, more reciprocity needs to be practiced: academics need to ask

111 Elders could, for example, work as oral storytellers and language teachers. Employing Elders in these ways is already being carried out at North American universities (Native Studies programs). Take, for example, the Native Studies Department at the University of Alberta. Its website has a Q&A section which explains if any Elders are involved in the department's and the university's activities: "Elders are involved on a regular basis with the Faculty of Native Studies. Elders sit on our Faculty Council and participate in most activities organized throughout the academic year: including guest lectures, conferences, special events and gatherings such as our semi-annual Tea & Bannock" ("Why Native Studies").

112 Providing a detailed analysis of the specific contexts of European institutions' (as opposed to North American institutions') offering permanent positions to Indigenous individuals from North America, would exceed the scope of this paper.

themselves if their findings, which promote their careers, might also benefit First Nations in any way. There are many possible ways of giving something back: it could mean a financial gift for a First Nations student project, a volunteer activity in a Native community, a funding application to support a visit from a Native representative, or it might mean a project work that is commissioned by a First Nation, by Métis, or by an Inuit community. The implementation, of course, is dependent on the Native agents. Again, the offer to give something back is potentially problematic, as it might be interpreted by Native contact persons as a patronizing gesture and thus, as a perpetuation of a subaltern status. These points should raise awareness and contribute to centralize Indigenous voices as primary sources – without assimilating them as mere appendix knowledge.

Acknowledgments

I would like to express my gratitude to the following people and institutions: I am indebted to Tahltan artist/curator Peter Morin and Haida artist/curator Kwiaawah Jones for patiently sharing their knowledge on Indigenous issues. I would like to thank Elder Lorraine Spence (Nicomen) from NVIT's Elders Council, Byron Robbie, and NVIT Dean of Education Dr. John Chenoweth (Okanagan First Nation) for their successful First Nations building[113] and shared information on AlterNative state-of-the-art education and training during a tour around NVIT. Thanks are due to the MOA, the RBCM, and the Bill Reid Gallery for knowledge sharing and critical thinking: Jennifer Kramer, Pamela Brown (Heiltsuk), Karen Duffek, and Martha Black. Thanks must also go to the UBC's First Nations Longhouse library facility. Words of thanks must go to the local guides, their tour offer and fry bread at the magnificent Lil'wat/Squamish Cultural Centre in Whistler. I am absolutely grateful to Heather Igloliorte (Inuit, Nunatsiavut Territory of Labrador) as an excellent and generous teacher. I also cordially thank visual artist Nadia Myre (Algonquin heritage) and Steven Loft (Mohawk-Jewish heritage) for offering workshops on Indigenous survivance and sharing the knowledge with the audience.

113 According to Mishuana Goeman, the core of First Nations building is storied land signifying geopolitical and cultural importance to First Nations (Goeman 2008: 23-33).

Works Cited

Alcoff, Linda Martín (2000): "What Should White People Do?" In: Uma Narayan/Sandra Harding (eds.), Decentering the Center: Philosophy for a Multicultural, Postcolonial, and Feminist World, Bloomington and Indianapolis: Indiana University Press, pp. 162-182.

Alfred, Taiaiake/Corntassel, Jeff (2005): "Being Indigenous – Resurgences against Contemporary Colonialism". In: Government and Opposition 40/4, pp. 597-614.

Blaeser, Kimberly (1996): Gerald Vizenor. Writing in the Oral Tradition, Norman and London: University of Oklahoma Press.

Carstarphen, Meta G./Sanchez, John P. (eds.) (2012): American Indians and the Mass Media, Norman: University of Oklahoma Press.

Cavender Wilson, Angela (1998a): "American Indian History or Non-Indian Perspectives." In: Devon A. Mihesuah (ed.), Natives and Academics, Researching and Writing about American Indians, Lincoln and London: University of Nebraska Press, pp. 23-27.

Cavender Wilson, Angela (1998b): "Grandmother to Granddaughter: Generations of Oral History in a Dakota Family." In: Devon A. Mihesuah (ed.), Natives and Academics, Researching and Writing about American Indians, Lincoln and London: University of Nebraska Press, pp. 28-36.

Champagne, Duane (1998): "American Indian Studies Is for Everyone." In: Devon A. Mihesuah (ed.), Natives and Academics, Researching and Writing about American Indians, Lincoln and London: University of Nebraska Press, pp. 181-189.

Clifford, James (2013): Returns. Becoming Indigenous in the Twenty-First Century, Cambridge: Harvard University Press.

Fixico, Donald L. (2003): The American Indian Mind In A Linear World. American Indian Studies & Traditional Knowledge, New York and London: Routledge.

Fixico, Donald L. (2013): Call for Change: The Medicine Way of American Indian History, Ethos, and Reality, Lincoln: University of Nebraska Press.

Goeman, Mishuana (2008): "From Place to Territories and Back Again: Centering Storied Land in the Discussion of Indigenous Nation-building." In: International Journal of Critical Indigenous Studies 1/1, pp. 23-34.

Mihesuah, Devon A. (ed.) (1998): Natives and Academics, Researching and Writing about American Indians, Lincoln and London: University of Nebraska Press.

Mihesuah, Devon A. (2005): So You Want to Write about American Indians? – A Guide for Writers, Students, and Scholars, Lincoln: University of Nebraska Press.

Nduka-Agwu, Adibeli/Hornscheidt, Antje Leann (eds.) (2013): Rassismus auf gut Deutsch. Ein kritisches Nachschlagewerk zu rassistischen Sprachhandlungen, Frankfurt am Main: Brandes & Apsel.

Phillips, Ruth B. (2011): Museum Pieces. Toward the Indigenization of Canadian Museums, Montréal: McGill-Queen's University Press.

Sarkowsky, Katja (2007): Alternative Spaces. Constructions of Space in Native American and First Nations' Literatures, Heidelberg: Universitätsverlag Winter.

Schilling, Vincent (2017): "The Word 'Squaw': Offensive or Not?" In: Indian Country Today. Digital. Indigenous. News March 23 (https://newsmaven.io/indiancountrytoday/archive/the-word-squaw-offensive-or-not-AOz_cjbLkEaxKaUEa6kejg/).

Simpson, Audra (2014): Mohawk Interruptus: Political Life Across the Borders of Settler States, Durham: Duke University Press.

Sium, Aman/Desai, Chandni/Ritskes Eric (2012): "Toward the 'tangible unknown': Decolonization and the Indigenous future." In: Decolonization: Indigeneity, Education & Society 1/1, pp. I-XIII.

Tuck, Eve/Yang, K. Wayne (2014): "R-Words: Refusing Research." In: Django Paris/Maisha T. Winn (eds.), Humanizing Research: Decolonizing Qualitative Inquiry with Youth and Communities, Thousand Oakes: SAGE Publications, pp. 223-248.

Tuhiwai Smith, Linda (1999): Decolonizing Methodologies: Research and Indigenous Peoples, London: Zed Books.

Vizenor, Gerald/Lee, A. Robert (1999): Postindian Conversations, Lincoln: University of Nebraska Press.

"Why Native Studies." University of Alberta, Faculty of Native Studies (https://www.ualberta.ca/native-studies/about-us/why-native-studies/).

Witt, Laurie Ann (1998): "Cultural Imperialism and the Marketing of Native America." In: Devon A. Mihesuah (ed.), Natives and Academics, Researching and Writing about American Indians, Lincoln and London: University of Nebraska Press, pp. 139-171.

“The Danger of a Single Story”: Addressing Contemporary Public Discourse and Protest Movements in American Studies Classrooms in Germany

Saskia Hertlein

The question of “Who Speaks and Who Is Heard/Hurt?” has taken up considerable space in public discourse in recent years, albeit in different shapes and forms. Contexts cover a wide range from voices identifying as having generally or historically been underrepresented to those claiming to have become marginalized – or forgotten – over time. Societal debates over representation as well as participation and voice are, however, by no means new or surprising phenomena, and neither are they specific to a particular country or part of the world. Nonetheless, since some of these discourses have significantly impacted their respective societies, a closer look at some exemplary phenomena from the past years is warranted.

American Studies has been exploring a great variety of issues ranging from campaigns like *#BlackLivesMatter* or *#I, Too, Am Harvard*, which testify to the many unresolved questions in U.S. race relations, to the past U.S. presidential election campaign, which resulted in Donald Trump’s victory, whose call to return to a putative previous state of greatness, *Make America Great Again*, resonated with a large enough number of voters afraid of status loss.

In Germany, some recent issues in public discourse may in part be reminiscent of the aforementioned examples from the U.S., even though their contexts differ, for example, the anti-racial profiling campaign by *Initiative Schwarze Menschen in Deutschland/Initiative of Black People in Germany*, continued activism surrounding the Oury Jalloh[1] case, or anti-Muslim activism, e.g. by right-wing organizations like *Pegida/Patriotic Europeans against the Islamization of the Occident*, or the recently established party calling itself *Alternative for Germany*, AfD.[2] Recent transnational campaigns that have preoccupied both Germans and Americans include *#MeToo*, which seeks to raise awareness of sexual harassment.

In view of the aforementioned issues concerning "Who Speaks and Who Is Heard/Hurt?", especially in the context of American Studies classroom discussions in Germany, this paper proposes to adapt the concept of a *Critical Diversity Literacy* by South African sociologist Melissa Steyn to German American Studies classrooms,[3] and combine it with U.S. English professor Paula M. L. Moya's suggestions put forth in *The Social Imperative* of how to transform schemata through close reading[4] in order to foster continuing transformation and heightened awareness in the classroom and beyond. This combination of approaches that is explored below is meant to provide students with a "diversity grammar" or "'reading practice' – a way of perceiving and responding to the social climate and prevalent structures of oppression"[5] (Steyn 2010: 20), a means and practice to decode both overt and covert hegemonies causing particular

1 Sierra Leone national Oury Jalloh died in police custody in 2005. Since then, the precise circumstances of his death have not been fully resolved, and the investigation as such has become subject to scrutiny, as well. Different perspectives on this issue can be found in German news media as well as on the website *In Remembrance of Oury Jalloh* (https://initiativeouryjalloh.wordpress.com). Bühl relates this case to lethal police violence in the U.S., e.g. to the death of Michael Brown in Ferguson, Missouri (2016: 66).

2 See e.g. Hirschmann (2017) for an approach to what he terms contemporary national populism polarizing Western societies.

3 Due to my own situatedness in American Literary and Media Studies, and Inter-American Studies in Germany, this proposal mostly refers to this specific context. I consider the disciplinary diversity of American Studies in Germany as a strength, but will leave concrete proposals for approaches to American Studies beyond my expertise to the experts in those fields.

4 "By close reading, [Moya means] the kind of intensive reading and re-reading that calls for a heightened attention to literary language and form, considering both as semantic structures that mediate authors' and readers' perceptions of the social world" (Moya 2016: 9).

5 In this quote, "reading" refers to a sociologist's approach to the concept. In the following part of this contribution, this sense of "reading" will be combined with Moya's proposal for a "close reading" to be understood in the literary studies sense.

individuals to speak, be silenced, or be heard/hurt. I will argue that an exploration of the concepts of exemplary markers of difference may not only serve as a tool to understand certain incidents and phenomena in the U.S., but also support a reevaluation of – sometimes hidden – discrimination and exclusion in Germany, in particular when considering issues connected to *race*, which have been difficult to address in Germany due to the notion's strong link to National Socialism.[6] The following considerations mobilize the concept of *diversity,* more specifically Steyn's *Critical Diversity Literacy* (Steyn 2010: 20), which develops France Winddance Twine's (2004) *Racial Literacy* further (the more precise analytical criteria for "reading diversity" will also be introduced below). Through a focus on diversity, then, I propose to take into consideration the many different potential forms of oppression that empower certain people to speak while silencing others, that render some heard and others hurt. However, this paper at the same time cautions against oversimplified comparisons or analogies that would neglect the very specific legacy that *Critical Diversity Literacy* calls for in its analysis of the individual phenomena.[7] Moreover, the

---

6 For an introduction to the difficulty of addressing race or racism in the German – or Western European – context, see e.g. Lentin (2008), Salem/Thompson (2016), Drews-Sylla/Makarska (2015), or Völkel/Pacyna (2017). Fatima El-Tayeb argues in *European Others*: "To reference race as native to contemporary European thought, however, violates the powerful narrative of Europe as a colorblind continent, largely untouched by the devastating ideology it exported all over the world. This narrative, framing the continent as a space free of 'race' (and, by implication, racism), is not only central to the way Europeans perceive themselves, but also has gained near-global acceptance" (2011: xv; qtd. in Wekker 2016: 4). Please refer to Balibar/Wallerstein (2002) for a connection of the concepts of *Race*, *Nation*, [and] *Class*, and Balibar's concept of "cultural racism." Moreover, a closer look at what Paula Moya and Hazel Markus term "conversations" ("interpretive frameworks, [i.e.] models, schemas, discourses, or scripts" Moya/Markus 2010: 5-6), reveals a number of attitudes that exist in Germany just the same, even if Moya's and Markus' examples are predominantly taken from the U.S. context.

7 The scholarly choice to deploy *racism* as a multifaceted concept may already be seen as an inclusive approach. However, especially because of the potential oversimplified dismissal of even noticing *race* or *racism* e.g. in the German context, *diversity* as an even more inclusive approach provides a suitable basis for addressing otherwise possibly overlooked issues and their intersections. Nonetheless, *race* and *racism* as well as other markers of difference, and discrimination deserve to be addressed directly in the discussions as well, as individual phenomena worthy of our attention and as having a specific quality, not only as aspects of *diversity*.

danger of *over*emphasizing as well as *under*emphasizing difference or perpetuating the very categories one wishes to expose as unfitting and discriminating is taken into account.

Overall, this paper argues that a recognition of the impact of different factors of diversity can be helpful as a first step towards (re-)evaluating both the issue of being represented at all, and the way in which different groups are represented. "Who Speaks and Who Is Heard/Hurt?", as this paper claims, depends first and foremost on the ability to recognize and articulate the presence or absence of different potential speakers or heard and hurt people; an informed view that allows for a certain distance may support this recognition.

While overly reductive classifications are problematic,[8] this paper also highlights the necessity to group people for some specific purposes in order to make clear that the phenomena at hand are not individual cases only. It remains a challenge to find an appropriate balance between necessary grouping and recognizing the individual.[9] Therefore, in the context of this paper, classifications are always proposed as a tool that needs to be supplemented by recognizing the individual, as well.

As a further aspect in the following proposal, German American Studies already implies a transnational endeavor, at least in the sense of considering the situatedness of the students and staff compared to their objects of study. Therefore, a conscious renegotiation of the field of inquiry in one's own

8 "Insofar as identities track social relations, they are contextual and subject to change in response to the transformation of social relations; identities come into being through the kinds of experiences people have, and they inform the way people interpret the social worlds they live in. Under this conception, identities are not reducible to social categories […], nor do they refer exclusively to people's subjective […] senses of self. Rather, *identity* refers to the complex and mediated way a multiply-situated and embodied human being looks out onto and interprets the social world she lives in" (Moya 2016: 28; original emphasis).

9 As e.g. Carolin Emcke highlights in *Against Hate* [*Gegen den Hass*] (2017), reducing people to single characteristics allows for targeting seemingly homogenous groups of people, e.g. in a racist manner, without recognizing the diversity and plurality inherent in humans and people's multitude of voluntary and ascribed roles, identities, and characteristics. While I acknowledge the danger of what Chimamanda Ngozi Adichie calls "The Danger of a Single Story" and value the diversity and plurality that to me as well is inherent in human beings, I would like to add the perspective that an overemphasis on individuality may equally be used to perpetuate problematic ideology and practice by reducing e.g. structural or institutional phenomena to allegedly individual incidents, underemphasizing their scope and impact.

situation may provide perspectives beyond the ones prevalent in one's own environment. This potential should not obscure the necessity to contextualize the studied phenomena, as otherwise they might be misread as only applicable to a different environment, externalizing issues that might also be included in the German or German American Studies context. Nonetheless, studying the phenomena and other approaches to them – like in U.S. American Studies or American Studies elsewhere as well as related approaches such as the ones proposed in *Inter-American Studies*[10] – might provide opportunities for critical reflection of what might be neglected or oversimplified in one's own environment. In the following, the proposed approach to a continuing reflection of American Studies research and teaching in Germany will therefore briefly be related to the field of inquiry – the different versions of American Studies – as well. Similarly to Moya's observation concerning the questions[11] she asks in *The Social Imperative*, "[t]hese questions are not new, but their answers have changed and taken on increased urgency in the context of a changing American [and German] society in which literacy about race and ethnicity [and more generally, a *Critical Diversity Literacy*] will be more needed than ever" (Moya 2016: 5). Therefore, this paper does not propose that all of the following aspects are entirely new, but rather that this combination might add another potential dimension to the task of teaching American Studies in Germany conscious of "Who Speaks and Who Is Heard/Hurt?".

The phenomena briefly referred to in the introduction may serve as reminders of the urgency of this question. While some incidents may have come as a surprise, tensions have been visible for a while. In Germany, right-wing

---

10 See e.g. Josef Raab's presentation entitled "Inter-American Studies: Why and Whither?" (2014) or his contribution "Thinking Beyond the Nation: The Project of Inter-American Studies" (2015), Wilfried Raussert's "Introduction" to *The Routledge Companion to Inter-American Studies* (2017), his "Selected Key Tropes in Inter-American Studies: Ways of Looking at Entangled Americas" (2015), or his joint "Introducción" with Yolanda Campos García and Margarita Ramos Godínez (2015) for recent assessments of an inter-American approach.

11 The "four theoretical questions with methodological implications [...] are: What is the power of a work of literature to affect a reader's perception of his or her world? How might a nuanced and insightful interpretation of a given text affect our perception of that text – and, by extension, of the world it represents? What is the status of 'close reading' within a literary critical landscape that includes quantitative formalism and cognitive approaches to literature? And, how important is it that literary critics maintain a focus on the individual literary text?" (Moya 2016: 5).

groups[12] have gained visibility for example via the Pegida marches starting in 2014 and AfD party politics, most obviously when comparably high numbers of refugees arrived in Germany in the summer of 2015, which some people refer to as a "refugee crisis," and its extensive media coverage.[13] Moreover, questions of belonging or the makeup of German society have become more prominent again. However, Germany has never been a monolithic society.[14] Continuing debates about the status of Islam in Germany,[15] but also regarding the divide between 'rich' and 'poor' in its various potential definitions, or East and West Germany, just to name a few, underscore that Germany is a heterogeneous society marked by many potential dividing lines.

---

12 See e.g. journalist Andreas Speit's analysis of German right-wing groups and their impact on public discourse (2017).

13 The term 'refugee crisis' refers to a rapidly growing number of refugees who reached Europe in 2015. Due to their high number and Europe's unpreparedness for such a scope, some people referred to the situation as a 'crisis.' In the broader context of these events, some right-wing groups gained momentum by conducting well-publicized demonstrations and by claiming that a 'wave' of refugees had reached Germany. In addition, Pegida or AfD also emphasized exclusive concepts of belonging and "the people" (explored in a historical perspective e.g. by Wildt [2017]), (re-)intensifying debates about a German '*Leitkultur*' as well. At the same time, this 'crisis' has also shown resilience and creativity, as e.g. Schiffauer, Eilert, and Rudloff point out in their survey of exemplary projects (2017). Hence, a simple evaluation of this complex situation does not seem appropriate. Nonetheless, in mid-2018, politicians of conservative Bavarian party *Christian Social Union* continue to use references to this 'crisis' to promote a number of policy proposals on a national level. Therefore, it is important to acknowledge that the concept of a 'refugee crisis' continues to have an impact on public discourse years after the EU-Turkey refugee agreement of March 2016 drastically reduced the numbers of refugees reaching Germany and other EU countries.

14 Offering a historical perspective, e.g. Jörn Rüsen highlights that German society derives from different tribes ("Stämme" [231]), (Rüsen/Völkel 2017: 230-231).

15 For instance, there have been public debates over the question whether Islam 'belonged to' Germany, whether headscarves may be worn in public service, how successful integration (sometimes understood as tantamount to assimilation) was supposed to look like etc. For a pre-"refugee crisis" impression of German public discourse on these issues, see e.g. Meier-Braun and Weber's edited volume from 2013. For an insightful exploration of the ethnicization of Islam in Germany, see e.g. Çakır (2014), for other European examples, see e.g. the different contributions in Drews-Sylla/Makarska (2015).

In the U.S., in addition to the prevailing exclusion of and discrimination against, for example, African or Asian Americans, 9/11 put 'Muslims,' that is, individuals presumed to be Muslim based on their appearance, in the center of public attention. Up until today, the heightened sensitivity towards a potential threat by people who are read as being 'Muslim,' which is conflated with being an Islamist terrorist, causes problematic situations.[16] Moreover, during President Obama's tenure, who had attempted to redefine the U.S.' "patchwork heritage [as] a strength, not a weakness" (Obama 2009), a number of African American initiatives (*#BlackLivesMatter, #I, Too, Am Harvard*) became active in society in order to bring incidents to public attention that might be considered discrimination, hate crimes, or police violence against African Americans. These incidents reveal the continuing challenges in the realm of racial justice U.S. society still faces despite hopes for substantial advances in light of the election of the first African American president or even claims of a "post-race era." In addition, some dividing lines that have become more visible in recent public discourse were already operative under Obama. As Jeff Chang points out,

> [w]ith Obama's election in 2008, everything changed. And nothing changed. We have entered a new era in U.S. history. But how do we describe this time?
> The divides remain. We live in an era in which the primary social schism is not that between so-called red states and blue states, but between those stuck on monoculturalism and a singular 'American way,' and those comfortable with demographic change and cultural difference; those fearful over the great America in danger of being lost forever, and those hopeful about the one being made anew; those stuck in black-and-white, and those living in color. Americans remain overly apocalyptic on the one hand and overly ardent on the other, identity-fatigued and post-racially euphoric.

16 Incidents in which people who are 'identified' as posing a potential threat based on a presumed 'Muslim' and Islamist appearance include an economics professor solving math problems on an airplane (cf. Reilly 2016). Literary and cultural explorations also highlight this interpellation of Muslims as Islamist terrorists, for instance Karan Johar's 2010 movie *My Name is Khan*, or more recently Lynn Gaspard's edited volume with the telling title *Don't Panic, I'm Islamic* (2017). In the latter, for example Arwa Mahdawi highlights the cultural panic surrounding Muslims in the West and the resultant hypersensitivity in her satirical contribution "A Personal Guide to Extreme Vetting: How to Distinguish an Acceptable Arab from a Terrorist in 6 Easy Steps," in which she provides her readers with lists of "Arabic words you should [not] be very alarmed by" (Mahdawi 2017: 6-7) and 6 steps, each with a satirical explanation as to why they are supposedly signs of potential danger (ibid: 4-5).

> Cultural desegregation *has* changed America. We can be seen as a happy rainbow country. Yet all our social indexes show rising rates of resegregation and inequity. In other words, there is a growing gap between what we see and what we think we see. For these conditions hide in plain view. Even as our image world expands at a profound rate, making us believe that every thing [sic] worth seeing is available to us, what sits in our blind spots may be more important than ever. There is also a growing gap between what we think and what we say. Blindness and denial – personal and systemic – often stop us from speaking at all about race (2014:10; original emphasis).

As Chang[17] highlights, despite hopes for substantial change towards more equity, inequity was a characteristic of U.S. society during the Obama presidency, as well.

While some of the other aforementioned incidents in the U.S. reached at least some public attention in Germany, the 2016 presidential election campaign culminating in Donald Trump's victory came as a surprise for a significant number of people. For American Studies in Germany, such a development may be regarded as an invitation to reconsider the discipline's potential and value for the public, having a diverse, long-term expertise in exploring the complexities of 'America,' which precludes both reductive simplifications of complex issues or mere surprise at the unexpected. In American Studies classrooms, some of the aforementioned issues may be discussed, providing both a chance for a deeper understanding of the particular situation in the U.S. and for a questioning of normalized phenomena in Germany, and for finding a vocabulary and grammar to address these issues.[18]

In this context, *schemas*[19] that everyone brings to the discussion can be revealed. In order to use this setting to expand one's schemas, active engagement is needed, which includes different aspects of scholarly work – from precise definitions to references to different modes of sensemaking. While fostering active engagement presumably constitutes the core of pedagogical practice in

17 Chang (2014) uses an inclusive understanding of *race*. For an overview of contemporary approaches to *race* and *racism*, which include a variety of inclusive conceptions, please refer to e.g. Bühl (2016).

18 As criterion 5 of *Critical Diversity Literacy* introduced below calls for.

19 "In social psychology [...] schema refers to the active organization of past experiences (physical and emotional) and past reactions (sensory-motor and cognitive-affective) through which a person apprehends and interacts with incoming stimuli. [...] Schemas thus have a temporal dimension characterized by evolution across time – they are anticipatory as well as retrospective, even as they orient a person's behavior in the present" (Moya 2016: 15).

American Studies classrooms in Germany, I would like to highlight its importance – especially an active engagement with differing opinions – in the context of being present in such a classroom. Of course, one might also engage with differing opinions while reading, or in digital settings, but in addition to the mediated approaches, the repeated practice of a direct rather than mediated engagement may provide students with additional aspects of understanding multiple perspectives. Moreover, such a direct engagement may also support the continuing development of their *diversity literacy* by added features such as affective aspects of debating opinions, or finding one's voice alongside dealing with potentially difficult situations in debating such potentially difficult issues.[20]

These discussions may, however, be based on the students' experiences from their solitary close readings. Literature or stories therefore can play a crucial role in developing schemas further, as Ben Okri argues when he states that "[s]tories are the secret reservoir of values: change the stories individuals and nations live by and tell themselves, and you change the individuals and nations" (Okri 1997: 112). Chimamanda Ngozi Adichie (2009) attributes a comparably crucial role to literature in her description of her own reading experience, concerning the expansion of her own "Single Story" of literature seemingly having to be about foreigners – as she read British and American literature – until she discovers African literature, e.g. by Chinua Achebe, featuring characters and lifestyles that she can relate to. Moya's observation does not contradict that

> [i]n the case of some works of literature written by racial, ethnic, or gendered 'others,' the effect can be (although it is not always the case) that readers are pulled in and given a kind of access to social worlds (and to worlds of sense) to which they might not otherwise be exposed, even if they live and work side by side with people of other races. The activity of reading good literature can expand the reader's horizon of possibility for encountering, living with, and loving characters [...] different from themselves (Moya 2016: 26).

Adichie's remark that the complete absence of any character that she felt she could identify with in her early readings should, however, also be discussed

20 Moya reminds us that "the number, content, and complexity of the schemas available to a person at any given time will have significant implications for that person's cognition, emotion, and motivation. [...] A person's ability to develop schemas is thus an essential aspect of survival and adaptability to changing circumstances" (Moya 2016: 20). The expansion of schemas in "number, content, and complexity" therefore appears to be a desirable teaching goal and may, as this paper claims, be reached best by a combination of methods ranging from solitary (close reading) to communal (classroom discussion) activities.

critically, since this reductive reading experience resulted in her misconception that "girls like [her]" were not included in literature, but that only "foreigners" qualified for that. Thereby, while some schemas were expanded during Adichie's described reading experience, other oversimplified ideas remained. Still, literature may be regarded as a good tool,[21] as Moya explains:

> [S]chemas are embedded into works of literature through a variety of narrative features, including focalization, narrative form, rhetorical devices, and historical and literary allusions. As part of the social world, literature is a system of social communication through which information, ideas, and norms are transmitted from author to reader, and among different communities of readers. As such, it is not unique. But neither is literature the same as every other system of communication. Literary texts are culturally consecrated; […] Moreover, literary texts work differently than do other forms of social communication. As aesthetic objects that engage readers deeply on the cognitive-affective level over a duration of time, works of literature make us think – but they also make us feel. They give us language, while also prompting us to form in our minds images that evoke a whole variety of associations. Literary texts have the potential to alter our perceptions and teach us how to interpret unfamiliar phenomena. They shape our cultural imaginaries and build for us schemas through which we interpret the social world (Moya 2016: 163-164).

Therefore, the engagement in direct conversation proposed above is meant as an addition, not a replacement of an engagement with literary texts.

Moreover, the selection, contextualization, and transfer of materials and concepts will need to be addressed from a *diversity literate* perspective,[22] fos-

---

21 This idea is countered e.g. by Adichie's linking of oversimplified reductions, hegemonic power, and literature in a self-reflexive story about an incident at one of her speeches "at a university where a student told [her] that it was such a shame that Nigerian men were physical abusers like the father character in [her] novel. [Adichie] told him that [she] had just read a novel called *American Psycho* […] and that it was such a shame that young Americans were serial murderers. […] It would never have occurred to [her] to think that just because [she] had read a novel in which a character was a serial killer that he was somehow representative of all Americans. And now, this is not because [she is] a better person than that student, but, because of America's cultural and economic power, [she] had many stories of America. [She] had read Tyler and Updike and Steinbeck and Gaitskill. [She] did not have a single story of America" (2009).

22 As Wilfried Raussert argues, the field of *Inter-American Studies* "requires a flexible positioning of scholars in the field" (2017: 4) and takes a rather inclusive approach to

tering an internalization of a *diversity literate* approach among the students. As indicated above, *race* can be used to exemplify factors based on which somebody is potentially in- or excluded from speaking or is heard/hurt, along with acknowledging the intersecting impact of other factors such as class, gender, or sexual orientation. In an approach seeking to widen schemas and foster *Diversity Literacy,* explicit methodological reflections aimed at avoiding oversimplified transfers or narrow selections of discussed terms and concepts may support students to become more aware of such issues. This awareness may include Moya and Markus' insight that "much of what affects our thinking about race lies below the surface of our awareness" (Moya/Markus 2010: 32), and therefore be of particular relevance to German and other public discourses in which terms like *race* or *racism* are mostly absent. In addition, the exploration of concepts like *race* and *racism* in other contexts than one's own might, through the multiplicity of perspectives thus provided, foster a deeper understanding of such concepts, expand one's schemas, as well as invite a reevaluation and diversification of a grammar and vocabulary suitable for discussing a variety of forms of oppression or discrimination.[23] These forms of discrimination or oppression may differ according to each given context, but may also serve as a reminder to reevaluate taken-for-granted ideas, e.g. the notion of Germany as a homogenous society.

The example of national minorities[24] may be added as another prominent issue for discussing "Who Speaks and Who Is Heard/Hurt?" in the German

---

the consideration of knowledge production in the Americas (cf. ibid: 6). In that sense, *Inter-American Studies* – along with other proponents of decolonizing and diversifying knowledge production – may contribute an important aspect of a *diversity literate* widening of schemas.

23 Cf. Steyn's sixth criterion of *Critical Diversity Literacy* explored below.

24 The German Federal Ministry of the Interior, Building and Community (2018a) [*Bundesministerium des Innern, für Bau und Heimat* 2018] provides some information about minorities in Germany in general and the minority groups: in the English version, an overview explains some general facts and defines and names the "[f]our officially recognized national minorities [...] in Germany: the Danes, the Frisians, the German Sinti and Roma, and the Sorbs" (*National Minorities*), and on the German website, some more details are provided on what is referred to as "*Nationale Minderheiten und deutsche Minderheiten*" ["National and German Minorities"; my translation]. Nonetheless, while this information may be available, the existence of such minorities is neither taught consistently throughout German secondary education – the curricula are designed by the respective federal states – nor made particularly visible in public discourse. Concerning the Sinti and Roma, antiziganism and a

context. German secondary school curricula tend to cover national minorities in some Anglophone parts of the world in some detail while they usually do not include the study of German national minorities. This neglect can be read as implying no need to preoccupy oneself with indigenous people in Germany, and might lead to the misconception that there were no indigenous people on what is German territory today, in the first place. Consequently, students may be misled to externalize indigenous people as an "American" – in whichever sense of the term – and Australian issue only. In addition, the question of "What is an American?" in a broader context, which has preoccupied American Studies in many different shapes and forms, may provide German students with further options to broaden their schemas in order to discuss identity and belonging in the German context also, especially as this question is not as easily solved as somewhat humorously explored by Gish Jen in the short story "What Means Switch," in which teenage protagonist Mona tells her Japanese friend that he "could become American" and "'d just have to switch, that's all" (2007: 185).

Just switching like the protagonist suggests frequently does not work in order to integrate – in whichever sense of the term – especially not when intersecting axes of discrimination and oppression are involved. In the late 1980s, sociologist Deborah King pointed out that the impact of racism, sexism, and classism was not merely additive if people were affected by more than one of these axes of oppression. Instead, these are "multiple jeopardies" (1988: 47), "[t]he modifier 'multiple' refer[ing] not only to several, simultaneous oppressions but to the multiplicative relationships among them […] the equivalent formulation is racism multiplied by sexism multiplied by classism" (ibid.). The differing quality and impact of the diverse intersecting components may also further complicate visibility or collective action in such a context. In the early 1990s, legal scholar Kimberlé Crenshaw drew on her experiences as a lawyer in formulating the theory of intersectionality, which enabled the analysis of subject positions that are vulnerable to multiple, 'intersecting' structures of oppression.[25] Her analysis focuses in particular on the intersection of race, class, and gender. Like King, Crenshaw points out that

conflation with class issues may complicate the public discourse even more while the mere existence of the other groups is usually not part of the German national discourse – especially not in the context of an allegedly 'homogenous' German society.

25 In the context of analyzing *racism*, however, as Salem and Thompson argue, "[i]t is important to note that racism does not simply intersect with other structures of domination such as gender, class, disability and so on, but is itself intersectional in its arrangements of signifiers and markers" (Salem/Thompson 2016: 13).

> [t]he effect of both these marginalizations [misogynist violence and racism] is that women of color have no ready means to link their experiences with those of other women. This sense of isolation compounds efforts to politicize sexual violence within communities of color and permits the deadly silence surrounding these issues (1991: 1282).

As Crenshaw highlights,

> current debates over representation continually elide the intersection of race and gender in the popular culture's construction of images of women of color. Accordingly, an analysis of what may be termed "representational intersectionality" would include both the ways in which these images are produced through a confluence of prevalent narratives of race and gender, as well as a recognition of how contemporary critiques of racist and sexist repression marginalize women of color (ibid: 1282-1283).

While it remains difficult to find a suitable balance between intragroup difference[26] and group building, finding a voice and a language to critically evaluate "Who Speaks and Who Is Heard/Hurt?" seems crucial. Especially if one takes the different potential impact factors into account, diversity concepts might be helpful in coming to terms with this challenge.

As a competence covering a variety of intersecting hegemonic social formations, Melissa Steyn proposes the concept of *Diversity Literacy*[27] as a tool to evaluate a critical diversity mindset. As a concept,

---

26 "The Danger of a Single Story" (Adichie 2009) or oversimplifying a multifaceted issue has been addressed in footnote 9, as well. In the context of intersectionality, Crenshaw reminds her readers that "[t]he problem with identity politics is not that it fails to transcend difference, as some critics charge, but rather the opposite – that it frequently conflates or ignores intragroup differences. In the context of violence against women, this elision of difference in identity politics is problematic, fundamentally because the violence that many women experience is often shaped by other dimensions of their identities, such as race or class. Moreover, ignoring difference *within* groups contributes to tension *among* groups, another problem of identity politics that bears on efforts to politicize violence against women" (Crenshaw 1991: 1242; original emphasis). While simplification may in some contexts make sense in order to reach a broader visibility for a marker of difference that is shared by a larger group of people, *Critical Diversity Literacy* underscores the importance of analyzing "how these systems of oppression intersect, interlock, co-construct and constitute each other" (Steyn 2007, qtd. in Steyn 2010: 20).

27 As Steyn argues concerning the self-reflexivity of this position, its "strength [...] is that it does not present itself as value free, aligning with the now well-established

"Diversity literacy" can best be characterised as a "reading practice" – a way of perceiving and responding to the social climate and prevalent structures of oppression. The analytical criteria employed to evaluate the presence of diversity literacy include the following: 1) a recognition of the symbolic and material value of hegemonic identities [...]; 2) analytic skill at unpacking how these systems of oppression intersect, interlock, co-construct and constitute each other; 3) the definition of oppressive systems such as racism as current social problems rather than a historical legacy; 4) an understanding that social identities are learned and an outcome of social practices; 5) the possession of a diversity grammar and a vocabulary that facilitates a discussion of race, racism, and antiracism, and the parallel concepts employed in the analysis of other forms of oppression; 6) the ability to translate (interpret) coded hegemonic practices; 7) an analysis of the ways that diversity hierarchies and institutionalised oppressions are mediated by class inequality and inflected in specific social contexts; and 8) an engagement with issues of transformation of these oppressive systems towards deepening democracy/social justice in all levels of social organisation (Steyn 2007; qtd. in Steyn 2010: 20).

As has been exemplified above, in American Studies classrooms in Germany, these criteria may be integrated concerning both the target culture in the U.S./the Americas and the German context. Students may benefit from this approach by learning not to distance themselves from the issues they study, for instance, by externalizing issues as mere "American issues." Instead, they use them in order to expand their perspective on German society, which they currently live in, as well. In that sense, Steyn's third point entails a recognition of "oppressive systems [...] as current social problems" (20) in Germany and other countries, rather than as "American" problems only. Moreover, complex concepts like *race*

---

argument in feminist, anti-racist, postcolonial and other emancipator [sic] scholarship that no research ever is value neutral, but that scholarship that claims value neutrality inevitably reproduces dominant ideologies. Rather, the research declares its social agenda up front. In brief, this particular stance towards diversity

- departs from a profound commitment to the values of democracy, social justice, equity and empowerment;
- recognises that the incorporation of people that have been marginalised should not involve a process of assimilation, but a transformation of the cultural milieu in order to bring about new social meanings and representations;
- rejects essentialised notions of identity, naturalised notions of race, gender, etc., and discourses which reify homogeneity;
- stresses that identity and difference are constructed within specific historical, cultural and power relations. (Carr 2000, Giroux 1997, Goldberg 1994, Steyn et al. 2003)" (Steyn 2010: 19).

or *racism* require a multidimensional approach, providing students with an understanding of both the impact of specific contexts and a more general understanding.

This multidimensional approach also includes a variety of methods, among them the proposed adoption of Moya's suggestion of close readings in order to expand one's schemas, but also a heightened attention towards active participation in classroom discussions. If taken seriously, however, this approach also involves a critical examination of the composition of our student groups as well as the representation and voice granted to diverse groups, in class content and interaction as well as in the professional field of German American Studies more generally. Encouraging our students to continuously work on their own *Critical Diversity Literacy* may thus foster a conversation about diversity and inclusion not only concerning issues that they perceive in society at large, but also in their immediate (learning) environment, and in turn, may contribute to lecturers' reevaluation of their own approaches, as well.

Students and staff may need to engage in a conversation that critically examines curricula as well as the field more generally, so that German American Studies may continuously reexamine the composition of its faculty and students at various levels, its research and teaching practices, as well as its research and teaching foci.[28] Such a diverse conversation might lessen the danger of overlooking the beam in one's own eye[29] like Wendy Wasserstein's character Professor Laurie Jameson in her play *Third*, for example, who understands herself as being very open, even declaring her "classroom [...] a hegemonic-free zone" (2008: 5). Her experience may serve as a warning since it takes her some time to discover that she may not have been as open and inclusive after all. In "The Danger of a Single Story," Adichie also emphasizes the necessity to be attentive and self-reflexive: While she notices her future roommate's "single story" of Adichie's life, Adichie realizes her own "single stories" both in an international setting and within her close vicinity when traveling across class boundaries. Oversimplifications, power, and prevailing hierarchies can, as both Wasserstein's and Adichie's contributions caution, mislead everyone, even those who perceive themselves to be immune against challenges of lacking self-reflexivity. Adichie remarks that "[p]ower is the ability not just to tell the story

28 Further potential sites for critical engagement with diversity and inclusion in German American Studies include e.g. the *German Association for American Studies*, which held a special conference "Diversity and/in the GAAS" at the *Amerikahaus* in Munich on October 20-21, 2017.

29 "And why beholdest thou the mote that is in thy brother's eye, but considerest not the beam that is in thine own eye?" (*The Holy Bible*, *KJV*, Matt. 7:3).

of another person, but to make it the definitive story of that person." The continuing expansion of schemas and a consistent *diversity literate* reading practice may therefore be helpful in overcoming "The Danger of a Single Story."[30] This self-reflexive approach may thereby contribute to a raised awareness among different actors in German American Studies and about the question of "Who Speaks and Who Is Heard/Hurt?" not only as observers and analysts of others, but also in our own contexts.

This long-term goal can only be achieved through a multidimensional approach. In addition to the development of *Critical Diversity Literacy*, a critical understanding of difference can complement raising awareness for the complexity that concepts like difference and diversity entail. Here again, American Studies students in Germany may benefit from their view from both the "inside" and the "outside": studying different phenomena within the framework of single dominant national (master) narratives may block one's view on implicit bias. If, however, one adopts a more inclusive and transnational or *inter-American*[31] perspective, mutual learning across multiple potential borders – students and instructors; theories, methods, materials, and approaches from a variety of potential backgrounds; other axes of difference – may support a more *diversity literate* view.

In terms of methodology, such a perspective necessitates the continuing creation of an appropriate language as well as an ever-expanding, inclusive reference framework both of theories and of examples. American Studies as an inherently interdisciplinary, transnational field does provide opportunities for such a multiplicity of perspectives and may therefore use its own patchwork quality as a strength, similar to what President Obama once said about America's "patchwork heritage" (Obama 2009). Contextualizing the issues raised, both in their original environments and in their reception in Germany by various potential groups seems crucial in order to foster a broader understanding of the issues at hand. A critical perspective thus entails a continuing reassessment and cautious, reflexive self-awareness, particularly towards issues of power and

30 As quoted in footnote 27, *Critical Diversity Literacy* is built upon self-reflexive foundational premises, trying to be conscious of unavoidable potential challenges within the concept.

31 According to Shelley Fisher Fishkin, a transnational perspective entails focusing on "multidirectional flows" and questioning "the naturalness of some of the borders, boundaries, and binaries" (2005: 22). Moreover, heightened attention is put on "*broadening the frame*" (Fisher Fishkin 2012: 31; original emphasis). Some aspects have also been investigated in the *Inter-American Studies* perspective (cf. footnotes 10; 22).

hierarchy that might not be overtly visible, but may result in ever-new forms of discrimination. As Moya concludes,

> [l]iterature by itself will never change the world or create racial [or diversity] literacy. But it nevertheless remains a highly powerful tool, and an important actor, in the ongoing struggle to imagine, as the Mexican poet and novelist Rosario Castellanos says, another way to be human and free (2016: 165).

In a similar fashion, Adichie remarks that "[s]tories matter. Many stories matter. Stories have been used to dispossess and to malign. But stories can also be used to empower, and to humanize. Stories can break the dignity of a people. But stories can also repair that broken dignity."[32]

As long as the question of "Who Speaks and Who Is Heard/Hurt?" remains relevant, more stories will be needed in order to break the power of potential "single stories." This diversity of stories, habitually read in an attentive and *diversity literate* close reading and negotiated among others in order to diversify one's schemas in multiple ways, may provide a chance to counter reductive tendencies, and expand both schemas and *diversity* grammar and vocabulary. In that sense, and conscious of Moya's reminder cited above that "[l]iterature by itself will never change the world or create racial [or diversity] literacy" (165), this proposed combination of methods and approaches may support the continuous development of more equitable answers to the question of "Who Speaks and Who Is Heard/Hurt?".

32 Stories' potential to malign is particularly pertinent when they are used in a reductive, stereotypical fashion, as Adichie argues: "The single story creates stereotypes. And the problem with stereotypes is not that they are untrue, but that they are incomplete. They make one story become the only story" (2009).

Works Cited

Adichie, Chimamanda N. (2009) "The Danger of a Single Story." TED-Talk, 18:43, October (https://www.ted.com/talks/chimamanda_adichie_the_danger_of_a_single_story/up- next?language=de#t-922).

Balibar, Étienne/Wallerstein, Immanuel M. (2002): Race, Nation, Class: Ambiguous Identities. Chris Turner (trans.), London: Verso.

Bühl, Achim (2016): Rassismus: Anatomie eines Machtverhältnisses, Wiesbaden: Marix Verlag.

Bundesministerium des Innern, für Bau und Heimat (2018a): "Nationale Minderheiten und deutsche Minderheiten." (https://www.bmi.bund.de/DE/themen/heimat-integration/minderheiten/minderheiten-node.html;jsessionid=4549D6C2D08ABE6916C63E4B64A7B8DA.2_cid295).

Bundesministerium des Innern, für Bau und Heimat (2018b): "Themen: Heimat & Integration." (https://www.bmi.bund.de/DE/themen/heimat-integration/heimat-integration-node.html).

Çakır, Naime (2014): Islamfeindlichkeit – Anatomie eines Feindbildes in Deutschland, Bielefeld: transcript.

Campos García, Yolanda M./Ramos Godinez, Margarita/Raussert, Wilfried (2015): "Introducción." In: Yolanda M. Campos García/Margarita Ramos Godínez/Wilfried Raussert (eds.), Cruzando Fronteras en las Américas. Las Dinámicas del Cambio en la Política, la Cultura y los Medios. Fronteras Simbólicas: Cine, Media, Musica. Crossing Boundaries in the Americas: Dynamics of Change in Politics, Culture, and Media, Guadalajara: Universidad de Guadalajara, pp. 9-17.

Chang, Jeff (2014): Who We Be: The Colorization of America. New York: St. Martin's.

Crenshaw, Kimberlé (1991): "Mapping the Margins: Intersectionality, Identity Politics, and Violence Against Women of Color." In: Stanford Law Review 43/6, pp. 1241-1299.

Drews-Sylla, Gesine/Makarska, Renata (2015): Neue alte Rassismen? Differenz und Exklusion in Europa nach 1989, Bielefeld: transcript.

Emcke, Carolin (2017): Gegen den Hass, Bonn: BpB Bundeszentrale für politische Bildung.

Federal Ministry of the Interior, Building and Community (2018): "National Minorities" (https://www.bmi.bund.de/EN/topics/community-and-integration/national-minorities/national-minorities-node.html).

Federal Ministry of the Interior, Building and Community (2018): "Topics: Community and Integration" (https://www.bmi.bund.de/EN/topics/community-and-integration/community-and-integration-node.html).

Fisher Fishkin, Shelley (2005): "Crossroads of Cultures: The Transnational Turn in American Studies – Presidential Address to the American Studies Association, November 12, 2004." In: American Quarterly 57/1, pp. 17-57.

Fisher Fishkin, Shelley (2012): "Mapping Transnational American Studies." In: Udo J. Hebel (ed.), Transnational American Studies, Heidelberg: Winter, pp. 31-74.

Gaspard, Lynn (ed.) (2017): Don't Panic, I'm Islamic: Words and Pictures on How to Stop Worrying and Learn to Love the Alien Next Door, London: Saqi.

Hirschmann, Kai (2017): Der Aufstieg des Nationalpopulismus: Wie westliche Gesellschaften polarisiert werden, Bonn: BpB Bundeszentrale für politische Bildung.

Jen, Gish (2007): "What Means Switch." In: Mary Frosch (ed.), Coming of Age in America: A Multicultural Anthology, New York: The New Press, pp. 175-194.

King, Deborah K. (1988): "Multiple Jeopardy, Multiple Consciousness: The Context of a Black Feminist Ideology." In: Signs 14/1, pp. 42-72.

Lentin, Alana (2008): "Europe and the Silence about Race." In: European Journal of Social Theory 11/4, pp. 487-503.

Mahdawi, Arwa (2017): "A Personal Guide to Extreme Vetting: How to Distinguish an Acceptable Arab from a Terrorist in 6 Easy Steps." In: Lynn Gaspard (ed.), Don't Panic, I'm Islamic: Words and Pictures on How to Stop Worrying and Learn to Love the Alien Next Door, London: Saqi, pp.2-7.

Meier-Braun, Karl-Heinz/Weber, Reinhold (eds.) (2013): Migration und Integration in Deutschland: Begriffe – Fakten – Kontroversen, Bonn: BpB Bundeszentrale für politische Bildung.

"Minorities under International Law." UN OHCHR, March 25, 2017 (http://www.ohchr.org/EN/ Issues/Minorities/Pages/internationallaw.aspx).

Moya, Paula M. L./Markus, Hazel R. (2010): "Doing Race: An Introduction." In: Hazel R. Markus/Paula M. L. Moya. (eds.), Doing Race: 21 Essays for the 21st Century, New York: W.W. Norton, pp. 1-102.

Moya, Paula M. L. (2016): The Social Imperative: Race, Close Reading, and Contemporary Literary Criticism, Stanford: Stanford University Press.

Obama, Barack (2009): Inaugural Address. The White House, January 30 (http://www.whitehouse.gov/blog/inaugural-address/).

Okri, Ben (1997): A Way of Being Free, London: Phoenix.

Raab, Josef (2014): "Inter-American Studies: Why and Whither?" fiar forum for inter-american research 7.3, 42:19 (http://interamerica.de/volume-7-3/raab/).

Raab, Josef (2015): "Thinking Beyond the Nation: The Project of Inter-American Studies." In: Yolanda M. Campos García/Margarita Ramos Godínez/Wilfried Raussert (eds.), Cruzando Fronteras en las Américas. Las Dinámicas del Cambio en la Política, la Cultura y los Medios. Fronteras Simbólicas: Cine, Media, Musica. Crossing Boundaries in the Americas: Dynamics of Change in Politics, Culture, and Media, Guadalajara: Universidad de Guadalajara, pp. 21-51.

Raussert, Wilfried (2017): "Introduction." In: Wilfried Raussert (ed.), The Routledge Companion to Inter-American Studies, pp. 1-12.

Raussert, Wilfried (2015): "Selected Key Tropes in Inter-American Studies: Ways of Looking at Entangled Americas." In: Wilfried Raussert/Brian Rozema/Yolanda Campos/Marius Littschwager (eds.), Key Tropes in Inter-American Studies: Perspectives from the Forum for Inter-American Research (fiar), Trier and Tempe: WVT Wissenschaftlicher Verlag Trier Tempe, pp. 1-11.

Reilly, Katie (2016): "Flight Delayed After Passenger Reports Professor's Math Equations as Suspicious." Time May 7 (http://time.com/4322154/airplane-professor-math-equations-suspicious/).

Rüsen, Jörn/Völkel, Bärbel (2017): "Historische Orientierung: Was uns wichtig war und wichtig ist: Eine kritische Bestandsaufnahme. Streitgespräch mit Jörn Rüsen am 9.5.16." In: Bärbel Völkel/Tony Pacyna (eds.), Neorassismus in der Einwanderungsgesellschaft: Eine Herausforderung für die Bildung, Bielefeld: transcript, pp. 213-53.

Salem, Sara/Thompson, Vanessa (2016): "Old Racisms, New Masks: On the Continuing Discontinuities of Racism and the Erasure of Race in European Contexts." nineteen sixty nine: an ethnic studies journal 3/1 (https://escholarship.org/uc/item/98p8q169).

Schiffauer, Werner/Eilert, Anne/Rudloff, Marlene (eds.) (2017): So schaffen wir das – eine Zivilgesellschaft im Aufbruch: 90 wegweisende Projekte mit Geflüchteten, Bielefeld: transcript.

Speit, Andreas (2017): Bürgerliche Scharfmacher: Deutschlands neue rechte Mitte. Lizenzausgabe, Bonn: BpB Bundeszentrale für politische Bildung.

Steyn, Melissa (2010): "Critical Diversity Literacy. Diversity Awareness in 12 South African Organisations." In: Melissa Steyn (eds.), Being Different Together: Case Studies on Diversity Interventions in some South African organisations, South Africa: iNCUDISA, pp. 15-42.

The Holy Bible, Authorized King James Version. 1997, Oxford and New York: Oxford University Press.

Twine, France W. (2004): "A White Side of Black Britain: The Concept of Racial Literacy." Ethnic and Racial Studies 27/6, pp. 878-907.

Völkel, Bärbel/Pacyna, Tony (eds.) (2017): Neorassismus in der Einwanderungsgesellschaft: Eine Herausforderung für die Bildung, Bielefeld: transcript.

Wasserstein, Wendy (2008): Third, New York: Dramatists Play Service.

Wekker, Gloria (2016): White Innocence: Paradoxes of Colonialism and Race, Durham: Duke University Press.

Wildt, Michael (2017): Volk, Volksgemeinschaft, AfD, Bonn: BpB Bundeszentrale für politische Bildung.

IV. Shifting Perspectives: Transatlantic and Genre-Crossing Reflections on White Normativity

Goethe Meets Baldwin: Notes towards a Comparative Perspective beyond Misappropriation

Derek C. Maus

I was exposed to the writing of Canadian-born journalist (and, later, politician) Michael Ignatieff, specifically his book *Blood and Belonging: Journeys into the New Nationalism* (1993) while researching my undergraduate honors thesis on Cold War Russian and American satirical fiction (a project that later morphed into my doctoral dissertation and my first book). Like me, Ignatieff is rather skeptical of nationalism, especially chauvinistic ethnic nationalism, and his multinational survey of its effects provided an extremely powerful lens through which my young self could interpret the world. The first half of the 1990s had, after all, been rife with various forms of nationalist conflict, including the Iraqi invasion of Kuwait and the subsequent Persian Gulf War, the series of wars fought in the former Yugoslavia, the Rwandan Genocide, the First Palestinian Intifada (and the stimulus for and backlash against it), Quebecois separatists nearly winning a second referendum about secession from Canada, the Iraqi Kurdish Civil War, and so many more.

In the absence of the overarching geopolitical rivalry of the Cold War, Ignatieff saw the 1990s being defined by a particular kind of localized/regionalized conflict: "[W]hat has succeeded the last age of empire is a new age of violence. The key narrative of the new world order is the disintegration of nation-states into ethnic civil war; the key architects of that order are warlords; and the key language for our age is ethnic nationalism" (1993: 5). While cataloguing and lamenting six of the most prominent examples of such nationalistic strife, Ignatieff also wrote about its apparent opposite – cosmopolitanism – in a way that resonated with me powerfully then, and still does today:

> Anyone whose father was born in Russia, whose mother was born in England, whose education was in America, and whose working life has been spent in Canada, Great Britain, and France, cannot be expected to be much of an ethnic nationalist. If anyone has a claim to being a cosmopolitan, it must be me. I wish I spoke more languages than I do, I wish I had lived in more nations than I have, and I wish that more people understood that expatriation is not exile: it is merely the belonging of those who choose their home rather than inherit it (ibid: 11).

I spent a good deal of the 1980s sleepless with nightmares about nuclear war. I even wrote to the Soviet embassy at some point in junior high, requesting more information about the nation that ostensibly justified the constant threat of nuclear apocalypse. The multicolored poster I received in return certainly informed me about the population and major exports of the Nagorno-Karabakh Autonomous Oblast, but it didn't bring me much clarity about whether these people were my friends or enemies. Partly out of adolescent idealism and partly out of terrified desperation, I gravitated towards any worldview that transcended the predominant binaries of the day. Thus, when I read Ignatieff's words about the possibility of productive expatriation, of choosing rather than inheriting my home, I was hooked.

The idea of defining myself in a way that transcends geographic, cultural, or linguistic boundaries had always appealed to me because I felt like I had grown up largely without such inherent limitations. I came into self-consciousness as a bilingual (and eventually quadrilingual) child of bilingual parents, having lived in both Germany and the United States before starting school. After my parents' divorce, I experienced a binary childhood and adolescence in which I generally spent the school year in one place (Little Rock, Arkansas) and the summer in another (Kansas City, Missouri). Although these two locations are not so radically different from one another in the grand scheme of things, for a ten-year-old trying to navigate the often-turbulent waters of social interaction, the peregrination from one context to the other required (re-)learning a completely different set of social codes every few months. Finally, my early experiences of race and class in Little Rock created a third level of what one might call partial or incomplete integration. My mother and I were unusual white "immigrants" into a lower-middle-class inner-city neighborhood undergoing "white flight" in the late 1970s, and I was part of the racial minority at almost every school I attended from 1978 to 1990. Rather than seeing these various layers of liminality as barriers to belonging (and, frankly, having had the privilege of ultimately not needing to assimilate in order to belong) within the various societies in which I have lived, I have come to think of them instead as the stimulus for the perspective with which I have tried to view the world, personally and professionally.

When I started graduate school in 1995 and began in earnest my professional development as a literary scholar, I had a relatively simplistic understanding of how and why ideals like multiculturalism and cosmopolitanism might become a part of my critical perspective. If nothing else, they provided a political/ philosophical explanation for why I was drawn to writing from countries to which I had not yet traveled, and by writers from ethnic, racial, and national backgrounds different from my own. They also aligned with my desire to produce scholarship that challenged the conventional critical wisdom about such things as the ostensible anti-Americanism of leftist writers during the 1930s (the subject of my MA thesis), or the inherently binary nature of the Cold War (the subject of my doctoral dissertation). Cosmopolitanism and multiculturalism seemed like the perfect antidotes to the ethnocentric bigotry and/or nationalistic provincialism regularly expressed during the 1990s by such "culture warriors" as Pat Buchanan, William Bennett, and Harold Bloom. Although I felt neither internal desire nor external pressure to reject my identity as a white man or as an American citizen of German national/cultural heritage, I also knew that I did not want those aspects of my identity to overdetermine the literature about which I could write. Using Ignatieff's words as my manifesto – "Cosmopolitans made a positive ethic out of cultural borrowing: in culture, exogamy was better than endogamy, and promiscuity was better than provincialism" (ibid: 11) – I pronounced myself a cosmopolitan multiculturalist and set out to culturally borrow whatever struck my fancy in constructing my scholarly identity!

Hold on, hold on … Before you decide to set that well-intentioned but horribly naïve version of me straight about privilege and appropriation, let me assure you that I have been fortunate to have several wise (and patient) friends, mentors, and colleagues in my life that have already undertaken that unenviable task. Thanks to their interventions, I not only processed the entirety of Ignatieff's comments about cosmopolitanism (rather than just cherry-picking the parts that best suited my desired self-image …), but also began to develop a more nuanced understanding of how to teach and to write critically about a wide range of literary works while remaining cognizant of the cultural and social power dynamics in which such interpretive acts take place.

The notion of being a scholar of "world literature" initially struck me as fairly uncomplicated. For longer than I care to admit, I thought that being a scholar of world literature was as simple as stepping outside the traditional American literature and British literature "tracks" of study available to most undergraduate English majors in the United States at that time. Not only had the broader implications of reading works in translation rather than in their original language not occurred to me yet, but I was also wholly unfamiliar with Goethe's 19th century concept of *Weltliteratur* – to say nothing of related concepts put forth by Diogenes, Immanuel Kant, Homi Bhabha, Franco Moretti, Judith Butler, Gayatri Chakravorty Spivak, Kwame Anthony Appiah, and Paul Gilroy, just to name a few (the waters of my undergraduate education had not been particularly muddied by post-colonial theory, as you might have guessed ...). I had considered enrolling in Comparative Literature instead of English for my doctorate at the University of North Carolina, but ultimately rejected that option because it seemed more predicated on linguistic comparisons than on the thematic ones in which I was primarily interested. Studying world literature – at least in my conception of it at the time – would allow me to glean knowledge found in works from other cultures and incorporate it into my own. What I imagined as the result of such study was a cosmopolitan intellect comparable to a World's Fair, with as many cultures as possible represented by at least a single exemplar in their respective pavilions.

With time and experience, the unsavory aspects of this metaphor have become clear to me. I recognized, for example, that every nation's opportunities for self-representation within a World's Fair were constrained by the values and desires of the society hosting those exhibits (e.g., the absence of a Soviet exhibit from the 1964-65 New York World's Fair). Every work of literature exists wholly independent of my (or any other critic's) desire to reframe it from my own vantage point. Such acts of reframing are understandable, since every reader is a unique individual who invariably brings his or her own background to a given text. An excess of cosmopolitan or multicultural zeal, however, can unwittingly appropriate a text by downplaying or otherwise erasing its cultural origins; Such a process is equally onerous and damaging when it arises from a desire to define a canon of ostensibly universal "Great Books," and when it stems (as it did with me) from the desire to transcend cultural distinctions in favor of an overarching humanism.

Thankfully, James Baldwin had left a trail of breadcrumbs that would eventually lead me to a better practice. I had read a considerable bit of Baldwin's writing as an undergraduate and had appreciated him both as a stylist and a contrarian, but it was not until years later that I understood what I now see as his profoundly necessary (and as-yet largely unheeded) advice to white people who wish to stop perpetuating and benefitting from America's racism. As anyone who has read even one of his works must surely recognize, Baldwin pulls few punches when it comes to speaking his mind about the causes of racism in America. As I revisited *Another Country*, *The Fire Next Time*, and various others of his occasional pieces, I started to notice that Baldwin calls out not just overt white supremacists like "Bull" Connor and James Eastland, but also white liberals who revel in the putative correctness of their attitudes while remaining ignorant of the ineffectuality or even harmfulness of their (in-)actions:

> People talk to me absolutely bathed in a bubble bath of self-congratulation. I mean, I walk into a room and everyone there is terribly proud of himself because I managed to get to the room. It proves to him that he is getting better. It's funny, but it's terribly sad. It's sad that one needs this kind of corroboration and it's terribly sad that one can be so self-deluded. The fact that Harry Belafonte makes as much money as, let's say, Frank Sinatra, doesn't really mean anything in this context. Frank can still get a house anywhere, and Harry can't. People go see Harry and stand in long lines to watch him. They love him onstage, or at a cocktail party, but they don't want him to marry their daughters. This has nothing to do with Harry; this has everything to do with America (Baldwin 1964: 74).

Baldwin wrote those words in 1964, but I felt their relevance in 1996 as much as I still feel it in 2017. When I hear presumably idealistic rhetoric that seeks to "raise awareness" or "promote tolerance" by studying literary works by members of "marginalized" or "historically underrepresented" groups, I almost invariably feel myself soaking in the metaphorical "bubble bath" of which Baldwin speaks. Although noble-sounding, these goals ultimately retain the inherent "othering" of such literatures and the authors who created them, much as cosmopolitanism and multiculturalism often retain a quasi-colonialist subjective privilege. They do nothing to dismantle or even to question the social hierarchy of power that dispatches Baldwin, Belafonte, Toni Morrison, and Percival Everett (to name a few) to the category of "black" artist, whether or not they wish to claim that label (and the presumptions that accompany it).

Had I read Ignatieff more attentively back in 1993, I would have noticed that he warned himself (and me) of the potential for privileged self-delusion that exists within cosmopolitanism:

> What has happened in Bosnia must give pause to anyone who believes in the virtues of cosmopolitanism. It is only too apparent that cosmopolitanism is the privilege of those who can take a secure nation-state for granted. […] [C]osmopolitans like myself are not beyond the nation; and a cosmopolitan, post-nationalist spirit will always depend, in the end, on the capacity of nation states to provide security and civility for their citizens.[…] At the very least, cosmopolitan disdain and astonishment at the ferocity with which people will fight to win a nation-state of their own is misplaced. They are, after all, only fighting for a privilege [that] cosmopolitans have long taken for granted (1993: 13-14).

Ignatieff claims (rightly, I believe) that what has long been celebrated as the open-mindedness of cosmopolitanism is actually a sheltered viewpoint that results from perceiving one's own relative security as a norm to which others would reasonably aspire. The political-cultural philosophy of American exceptionalism is wholly predicated on such a perception, as is white supremacy; Both protect the ostensibly natural "rights" of the favored nation/race with great violence, figurative and literal.

Overly simplistic diversity rhetoric that espouses "color-blindness" or a "post-racial" mindset relies on a similar self-delusion that willfully ignores the persistence of systemic inequalities that are not addressed meaningfully by such concepts. As Baldwin puts it in *The Fire Next Time* (1963), "White Americans find it as difficult as white people everywhere do to divest themselves of the notion that they are in possession of some intrinsic value that black people need, or want" (108). I am not suggesting that extant discourses of cosmopolitanism and diversity are the ethical equivalents of chauvinistic nationalism and white supremacy, but rather that they are perhaps less of a remedy than I (and others) have presumed them to be.

Baldwin's complaint about white Americans' response to racism tellingly echoes Ignatieff's indictment of cosmopolitanism:

> Most people guard and keep; they suppose that it is they themselves and what they identify with themselves that they are guarding and keeping, whereas what they are actually guarding and keeping is their system of reality and what they assume themselves to be. One can give nothing whatever without giving oneself – that is to say, risking oneself. If one cannot risk oneself, then one is simply incapable of giving. And, after all, one can give freedom only by setting someone free. This, in the case of the Negro, the American republic has never been sufficiently mature to do. White Americans have contented themselves with gestures that are now described as "tokenism." [...] [T]he sloppy and fatuous nature of American good will can never be relied upon to resolve hard problems (ibid: 100-101).

The institutionalized forms of cosmopolitanism and multiculturalism in which I have participated as a scholar and teacher of world literature are far from immune to charges of "tokenism," so I take Baldwin's accusations of "sloppy" and "fatuous" practice to heart. Fortunately, he also offers a provocative solution that can form the basis for a pedagogical and scholarly praxis that moves beyond such superficiality:

> The only way [the white man] can be released from the Negro's tyrannical power over him is to consent, in effect, to be black himself, to become part of that suffering and dancing country that he now watches wistfully from the heights of his lonely power and, armed with spiritual traveller's checks, visits surreptitiously after dark (ibid: 110).

This act of "consent[ing] [...] to be black" is importantly distinct from the more touristic impostures that pervade brands of cosmopolitanism and/or multi-culturalism that refuse to acknowledge and to step outside the presumption of "security and civility" that Ignatieff mentions. It goes well beyond the temporary empathy of "walking a mile in the shoes" of the putative Other, a non-binding performance of "risking oneself" that always allows the subject to return to the sanctity and safety of his or her starting point:

> [W]hen we talk about what we call "the Negro problem" we are simply evolving means of avoiding the facts of this life. Because in order to face the facts of a life like Billie [Holliday]'s or, for that matter, a life like mine, one has got to – the American white has got to – accept the fact that what he thinks he is, he is not. He has got to give up, he has got to surrender his image of himself, and apparently this is the last thing white Americans are prepared to do (Baldwin 1964: 74).

It is this act of "surrender" of one's self-image that I believe can transform the study of literature from individuals and nations different from oneself from a shallow tokenism into a meaningful act of humanist solidarity. I do not believe that such "surrender" requires either negation of one's identity or uncritical acceptance of the values of all other cultures, the two anxieties that seem to trouble multiculturalism's fervent opponents within and outside academe. It does, however, require a difficult and potentially uncomfortable process of discarding the presumption that what is important or desirable in others is defined solely by what is either "universal" or in some other way comprehensible through the lens of one's own existence; In Baldwin's terms, one must be willing to drop the "guard" on one's "system of reality" for more than just a fleeting moment. Metaphorically speaking, it means disembarking from the air-conditioning, plush seats, and tinted windows of the tourist-bus and "risking oneself" among the locals on their own terms as much as possible.

In his *The Idea of World Literature: History and Pedagogical Practice* (2006), John Pizer articulates a scholarly and pedagogical model that squares with Baldwin's ideas without referring to them directly. Pizer begins by stating his belief "that one of the fundamental desiderata of a World Literature course should be the inculcation of an *appreciation* for the nuances of alterity, of a belief that life and literature outside the United States are inscribed by unique linguistic/cultural matrices perhaps no longer defined at the national level, but capable of being glimpsed through the filter of the subnational-transnational dialectic" (2006: 15, original emphasis). He claims that this "filter" originates with Goethe's concept of *Weltliteratur,* which Pizer asserts is "imbued by a strong openness to […] the cultural and linguistic Other. It is temporally oriented toward the future rather than the past" (ibid: 110). Pizer spends the remainder of his book tracing the historical development from that foundation to a contemporary metatheoretical pedagogy of world literature that would "educate beginning students in the complex diversities of the globe's cultures while concomitantly highlighting their universal elements" (ibid: 94).

From my own perspective, one of the most useful expressions of this "subnational-transnational dialectic[al]" approach is Pizer's application of sociologist Roland Robertson's theory of "glocalization" to the practice of literary study. Pizer not only expands on Goethe's ideas, but also on those of David Damrosch, whose *What Is World Literature?* (2003) remains a core text for contemporary scholars seeking to extricate the practice of *Weltliteratur* from the culture wars' obsession with canonical *Wertliteratur* ("worthy literature"). Pizer summarizes Damrosch's "ideal reading" practice as simultaneously "driven by a pleasure in the difference of foreign works from one's contemporary cultural framework, a gratification in their similarity, and an exploration of 'what is *like-but-unlike* – the sort of relation most likely to make a productive change in our own perceptions and practices'" (ibid: 84). Pizer acknowledges that both the unavoidable loss of nuance that accompanies translation and the inherent difficulty of acquiring native proficiency in foreign languages/cultures tend to constrain the study of "the universalities and particularities of human experience that can be gleaned and critically pondered from all worthwhile texts" to a "nodding, indeed superficial, acquaintance" (ibid: 109). However, he also believes that:

> a *Weltliteratur*-driven reading of contemporary "glocalized" literature must mediate among national, local, and universal contexts of place. Such a reading must show how discrete localities are imaginatively but realistically linked and transformed through discursive networks enabled by contemporary telecommunication technologies. This "two-dimensional" reading will indicate where the global and local are enmeshed, but will also demonstrate where the processes of globalization and uniformity are resisted and contested (ibid: 118).

If this complex "mediat[ion]" among different layers of context can be accomplished without presuming either the particular or the universal to be a desirable norm, then Baldwin's necessary act of self-surrender remains possible.

By inclination and by later training, I am a comparatist; As such, I value the inclusively dualistic perspective fostered by both cosmopolitanism and multiculturalism at their ethical best. I encourage my students to interpret literature stereoscopically, to see the "global" with the left eye and to see the "local" with the right. I also strive to achieve such a perspective in my daily life and in turn to apply it to both my teaching and my scholarship. By doing so, I hope to navigate between an interpretive Scylla and Charybdis. On one side, we find overly simplistic readings that merely "honor" or "sample" local variations without also seeking to understand how and why they matter to a text's reception outside its originating culture; On the other resides a canonizing impulse that assigns value to a work exclusively on the basis of its potential to transcend spatial and temporal borders. The narrow path between these two options involves remaining receptive to the unfamiliar without prejudging it – either positively or negatively – because of its alterity. In this way, the Other ceases to be defined either in opposition to the Self or as a desirable exoticism to be appropriated into it; Instead, both Self and Other become voices within a grand-scale and often halting conversation whose cognates, untranslatables, neologisms, and double-entendres all demand consistently mindful interpretation.

I am a white American, both by the accident of my birth and by my acculturation over the course of more than four decades. Because of the privileges it affords me, I strive to ensure that this identity is *only* the starting point for my subsequent investigations. There is *nothing* about my own experience of being American (or white, or male, or Southern, or second-generation German American, or any other group identity marker) that is definitive in terms of American-ness, even if many aspects of it are relatively representative. The first step towards a productively cosmopolitan surrender of the privileged self is to drop the presumption that any part of my identity – whether assigned, assumed, or insisted-upon – must invariably prescribe my relationship to others and vice-versa. As I tell my students, each of us can bring his or her personal experiences and values to bear productively on a text, provided that those experiences do not become a source of confirmation bias that imparts rigid expectations about what kind of literature is worthy of attention or exertion.

Instead of adopting the unchallenging cultural relativism that afflicts much of contemporary cosmopolitanism and multiculturalism, I have tried to adopt a "relativity of diversity" that incorporates the aforementioned ideas of Ignatieff, Baldwin, Goethe, and Pizer. Einstein redefined each individual's position in the physical cosmos by noting that the universe *inherently* appears different depending on the condition of the observer – a fact that comments neither on the intrinsic value of the observer, nor on that of the universe. Transmuted into literary terms, such a viewpoint seeks to understand how the idiosyncratic conditions of the observer (writer and/or reader) affect the universe (the written about), and to encourage constant inquiry into the changing nature of both the observer and the observed. If I have done my job well, my students or readers will become aware of how each new experience they incorporate into themselves fundamentally disrupts any inflexible or essentialist aspects of identity, but without eradicating that identity in the process. I do not cease being a white man of German American cultural background because I read and write about novels by Gish Jen, Colson Whitehead, Jhumpa Lahiri, or Sherman Alexie, but with each passing expansion of worldview, any potential rationalizations for either the superiority of this background or its incompatibility with others become increasingly untenable and, hopefully, undesirable. I have no wish to tell others who or what they should be, but I aspire to use discussions of literature to ask questions about the ways in which both individuals and groups make (and recognize) these distinctions to begin with. In the end, I seek means of interconnecting the otherwise disparate dots of humanity as possible without homogenizing them in the process.

Works Cited

Baldwin, James (1962): Another Country, New York: Dial Press.

Baldwin, James (1963): The Fire Next Time, New York: Dial Press.

Baldwin, James (1964): "The Uses of the Blues." In: James Baldwin/Randall Kenan (eds.), The Cross of Redemption: Uncollected Writings, New York: Vintage International, pp. 70-81.

Damrosch, David (2003): What Is World Literature?, Princeton and London: Princeton University Press.

Ignatieff, Michael (1993): Blood and Belonging: Journeys into the New Nationalism, Toronto: Viking Canada.

Pizer, John (2006): The Idea of World Literature: History and Pedagogical Practice, Baton Rouge: Louisiana State University Press.

Notes from the Margin: Academic White Spaces and the Silencing of Scholars of Color

Kimberly Alecia Singletary

1 July 2009 (Berlin, Germany)

Here starts my first major trip for my research abroad. It's a bit nerve-wracking ... This will be the two weeks in which I need to establish my project and start breaking ground as a scholar. Yikes.

2 July 2009 (Berlin, Germany)

[The airline] has lost my luggage. Mom and Dad fly with thirteen bags and a crocodile and all of their things make it safely. I have one bag and it's gone ... The front desk guy, Lars, was so apologetic. He tried to upgrade me, but the hotel is fully booked. I do get free breakfast for four days just because of my airport trouble. I imagine, though, that hoarding croissants in Ziploc bags may be a bit of an issue.

10 July 2009 (Berlin, Germany)

The bus driver wouldn't open the door this morning when he was stopped at the traffic light. He signaled that the light was going to change so he couldn't open the door, but it was for a second that I understood the irritation that minorities feel here. You can't tell if he would have stopped for someone who looked German. In the US, where so much of the working class are of color, you don't have the same questions – they didn't want to stop because they didn't feel like it. But here, where the people in upper, middle, and lower classes are equally likely to be white, it's difficult to tell. It keeps you guessing, always unsure.

June 2010 (Berlin, Germany)

I went to church today with [my landlord] who is the former minister of a Lutheran church at the Ruhleben stop. He and his wife own apartments they rent out near Innsbrucker Platz. It was a nice service. I sat next to two white German women. I greeted them in German and they didn't respond. When it came time for Communion, I sat in my seat but made space for the women to get by me. I smiled at them. They looked at me with blank faces. When I turned to face the front, one said to the other, "Don't worry. I'll watch your purse. You know how they are." Then one got up to take Communion and the other sat glaring at me like an angry Chihuahua. When [the minister] came to the back of the church after service, he greeted the women, whom he knew, turned to me and asked me, "Did you understand everything?" "Oh yes," I replied. "The German wasn't that hard at all." Then I turned to the women, who were listening to our conversation with their mouths open, said goodbye, and walked out of the church.

When I undertook my first major research trip to Germany in the context of my dissertation, I had actually already completed several other research- or work-related trips across Europe and Asia. As a rhetorician analyzing images of Blackness in the German public sphere and as a Woman of Color traveling alone, I was apprehensive as to how my racialized and gendered body would be read when my nationality was not apparent and interested in whether my color rendered me as much of a non-entity in Germany as it did in the United States. I arrived in Berlin in 2009, spending the first four days at the Hilton, then three days in Frankfurt to meet with researchers studying Black German identity, and finally traveling back to Berlin to find archives and speak at an academic conference. The experience at the hotel was exceedingly pleasant, but I understood that it was dependent upon the (mis-)perception that I had the money to pay for a mid-luxury hotel in one of Berlin's best neighborhoods. The only color that mattered was green (for what it is worth, I used my aunt's timeshare). Of my days in Berlin, I was almost exclusively located in academic spaces where I felt the absence of other Brown people profoundly; The only Person of Color I saw in the hotel was a Filipina staff member who advised me to watch television to learn everyday German when she saw me reading the *Frankfurter Allgemeine* alongside my German dictionary. Apart from my brief contact with her, I spent several days in isolation, conducting research, watching TV, and reading gossip magazines (to pick up on slang, of course). I often went to the Film and Television Museum in the *Sony Center*, which was where Fortuna, the

Cameroonian security guard "pointedly said hello to me," as I wrote in my field notes in 2009, and asked why I wanted to watch movies instead of talking to actual Black people. So I met him after his shift in Wedding, an immigrant-heavy neighborhood once derided by locals, but now gentrified and lauded as a hot new neighbourhood. We walked to a non-descript "African shop" and once inside, passed two kids minding the register at the front of the store and into a space in the back with several African men and one Afro-German woman. Up until that moment, I had had little chance to interact with People of Color in Germany in or outside of academia.

In my class-privileged position as a foreign researcher, Afro- or Asian-German subjects were wholly unrepresented in the academic spaces I frequented. At a university or in an archive, I was marked as a student simply by my gym shoes and bookbag. I claimed ownership of the racially exclusive intellectual space of the university library without thinking of the ways that the library might be a site of tension and representative of the paucity of racial and ethnic diversity in German higher education (Ha 2016). My national privilege, marked by letters of introduction from my R1, well-funded, private university, created a space in which my racial difference was sublimated when buttressed against my citizenship in one of the world's most powerful nations. When I moved out of the academic space where I had control over my movement and level of engagement, however, my privilege vanished. I became yet another not white person in Germany, working in isolation, trying to convince the bus driver to open the door, arguing that race impacted one's career, and waiting for someone to notice and befriend me. The academic space was one in which I navigated with ease, as long as the visual markers of my American identity – paraphernalia with the university logo, for instance – were easily identifiable.

The emotional significance of feeling 'raceless' in Europe cannot be discounted. It is not that race does not matter in the EU, I would posit, but that the de-emphasis on race and the explicit emphasis on national belonging create moments where Americans of Color can enjoy privileges theretofore restricted from them in the States. Europeans of Color, however, fight every day to be seen as European (cf. Sow 2009; Weheliye 2009). The deliberate use of my body as a site of inquiry underscores how race permeates even the most quotidian encounters and therefore requires researchers to address the impact skin color has on perception, cognition, and the production of scholarship.

My experiences in Berlin and the United States, recounted in reflections from my field notes and diaries, took place in a racial vacuum, whereby whiteness was so overrepresented that the lack of People of Color represented in the faculty, student, and administrative ranks at the university level seemed unremarkable. I became a fish without any knowledge of the fishbowl in which I swam. The university as an institution can often reflect what sociologist Elijah Anderson calls *white spaces*, whose signature "feature is their overwhelming presence of white people and their absence of black people" (Anderson 2015: 13). Academic white spaces persist in academic programs where marginalized communities are understood as case studies rather than collaborators, in major field publications that make overtures to the importance of race-related research without deeply engaging with it, or with the absence of minority faculty in supportive long-term, well-funded teaching and research positions (cf. Johnsrud/Des Jarlais 2010; Pittman 2010). As Kien Nghi Ha noted in relation to a 2013 survey for the University of Osnabrück, when studies on racial animus examine parts, but not the entirety of a university population, they are broaching, but not honestly addressing, the problem of racial exclusion (Ha 2016). Academic white spaces are further exacerbated by an overwhelming adherence to the notion of colorblindness, which further silences People of Color by positing that color – and, by extension, one's daily lived experience as a racialized Other – does not matter or is less important than gender, class, or sexuality.

This paper examines the potential impact academic white spaces can have on an institution's ability to produce scholarship that accurately reflects the 21st century reality of globally connected, racially diverse nation-states. As Scholars of Color and their allies work to create scholarship that more accurately reflects the increasing racial, ethnic, and class diversity of Europe and North America, they must also address the university's position as a white space that silences Other narratives through an adherence to colorblindness and race neutrality, even as the university purports to educate students about racial and social justice.

Imagining Difference

13 November 2013 (Berlin, Germany)

On the U-Bahn. FOUR Black people (affiliated with the U.S. military) squeezed in next to me. A male-presenting Afro-German boy sat across from us ... "Can I ask you guys a question?" he said. "You're all American and Black right? You're proud Black people? Well I am half Black and I call myself Black but the kids at school call me white. Am I Black to you?" ... The soldiers and their wives all agreed: "You look Black to me." ... I told him, "Don't let anybody define who you are. You define who you are." You should have seen this kid's face.

That words of affirmation from strangers connected only by their membership in the African Diaspora could have given a young man a look of joy is illustrative of how Othered identities can be smothered by rigid beliefs about racial phenotype, nation, and belonging. While scholars may understand that identity is inextricably intertwined with larger societal forces and racialized hierarchies of power, that teen in that moment was concerned with his everyday lived experience, not abstract theories. He had run into a common problem mixed-race and/or fair-skinned People of Color face: the insistence that one must swear allegiance to one racial group lest one's authenticity or commitment to group advancement be questioned (cf. Elam 2011; Sundstrom 2001; Zack 1994). For the boy to be part of the German body politic according to his friends, he had to de-racinate himself so as to be normalized. His friends applied a schoolyard version of the *jus sanguinis* principle (citizenship or "right" by blood), which steered German citizenship for centuries, and thus undermined the *jus soli* principle (citizenship or "right" by soil) enacted in Germany in 2000. Although they recognized his German identity, for him to claim his right by blood, he could not claim any other identity. The stubborn misconception that real Germans are white is a forceful erasure of Germans of Color akin to the one drop rule in the United States: any ounce of blackness excluded one from the juridical, financial, and social privileges bestowed upon whiteness. The equation of whiteness with national belonging, however, is not simply a German problem (cf. Maria Alexopoulou's contribution in this volume).

The U-Bahn interaction is indicative of what Fatima El-Tayeb identifies as a continent-wide imaginary in which, "Europe continues to imagine itself as an autonomous entity, simultaneously part and whole of the dialectic of progress, untouched by race matters, occasionally wizened but fundamentally unchanged by its contact with various Others who remain forever outside [...]" (2011: 13). She points to a failure at the societal and individual level in the inclusion of racial Others as equally European. Professors and institutions of higher learning are uniquely positioned to lead the charge in challenging Europe's mischaracterization of itself, yet universities are often the last space for challenging racial and ethnic inequality despite anti-racist activism on campus. As El-Tayeb argues, the white European imaginary is fueled by white citizens' assumptions that Europe is "a colorblind continent in which difference is marked along lines of nationality and ethnicized Others are routinely ascribed a position outside the nation, allowing the permanent externalization and thus silencing of a debate on the legacy of racism and colonialism" (2011: 14). Color still is considered notable because racialized minorities continue to be treated as if they do not have a rightful claim to their German identity, pitting colorblind ideology against those who still struggle to be recognized as equal and full citizens. As Michelle Wright opines:

> [W]hite Germans insistently and consistently misrecognize Afro-Germans as *Africans*, or Others-from-Without [...] Afro-Germans must confront a racist discourse directed at Africans, rather than Afro-Germans [...] [T]here is no such thing as an *anti-Afro-German* discourse, only *anti-African* discourse, raising the question of how one, as an Other-from-Within, should respond to discourse that posits one as an Other-from-Without (2003: 298).

White citizens of the imaginary white nation-state have held steadfastly to the falsehood that theirs is a country untouched by the influence of racial Others or unchecked racism, which makes producing race-centered scholarship feel almost Sisyphean. One rolls the stone equating racial Otherness and national belonging up the mountain of skepticism and denial only to see it roll back down with each new *Microsoft Word* document, conference paper, or inter-departmental meeting.

Choosing Ignorance

Spring 2017 (Portland, Oregon)

The worst interview I ever had was in 2016, at a university where the white female administrative dean told me I blamed white people for racism within the first five minutes of our meeting. She used her own knowledge of Germany, gleaned from her work as a scientist overseas decades before, to explain that Germans, not Americans, had the problems with race. I did not get the job. That interview was quickly followed by another in California, in which, after hearing my job talk on contemporary blackness, a senior white male researcher asserted that African slaves and white masters had higher consciousness and knew their place in society, insinuating that slaves were content with their ritual brutalization. I didn't get that job, either.

I am questioning whether I should stay in academia. The majority of the people with hiring power who I've encountered seem to engage with my work only to tell me I am being unfair to white people. Last week, a committee member asked me about "holes in my knowledge" because I focused "so much on identity." As if whiteness were a neutral designation. I had to remind them that I am as well versed in Aristotle as I am in Malcolm X. And then – I was ashamed as soon as the words left my mouth – I said, "My work is scary, but I'm not. I'm a very supportive and invested colleague." I so desperately need to keep medical insurance that I threw that Hail Mary and demeaned myself in the process. I didn't get that job, either.

In truth, spaces of higher learning often celebrate diversity while actively silencing People of Color. Scholars of Color doing race-related work are often at a disadvantage, spending precious energy reminding their audience how and why race matters, or addressing ill-informed challenges from those convinced that their outsider opinion is not just another perspective, but the only valid perspective. Rejecting the assumption that predominantly white colleges and universities (PWCUs) are spaces of racial tolerance and diversity, William A. Smith, Tara J. Yosso, and Daniel G. Solórzano discuss what they call "racial battle fatigue" as it relates to the emotional and physical distress Faculty of Color experience as a result of dealing with a constant barrage of racialized and/or racist comments and behavior from colleagues and students (Smith et al. 2006; cf. Collins 2015).

Racial fatigue is a common emotion in white academic spaces in which there are few – if any – Scholars of Color present. One learns to be prepared for the inevitable questions: "But what about *larger* categories of gender or class or sexuality?" or "Will identity politics help students in the real world?" or "Surely you don't mean *all* white people?". What those questions reveal is a willful ignorance in favor of centering the perspective of whiteness in a discussion about racial Otherness. When the Black Lives Matter Movement (BLM) began to take hold across North America and parts of Europe, there were those who argued that BLM was racist against non-black people; They intentionally misread the plea "Black Lives Matter (too)" as "Black Lives Matter (more)" despite claims to the contrary. The All Lives Matter countermovement was at best, understood as naivete on the part of those who espoused a politics of colorblindness and at worst, an attempt to undercut cries for social justice and attention to police brutality for Black people worldwide. When 2016 U.S. Presidential candidate Hillary Clinton proclaimed that all lives mattered at a campaign event, it was so damaging to her public perception that she was forced to issue an apology (Keith 2015). Clinton reflected a mindset that regarded race-centric conversations as destructive to societal goodwill rather than ameliorative. Charles Mills argues that white ignorance is difficult to combat because of its manifestation in two forms:

> general skepticism about nonwhite cognition and an exclusion from accepted discourse of nonwhite categories and frameworks of analysis. Thus, a double handicap will result – people of color will be denied credibility and the alternative viewpoints that could be

developed from taking their perspective seriously will be rejected as a priori wrong (2015: 22).

In Mills' view, white ignorance stems from a whiteness-centered mindset that influences a "prism of perception" as it relates to People of Color (2015: 218). That institutions of higher learning bolster an *All Lives Matter* mindset is technically an improvement from the *White Lives Matter* mindset on which many were founded. The case remains, however, that prisms of perception have created moments of professional crisis for scholars of race, and in particular, minority scholars of race.

In 1984, Robert Staples noted that Blacks in the intellectual class have little choice other than presenting themselves as non-threatening or becoming "superior" performers in order to gain the same professional outcomes as whites (Staples 1984: 4). The problem seems to have become even more acute decades later. The prevalence of racism and sexism on campus has led Black professors to feel as if their white colleagues were constantly negatively evaluating them as if they were producing less intellectually rigorous work – especially if that work was race-related. "[…] [B]lack professors who perceive the climate at their institutions as hostile are resisting stereotypes and attempting to ensure their advancement by being without fault and overachievers" (Griffin et al. 2011: 514). Unable to control the societal narrative that degrades Black-centered and Black-produced scholarship, Scholars of Color are often forced to hold themselves to unrealistic expectations of perfection. They ultimately are reliving Frantz Fanon's assertion from 1952 about the European's "fixed conception" of Blackness: "This may be the reason for the strivings of contemporary Negroes: to prove the existence of a black civilization to the white world at all costs" (Fanon 1952: 34). Scholars of Color, then, are put into an untenable situation: continue doing research that benefits their communities while overworking themselves to be seen as equally capable as their white colleagues or avoid race-related research, opting instead for a colorblind approach that treats whiteness as a race-neutral category at the expense of People of Color.

Visible Indifference

Colorblindness makes it possible to erase People of Color from discussions directly impacting their lives, such as in the would-be 2012 German production of *Clybourne Park*, which made international news when the U.S. playwright revoked German privileges to perform the play after learning that the German production was going to use white actors in blackface because they saw "no logical reason why we should cast an 'Afro-German'" and maintained that skin color was irrelevant since they planned to "experiment with makeup" (Rothman 2012). *Clybourne Park* was based on Lorraine Hansberry's 1959 play, *A Raisin in the Sun*, about a black family in Chicago. What might have been the perfect opportunity for Black Germans to infuse an American play about race with a German sensibility was avoided in favor of a colorblind policy that inexplicably advocated blackface. The use of blackface as a comedic or dramatic trope is not exceptional in 21st century Germany. A 2012 production of *I'm Not Rappaport* did not star a Black actor in the role of the Black character, as in the original 1984 play, but a white man in blackface. In 2011, German comedian Martin Sonneborn was criticized for blacking his face as he pretended to be U.S. President Barack Obama and defending his actions as non-offensive. Blackface has cropped up in German UNICEF campaigns, at soccer matches, plays, films, and in political messaging; Artifacts featuring blackface have been used uncritically in research and classrooms. The widespread use of such an offensive cultural practice indicates not just an unwillingness to give up a tradition roundly condemned by Black Germans and others under the guise of "innocent fun" but also a desire to treat Blackness as a means to an end.

The 2009 documentary *Schwarz auf Weiß* (Black on White) by white German journalist Günter Wallraff illustrates the ease with which People of Color are silenced in the name of colorblind intellectualism. Wallraff creates the character of and spends a year as "Kwami Ogonno," a Somalian asylum seeker complete with brown makeup and an Afro wig. As "Kwami," Wallraff encounters racist situations that would benefit from a concurrent conversation on the Black German experience. Yet no Black German scholars, activists, and/or journalists were interviewed; That omission, coupled with Wallraff's representation of an asylum seeker – an extremely politically charged category – made it impossible to divorce Blackness from foreignness. Furthermore, "Kwami's" costuming and mannerisms were so absurd that the film became less a referendum on white ignorance than a modern-day minstrel show in which a white man's idea of Blackness came to be more important than Black people's testimony about racism in Germany. Wallraff maintained that only he could properly tell the story of racism against Black people, thus illustrating Mills' concept of the double handicap of race. Henning Hoff writes that "[Wallraff] says a black person could not have made the same film because the transformation of his identity from a white German to a black German is an integral part of the story and resonates more powerfully with white audiences" (Hoff 2009). Wallraff told a story of Blackness through the prism of whiteness and implicitly reinforced the assumption of Germanness as inherently white while presenting Afro-Germans as insufficient experts on their own lives.

Although he claimed to have had the idea to go undercover as a Black man long before he knew of similar projects in the United States, Wallraff's film is strikingly similar to the work of John Howard Griffin, a white American journalist who medically darkened his skin to live as a Black man in the Deep South in the 1950s and wrote a book about his experiences, *Black Like Me*. The most significant difference between the two, however, is that Griffin includes the voices of Black people in his work, making Wallraff's exclusion of Black voices nearly 50 years later that much more egregious. Wallraff's refusal to include the experiences of Black Germans in a significant manner suggests a containment of Blackness, where inclusion is based on permission and granted through a restricted number of avenues. By speaking for instead of with Black Germans, Wallraff makes it impossible for Blacks to be identified by anything other than their Blackness. His insistence that his 365-day artificial existence as a Black man was equal to, if not more informed than, a lifetime of Blackness is a mindset common in academia: that years of study are equal to years of lived experience.

Objectifying Otherness

11 June 2009 (Berlin, Germany)

At the conference, I was the only POC on my panel. One woman couldn't understand what I meant when I said famous people in the United States were understood ... [in respect to] their various ethnicities. She [said she] saw them only as famous people. But that is a European thing. They say they don't see color, but color is in every description, every facet of society.

The same white German woman challenged me multiple times, while at the same time frowning, shaking her head, and trying to interject how her understanding of U.S. culture from her studies was significantly different from my recounted lived experience, intimating that I needed to better understand Americans of Color and their representation. Questions struck me from all sides, questions that focused not on my paper but white audience members' beliefs that I was not giving an accurate reading of representations of Americans of Color in the U.S. media.

In academia as in art, an adherence to colorblindness as a driving ideology only feeds the tumor of racial imbalance. Colorblindness allows others to justify racial inequality by pretending that racial antagonism does not influence the everyday lives of People or Color or by ignoring how structural racism creates barriers to entry for racial minorities in myriad sectors of society, including the professorial ranks (Haney Lopez 2011). It opens the door for what George Yancy's describes as *white hubris*, in which whites take it upon themselves "to define the 'reality' of non-whites in categorical terms, stipulations, conditions, and appellations predicated upon white power and privilege" (Yancy 2008: xix). When the *Journal* of *Political Philosophy* published a symposium on the Black Lives Matter movement, they did not include one Philosopher of Color – not even the philosopher who had recently published a book on Black Lives Matter as it related to the philosophical tradition (Lebron 2017). While one's race does *not* determine the quality of her or his scholarship, the absence of in-group members who can deepen and expand existing scholarly conversations perpetuates the exclusion of Others from necessary conversations on race and belonging. These dynamics all but determine that a department will create an echo chamber of similar voices and perspectives while purporting to present students with a diverse education. Equally problematic is the reliance on the invitation of guest speakers who can speak to the intersectionality of race in all manner of disciplines while the department ranks have few, if any, professors who reflect that diversity. Unlike permanent faculty who advise students, direct theses, and shape the conversation on race at a university over the course of many years, guest speakers do not mold or maintain a university's commitment to racial diversity.

Colorblindness gives the Academy an "out" because it allows its members to avoid challenging racial imbalances within its own ranks. In order to transform the white space into what I call a "multicolored space," which would better reflect the multitude of colors and ethnicities in Germany and the United States, race must come to the fore in conversations on hiring practices, scholarship applications, and student admissions. Race matters in discussions on urban planning, medical care, biology, LGBTQ outreach, women's studies, politics, history, media studies, philosophical inquiry, engineering, and myriad other topics under investigation at the university level. As Philomena Essed argues, when discussions of racial and ethnic dominance are eschewed, conversations on diversity are more likely to lead to:

> (1) the objectification of 'otherness,' (2) the institutionalization of the idea that the management of ethnic diversity is the main problem of race and ethnic relations, and (3) the marginalization of elements in society perceived as 'ethnic' (Essed 1994: 244).

For scholars in the humanities and social sciences whose work focuses on the importance of understanding how gendered racializations impact citizens' worldviews, the dangers Essed outlines are important to heed. Colorblindness ultimately does more harm to than good to People of Color, whose racial and ethnic backgrounds are often integral to navigating their professional and social lives. Eduardo Bonilla-Silva argues that colorblindness "otherizes softly," creating opportune moments for covert discrimination and that a "racial grammar" normalizes white domination to the point that it remains invisible to many, if not most, people. Just as we learn the grammar that structures our sentences and patterns of speech, "[...] racial grammar provides the 'deep structure' or the 'logic' and 'rules' for proper composition of racial statements and [...] what can be seen, understood, or even felt about racial matters" (Bonilla-Silva 2012: 2). The university campus is a prime environment to see how racial grammar creates a separate and unequal learning space.

Grammatically Incorrect

Beloved U.S. college traditions established before wide-scale integration necessitated separate minority equivalents. The Black Greek tradition (National Pan-Hellenic Council, often also known as "Pan dash Hell"), Black homecoming, minority collegiate beauty pageants, among other minority-focused traditions, were created – and continue – because of white resistance to minority participation. Universities considered "PWI" – predominantly white institutions – "have a history, demography, curriculum climate, and a set of symbols and traditions that embody, signify, and reproduce whiteness" (Bonilla-Silva 2012: 10).[1] Collegiate Indian mascots, widely considered racist and disrespectful by Native Americans, also belie a university's presentation of itself as a welcoming space for all Students of Color (Black 2002). Inadequate attention to issues of racial exclusion and discrimination have led Students of Color to a national fight for safer campus environments. Protests at the University of Missouri-Columbia (Mizzou) began when graduate student Jonathan Butler, alleging the school had not done enough to protect Students of Color from racially fueled attacks, went on a hunger strike (Butler 2015). Student groups and a large portion of the Division I football team soon followed

1 Often used in comparison to historically black colleges and universities (HBCUs), the term Predominantly White Institutions (PWIs) is a general shorthand for discussing schools with low or non-existent minority student populations. In his article, Bonilla-Silva uses the term HWCUs (historically white colleges and universities).

suit, forcing Tim Wolfe, the university president, to resign; Inspired by Mizzou, students at other schools began protesting at their universities under the hashtag #blackoncampus (Chung/Payne 2015).

While universities have opened their doors to People of Color, their racial grammar does not always cultivate white empathy with marginalized populations. As Bonilla-Silva writes, white empathy for injustices such as voter disenfranchisement, child abduction, or racial violence does not easily extend to People of Color. "[...] [H]uman solidarity, unfortunately, does not happen because these horrid things are not processed by whites as they are for folks of color. In short, these things are, for whites, ungrammatical" (Bonilla-Silva 2012: 11). The conditions that make empathy across racial difference legible, or grammatical, are cultivated when universities focus on race-based policies that recognize and reward racial and ethnic diversity. To recognize the need for diverse conversations without concentrating on how those conversations will be fostered, enacts violence against marginalized communities. Such was the case at the University of Bremen, where the research group "Black Knowledges" – consisting solely of white scholars – applied for a grant to implement a Black Studies program, which 75 Black German organizations and community members from Germany and Austria, as well as allies worldwide, condemned in an open letter. The envisioned program was comprised entirely of white professors, postdocs, and graduate students. Those who signed the open letter argued that the grant applicants implied that Black Studies was not a serious avenue of study because none of the envisioned program's personnel had experience teaching Black Studies. They also criticized the lack of input from Black German scholars in designing the program and grant applicants' decision to name Black scholars and activists as "current" or "prospective" partners without their permission, which the petitioners perceived as particularly frustrating given what they noted as a tendency of Black Germans to leave the Academy altogether because of an inability to find teaching and research positions. They wrote:

> [...] the current hiring practice amounts to little more than thinly disguised affirmative action for white academics, while German academia continues to systematically exclude Black scholars and other scholars of color due to the absence of legal mechanisms in Germany to ensure that underrepresented and marginalized groups participate in the life of universities as students and faculty [...] Black Germans can serve as the 'raw resource' or 'native informants' for white academics but are not permitted to act as scholars in their own right (Community Statement 2015: 4).

The "Black Knowledges" research group at the University of Bremen assumed their learned knowledge as white people was more valuable than the lived and learned knowledge of Black German citizens, residents, activists, and researchers. In staffing what would have been one of the first major black studies initiatives in Germany entirely with white people, the grant applicants attempted to maintain an academic white space.[2] The colorblind racism they enacted was just as violent as an encounter in which racial animus was clearly communicated.

Conclusion

30 Sept 2013 (Berlin, Germany)

A woman looked at me after I left the bathroom stall, [went in and] walked out, stood next to me at the sink, wet some toilet paper, walked back into the stall and washed the (clean) seat off.

The contemporary moment is one in which racism, as U.S. professional basketball player LeBron James noted after his house was vandalized with racial epithets, can no longer reliably be expected to be accompanied by men in hoods screaming angry words. Said James, "[...] [W]e know people hide their faces and will say things about you – when they see you, they smile at your face" (Davis 2017). Those things they say, however, may not always sound racially antagonistic. They may be gentle words and soft smiles of welcome that conceal an unwillingness to acknowledge feelings of racial animosity or to admit that race matters in every way, every day, for People of Color. Colorblindness creates what Bonilla-Silva calls "racism without racists," whereby whites express racial tolerance and an abhorrence of violent racist acts, but maintain that minorities retain characteristics that make them unsuitable citizens such as an unwillingness to assimilate to "normal" (read: white) cultural practices; inferior linguistic or intellectual ability; poor child-rearing skills; and/or laziness (Bonilla-Silva 2014). Often people assume that they are too decent to be racist, allowing themselves a wide berth to criticize or condemn Others because they are good moral actors in other respects. Moral licensing allows one to justify racially biased statements or actions because of existing charity work, friendly interactions with minority colleagues, or other deeds that illustrate one's

2 The grant application was denied and the "Black Knowledges" research group ultimately disbanded as a result of the criticism their attempt received.

assumed goodness. Moral licensing is particularly acute in academia where intellectual interest can often be confused with activism.

There often is an implicit assumption that Scholars of Color insert race in their work in order to subvert the canon in an exercise of intellectual whimsy or navel gazing. That assumption is both a form of silencing and a means to keep whiteness unnamed and at the center of intellectual discussion. If race talk is acceptable only in acts of clear racial violence, it becomes impossible to discuss everyday racism [*Alltagsrassismus*]. The indignities that happen in moments when no one is looking – being mistaken for staff instead of an invited guest or being arrested for trying to access one's own home – are those that explicate the need for societal change and intellectual discussion as much as a racially motivated attack in a public space.[3] "[...] [S]lights that may appear minor or inconsequential [...] have a debilitating effect when aggregated," write Griffin et al. (2011: 516). The university has a unique opportunity to lead the way in helping the academic and local community understand how avoiding race perpetuates the mistreatment of racialized Others. Academics have the opportunity to leave behind the assumption that their knowledge of racial hierarchies of power and privilege means that they are immune to implicit racial bias.

Scholars of Color in the United States and Germany are more likely to work in institutional white spaces, and – depending on the location of their university or research site – live in white spaces that "reinforce a normative sensibility in settings in which black people are typically absent, not expected, or marginalized when present" (Anderson 2015: 10). White spaces do not simply impact one's psyche. They also can have tangible effects on one's ability to succeed professionally, leading Scholars of Color to isolate themselves from colleagues, put pressure on themselves to publish prolifically so as to avoid scrutiny, or leave the Academy altogether (Griffin et al. 2011). On its own, a white space does not necessitate racial disharmony or discrimination. Yet a university that does not boast ethnic and racial diversity cannot be considered a space that accurately reflects a 21st century sensibility. Speaking about diversity is not the same as creating enduring opportunities for people to speak. To ignore the impact race has on one's ability to easily and successfully navigate interpersonal and professional relationships is to cede both the university's function as a center for scientific inquiry and its role as a leader in political and social thought. Uni-

3 For example, Black American Harvard University professor Henry Louis Gates, Jr. was arrested on suspicion of trying to break into his own home in 2009 (Seelye 2009), and businesswoman Melody Hobson was mistaken as kitchen help for an event she organized in 2014 (Matthews 2014).

versities must work more closely with scholar-activists and students from underrepresented groups and encourage all professors to create forward-thinking curricula with the aim of transforming the academic white space into a multicolored space that encourages dialogue across the boundaries of racial difference.

Works Cited

Anderson, Elijah (2015): "The White Space." In: Sociology of Race and Ethnicity 1/1, pp. 10-21.

Black, Jonathan (2002): "The Mascotting of Native America." In: The American Indian Quarterly 26/4, pp. 605-622.

Bonilla-Silva, Eduardo (2012): "The Invisible Weight of Whiteness: The Racial Grammar of Everyday Life in America." In: Michigan Sociological Review 26/1, pp. 1-15.

Bonilla-Silva, Eduardo (2014): Racism Without Racists: Colorblind Racism and the Resistance of Racial Inequality in America. 4th edition, New York: Rowman and Littlefield.

Butler, Jonathan (2015): "Letter to the Editor: Dear Chancellor Loftin: It isn't enough." In: The Maneater October 14 (http://www.themaneater.com/stories/2015/10/14/dear-chancellor-loftin-it-isnt-enough/).

Chung, Maya/Payne, Amber (2015): "Inspired by Mizzou Protests, Students Across Country Focus on Being #BlackonCampus." In: NBC News November 11 (http://www.nbcnews.com/news/nbcblk/inspired-mizzou-protests-ithaca-students-rally-demand-resignation-n461646).

Collins, Sam P. K. (2015). "Black People aren't Making Things up: The Science behind 'Racial Battle Fatigue'." In: Think Progress June 19 (https://thinkprogress.org/black-people-arent-making-things-up-the-science-behind-racial-battle-fatigue-9726fcebc938/).

"Community Statement: 'Black' Studies at the University of Bremen." January, 2015 (http://www.fb10.uni-bremen.de/inputs/pdf/CommunitystatJanuaryackStudiesBremen_engl_Undersgnd415.pdf).

Davis, Scott (2017): "LeBron James Issues Powerful Response to His House Being Vandalized with Racist Graffiti: 'Racism will always be a part of the world, a part of America.'" In: Business Insider May 31 (http://www.businessinsider.com/lebron-james-house-racist-graffiti-powerful-response-2017-5).

Elam, Michelle (2011): The Souls of Mixed Folk: Race, Politics, and Aesthetics in the New Millennium, Stanford: Stanford University Press.

El-Tayeb, Fatima (2011): European Others. Queering Ethnicity in Postnational Europe, Minneapolis: University of Minnesota Press.

Essed, Philomena (1994): "Making and Breaking Ethnic Boundaries: Women's Studies, Diversity, and Racism." In: Women's Studies Quarterly 22/¾, pp. 232-249.

Fanon, Frantz (2008[1952]): Black Skin, White Masks. Richard Philox (trans.), New York: Grove Press.

"German Play, 'I'm Not Rappaport' Angers Critics with Blackface Actor." In: Huffington Post January 10, 2012 (http://www.huffingtonpost.com/2012/01/10/german-play-im-not-a-rapp_n_1197160.html).

Griffin, Kimberly A./Pifer, Meghan J./Humphrey, Jordan R./Hazelwood, Ashley M. (2011): "(Re)Defining Departure: Exploring Black Professors' Experiences with and Responses to Racism and Racial Climate." In: American Journal of Education 117/ 4, pp. 495-526.

Ha, Kien Nghi (2016): "Weiße Parallelgesellschaft oder wie rassistisch ist die Universität?" In: Migazin May 10 (http://www.migazin.de/2016/05/10/weisse-parallelgesellschaft-oder-wie-rassistisch-ist-die-universitaet/).

Haney Lopez, Ian F. (2010): "Is the 'Post' in Post-Racial the 'Blind' in Colorblind?" In: Cardozo Law Review 32/3, pp. 807-831.

Hoff, Henning (2009): "Blackface Filmmaker Sparks A Race Debate in Germany." In: Time November 18 (http://content.time.com/time/world/article/0,8599,1940290,00.html).

Johnsrud, Linda K./Des Jarlais, Christine D. (2010). "Barriers to Tenure for Women and Minorities." In: The Review of Higher Education, 17/4, pp. 335-353.

Keith, Tamara (2015): "Hillary Clinton's 3-Word Misstep: 'All Lives Matter.'" In: NPR June 24 (http://www.npr.org/sections/itsallpolitics/2015/06/24/417112956/hillary-clintons-three-word-gaffe-all-lives-matter).

Lebron, Chris (2017): "An Open Letter to the Editors of the Journal of Political Philosophy; or How Black Scholarship Matters, Too." In: Politicalphilosopher May 24 (https://politicalphilosopher.net/2017/05/24/an-open-letter-to-the-editors-of-the-journal-of-political-philosophy-or-how-black-scholarship-matters-too/).

Matthews, Cate (2014). "Being Mistaken For Kitchen Help Just Made This Exec More Outspoken." In: Huffington Post May 23 (http://www.huffingtonpost.com/2014/05/23/mellody-hobson-color-blind-color-brave-ted-talk_n_5380095.html/).

Mills, Charles W. (2015): "Global White Ignorance." In: Matthias Grass/Linsey McGoey (eds.), Routledge International Handbook of Ignorance Studies, London and New York: Routledge, pp. 217-227.

"Outrage at German Comedian's Blackface Obama Billboard." In: Daily Mail September 17, 2011 (http://www.dailymail.co.uk/news/article-2038301/German-comedian-Martin-Sonneborns-blackface-Obama-billboard-causes-outrage.html/).

Pittman, Chavella T. (2010). "Race and Gender Oppression in the Classroom: The Experiences of Woman Faculty of Color with White Male Students." In: Teaching Sociology 38/3, pp. 183-196.

Rothman, Lily (2012): "Blackface 'Clybourne Park' Production Cancelled in Berlin." In: Time October 18 (http://entertainment.time.com/2012/10/18/blackface-clybourne-park-production-cancelled-in-berlin/).

Seelye, Katharine Q (2009). "Obama Wades Into a Volatile Racial Issue." In: New York Times July 23 (http://www.nytimes.com/2009/07/23/us/23race.html?_r=1&hp>June8,2017/).

Smith, William A./Yosso, Tara J./Solórzano, Daniel G. (2006): "Challenging Racial Battle Fatigue on Historically White Campuses: A Critical Race Examination of Race-Related Stress." In: C.A. Stanley (eds.), Faculty of Color Teaching in Predominantly White Colleges and Universities, Bolton and Massachusetts: Anker Publishing, pp. 299-327.

Sow, Noah (2009): Deutschland Schwarz Weiss: Der alltägliche Rassismus, München: Goldmann.

Sundstrom, Ronald R. (2001). "Being and Being Mixed Race." In: Social Theory and Practice 27/2, pp. 285-307.

Squires, Catherine R. (2014): The Post-Racial Mystique, New York: New York University Press.

Staples, Robert (1984): "Racial Ideology and Intellectual Racism: Blacks in Academia." In: The Black Scholar 15/2, pp. 2-17.

Weheliye, Alexander G. (2009). "My Volk to Come: Peoplehood in Recent Diaspora Discourse and Afro-German Popular Music." In: Darlene Clark Hine/Tricia Danielle Keaton/Stephen Small (eds.), Black Europe and the African Diaspora: University of Illinois Press, pp. 161-179.

Wright, Michelle M. (2003): "Others-from-within from Without: Afro-German Subject Formation and the Challenge of a Counter-Discourse." In: Callaloo 26/2, pp. 296-305.

Yancy, George (2008): Black Bodies, White Gazes: The Continuing Significance of Race, New York: Rowman and Littlefield.

Zack, Naomi (1994). Race and Mixed Race, Philadelphia: Temple University Press.

Transatlantic Postcolonial (T)Races in the Classroom: From Defoe's Desert Island to Larsen's *Quicksand* and *Black-ish* Suburbia

Elahe Haschemi Yekani

Countering an understanding of race as an essentialist constant is one of the most pressing tasks in the university classroom. Classrooms that in Germany, too, have never been homogenous, but too often relegate racialized markers of difference to other geographical settings. Fatima El-Tayeb, for instance, speaks of a particular European "'invisible' racialization" that relies on "visual markers that construct nonwhiteness as non-Europeanness" and at the same time suppose "a discourse of colorblindness that claims not to 'see' racialized difference" (2011: xxiv). Thus, "race is not mentioned yet referenced implicitly as a marker of not belonging, a strategy that relies on a shared iconography that remains unspoken" (El-Tayeb 2011: xxvi). As someone who teaches postcolonial literatures at a German university, I hope to promote an understanding of how race came and comes to matter and acquires different meanings historically and geographically, not only in the global Anglosphere but also in the German here and now. In the following, I want to trace race not as a given category of difference, but as a discursive and aesthetic formation that we need to reconstruct from its specific times and places of emergence. For this purpose, I will discuss Anglophone examples, but want to stress that such a contextual understanding of race also implies addressing racialized hierarchies closer to home as this collection as a whole suggests. I hope that the critical insights from such a historicizing perspective on how we might continue to use race as an analytical category in English and American Studies might become apparent by charting quite diverse territory in this paper, from a longer reading of Daniel Defoe's *Robinson Crusoe*, to two shorter vignettes on Nella Larsen's *Quicksand,* and finally to TV's *Black-ish* suburbia.

Introduction – A Long History of Postcolonial Studies, or: How Race Comes to Matter across Time and Place

Before turning to the specific readings, I want to elaborate on how a long transatlantic history of the framework of postcolonial studies might benefit the suggested contextualizing of race. Following Edward Said's focus on "imaginative geography" in *Orientalism* (2003 [1978]), postcolonial studies for a long time has been concerned with the national orderings and re-orderings that resulted from colonial contact, the high imperialism of the 19th century and postcolonial nation formation in the 20th century. More recently, however, by challenging the temporal logic of *before* and *after* colonialism as well as of Western teleological progress narratives, postcolonial studies is increasingly rejecting such linear timelines and one might provocatively ask whether we have ever been non-postcolonial (cf. Ganguly 2004, McClintock 1995). Medievalists like Jeffrey Cohen (2000) and Geraldine Heng (2011a; 2011b; 2015) speak of the postcolonial Middle Ages, 18th century scholars like Daniel Carey and Lynn Festa (2009) examine the European enlightenment as always already in entanglement with its assumed Others. This is a perspective I also advocate in my forthcoming book, in which I describe an entangled literary history of the rise of the British novel with the earliest written testimonies of Black Transatlantic writers. This framework rests on a threefold argument on how readings informed by a postcolonial epistemology of race can work as an intersectional method in literary and cultural studies that would also benefit a critical pedagogy of race.

First, as stated, there is a need to extend time frames to move away from a presentist and linear understanding of the postcolonial. Second, for this purpose we can build on a methodological strength of postcolonial studies, namely, the de-essentializing of categories of difference – and I will align this approach with the research paradigm of intersectionality at the close of the paper. Third and finally, it is specifically a comparative transnational perspective that can abet such a project. My motivation is decidedly not guided by the assumption that race is the only or most central category of postcolonial enquiry but rather, in line with critics like Paul Gilroy (2001) and Robert Stam and Ella Shohat (2012), I believe that a diachronic and transatlantic interrogation of race can illustrate the benefits of a long history of the postcolonial particularly well. This also motivates the choice to analyze three very different racial topographies as well as genres: from the encounters of so-called "proto-racial" multiple differences in Daniel Defoe's 1719 *Robinson Crusoe*, to the ossification of the color line in Nella Larsen's 1928 novel *Quicksand*, to the self-reflexive post-racial suburban humor of the contemporary US TV sitcom *Black-ish*. Clearly, these readings can therefore not be fleshed out in too much detail in this paper, but nevertheless hopefully show that it is precisely the multiplicity of genealogies of how race comes to matter differently across time and space that accounts for a truly postcolonial epistemology of thinking race intersectionally. In Germany, the term *Rasse,* because of its associations with Nazism and eugenics, is tabooed and ethnicity often becomes its euphemistic placeholder. Hence, while we must obviously caution against essentialist and biologist discourses, racializing markers impact social stratification in Germany to a large degree, and it is fundamental to understand how inequalities are based on processes of physical and/or cultural differentiation. Teaching race as a relational and analytical concept in English and American studies can thus also inform ways of making sense of race in Germany both in a historical and contemporary perspective (for further explorations of these issues, see Alexopoulou, Obute, and Singletary in this volume).

Daniel Defoe: *Robinson Crusoe* (1719)

Ever since Watt's seminal 1957 *The Rise of the Novel*, *Robinson Crusoe*[1] has been widely understood as the first depiction of a modern economic subject that partakes in an increasingly imperial capitalist order which becomes foundational of both a national understanding of Englishness and the modern English novel itself. Not being content with the "middle station of life" (Defoe 2003: 6) and due to his limited prospects as the third son, Robinson Crusoe sets off to his well-known ill-begotten adventures. Fleeing from parental expectations, it is specifically travelling and transcultural contact that leads to the transformation of Robinson Kreutznaer, whose father is a German immigrant from Bremen after all, to become the English entrepreneur Robinson Crusoe.

Crusoe's self-reflexive musings mainly revolve around his "original sin" (Defoe 2003: 154) of disobeying his father and around religious introspection on

1 The title reads in full: *The Life and Strange Surprizing Adventures of Robinson Crusoe, Of York, Mariner: Who lived Eight and Twenty Years, all alone in an un-inhabited Island on the Coast of America, near the Mouth of the Great River of Oroonoque; Having been cast on Shore by Shipwreck, wherein all the Men perished but himself. With an Account how he was at last as strangely deliver'd by Pyrates. Written by Himself.*

providence.[2] Nevertheless, the text is also appealing because of its adventurous encounters with Otherness. Thus, the extent to which this 18th century imagination of 17th century[3] colonial Others translates into a language of race, has come under scrutiny lately. I follow Roxann Wheeler's findings in her comprehensive study, *The Complexion of Race,* where she speaks of "residual proto-racial ideologies" (2000: 9) that are articulated through and with other contemporary markers of difference that include Christianity, civility, and rank (cf. Wheeler 2000: 7). This in turn helps account for the, again in Wheeler's terms, "situated multiplicity" (2000: 45) of race at the time. Accordingly, while I use the term 'race'[4] in relation to an 18th century source, I do so in what Jeffrey

---

2 Helga Schwalm calls providence and deliverance the leitmotifs of the Puritan spiritual autobiography to which *Robinson Crusoe* generically is indebted (cf. 2007: 240-241).

3 This is the common reconstruction of the timeline: Crusoe is born in 1632. The plot starts when he is 18 in 1650. One year later in 1651, he embarks on his first journey; He is shipwrecked in 1659 when he is 27 (in the text it says 26 which does not add up). After 28 years on the island, he leaves in December 1686, and after travelling again to Lisbon to sell his Brazilian plantation, he returns to England in 1687 after 35 years of absence. Crusoe finally gets married and has three children, but returns to travelling to the East Indies in 1694 (at the age of 62) when his wife (conveniently) dies, giving him the opportunity to return to his "colony" (RC 240). This is a foreshadowing of the plot of the second part, *The Farther Adventures of Robinson Crusoe; Being the Second and Last Part of his Life, and of the Strange Surprizing Accounts of his Travels Round Three Parts of the Globe. Written by Himself,* which was published immediately following the first instalment in 1719. This supposedly last part of Crusoe's tale, which ends in January 1705 with Crusoe's retirement at the age of 72, is then followed by yet a third and final book which was published in 1720, called *Serious Reflections during the Life and Surprising Adventures of Robinson Crusoe: with his Vision of the Angelick World. Written by Himself.*

4 Cf. Wheeler's helpful explanations on the term 'race' in 18th century usage: "Until the very end of the century, variety, not race, was the scientific term of choice to designate different groups of people. […] In its most common usage, *race* simply meant a group. […] Conventionally, race meant family lineage, and it could apply generally to 'the race of man' (as distinct from animals); to a subgroup of people, such as the Irish race; or even to nonhuman objects, such as the vegetable race. Unlike today in Britain or the United States, race was not primarily a characteristic of minority populations. During the late eighteenth century, the word *race* was used by some writers in recognizably incipient forms of its modern sense – denoting a fairly rigid separation among groups. At this time, skin color was the most typical way to differentiate 'races'" (2000: 31).

Cohen calls a deliberately "untimely" (2000: 2) manner. Race is thus deployed as a heuristic category, precisely to challenge those postcolonial readings that all too quickly read racialized literary characters such as Friday (but also Othello and Oroonoko) within a framework of scientific racism and the biologized black and white binary.[5] Bearing in mind such a heuristic understanding of race as emerging in a web of multiple meanings, we can, following Defoe's plot, I believe assert that English national identity is gradually racialized as white in *Robinson Crusoe* through manifold demarcations: from Muslimness in North Africa and the Blackness of the West Africans, to the nudity of the natives, and the 'barbarity' of the cannibals in the Caribbean. But, Englishness is also consolidated as benevolent against the so-called "'black legend' of Spanish cruelty" in inner-European national colonial rivalries (Boulukos 2008: 14; cf. Defoe 2003: 136). Hence, while race is not considered a biological given yet, that does not amount to the absence of racializing color-codings.

What is more, in the early 18th century, not only race, but also slavery[6] is still a category of multiplicity. George Boulukos explains that "[u]ntil at least the mid-eighteenth century the terms 'slave' and 'servant' could still be used interchangeably for English indentured servants and African slaves in metropolitan discourse, and likely also in the colonies" (2008: 119). Being himself enslaved by Corsairs in Morocco on his third voyage,[7] Crusoe spends two years of isolation in North Africa. Eventually Crusoe meets the young boy Xury. In contrast to the Moroccan Muslims, who are referred to as "Moors," Xury is labelled a "Maresco" (Defoe 2003: 18), which denotes Spanish, that is European, Muslims at the time. The Spanish were perceived, as mentioned, if not as Black, then definitely as Black*er* than the English. While initially sharing

5 Carey (2009), for instance, criticises Hulme in this regard and emphasises a broader spectrum of servitude in the 18th century (cf. also Boulukos 2008: 76-77).

6 Cf. Hartman's elaborations on the history of slavery: "The very term 'slavery' derived from the word 'Slav,' because Eastern Europeans were the slaves of the medieval world. At the beginning of modernity, slavery declined in Europe as it expanded in Africa, although as late as the seventeenth and eighteenth centuries, it was still possible to purchase 'white' slaves – English, Spanish, and Portuguese captives in the Mediterranean ports of North Africa. [...] It was not until the sixteenth and seventeenth centuries that the line between the slave and the free separated Africans and Europeans and hardened into a color line" (Hartman 2008: 5).

7 Hence, in many ways, like in William Shakespeare's *The Tempest* (1610-11), it is in fact the Mediterranean rather than the Atlantic that is the first location of encounters with slavery and Otherness in the novel. Thus, the 'Old' and 'New World' are symbolically linked in Shakespeare and Defoe.

the same fate, Crusoe can finally escape captivity and takes the young boy with him on condition that he serve him from then on (cf. Defoe 2003: 20-21). Despite being European, Xury's non-Christianness makes him a legitimate item of exchange in Crusoe's capitalist ventures in which religion, nationality, and racializing discourses are entangled. He sells Xury to the Portuguese Captain for 60 Pieces of Eight, 20 less than he gets for his boat. But, there is a marked temporal difference to chattel slavery, as Xury's servitude is finite – should he convert to Christianity after ten years.[8] Crusoe, despite being implicated in the already quite institutionalized transatlantic trade,[9] always highlights the singularity of his endeavors which is of course also a literary strategy of establishing individuality. In the logic of the narrative, this is not an established global framework of seafaring, trade, and slavery, but rather, Crusoe's unique (life) journey in finding his identity (and eventually repenting his filial sins). In this way, the novel itself narrates the co-existence of forms of race-independent indentured servitude and the rise of the transatlantic plantation economy which increasingly racializes slavery. Free again, Crusoe is subsequently taken to Brazil by the Captain and immediately invests in a sugar plantation actively seeking African enslaved labor for himself. Finally, shipwrecked on the supposedly uninhabited Caribbean island on the mission to buy more slaves, Crusoe lives self-sufficiently for more than 25 years before meeting another human being again.

Upon eventually saving a man native to the islands from the cannibals who visit from the neighboring shore for their gruesome rituals, the man supposedly willingly submits to Crusoe (cf. Aravamudan 1999: 74): He "laid his head upon the ground, and taking me by the foot, set my foot upon his head; this, it seems was in token for swearing to be my slave for ever" (Defoe 2003: 161). Friday is

8 In this passage, Xury himself seems happy to consent to this transaction: "[H]e would give the boy an obligation to set him free in ten years, if he turn'd Christian: upon this, and *Xury* saying he was willing to go to him, I let the captain have him" (Defoe 2003: 29). I will come back to such ostensibly non-realist elements of the story that can be read as wish-fulfilment and that are repeated in Friday's entirely consensual subjugation.

9 In fact, the plot of *Robinson Crusoe* coincides with the massive European colonial expansion of the British, Dutch and French during the mid-17th century sugar boom in the Caribbean which led to these 'new' colonial powers increasingly supplanting Spanish and Portuguese dominance in the transatlantic trade in people and goods (cf. Barrett 2014: 22-26). It is therefore also no coincidence that the economic rivals from (Catholic) Southern Europe are delineated in the mentioned colour-coded derogative language at the time.

quickly turned into an object of both subjugation and instruction in the familiar modes of Christianity and enlightened education. In stark contrast to Xury's temporally limited services, it also appears evident that Friday is permanently bound to him. Initially, not sharing a language, it is the body of the native that is read as communicating submission unambiguously. The initially quoted ritualistic subjection of bowing down in front of the Englishman is repeated once more in the text:

> At last he lays his head flat upon the ground, close to my foot, and sets my other foot upon his head, as he had done before; and after this made all the signs to me of subjection, servitude, and submission imaginable, to let me know how he would serve me so long as he liv'd; I understood him in many things, and let him know, I was very well pleas'd with him; In a little time I began to speak to him, and teach him to speak to me; and first, I let him know his name should be *Friday*, which was the day I sav'd his life; I call'd him so for the memory of the time. I likewise taught him to say *Master*, and then let him know that was to be my name […] (Defoe 2003: 162-163).

In the hyperbolic alliteration of "subjection, servitude, and submission" Crusoe reads into the described gesture indefinite service – despite the still common 18th century system of temporary servitude that he himself had experienced earlier. While the second scene of subjection describes the adoption of the title "Master" by Crusoe and the seemingly random naming of the Carib by the day of the week that marks their encounter, it is in the earlier scene that Friday is called a "slave" explicitly for the first and only time in the novel. Mostly he is referred to as "my man" and a "servant" (Defoe 2003: 220) and there is now a debate between 18th century and postcolonial scholars about the status of Friday's subjugation in relation to the master-slave dialectic. Before going into more detail, it is instructive to revisit the famous description of Friday's appearance, which I will quote at some length because of its significance for an understanding of the mentioned emergence of racialized difference:

> He was a comely, handsome fellow, perfectly well made, with straight strong limbs, not too large; tall and well shap'd, and, as I reckon, about twenty six years of age. He had a very good countenance, not a fierce and surly aspect, but seem'd to have something very manly in his face, and yet he had all the sweetness and softness of an *European* in his countenance too, especially when he smil'd. His hair was long and black, not curl'd like wool; his forehead very high, and large, and a great vivacity and sparkling sharpness in his eyes. The colour of his skin was not quite black, but very tawny; and yet not of an ugly yellow nauseous tawny, as the *Brasilians*, and *Virginians*, and other natives of *America*

are; but of a bright kind of a dun olive colour, that had in it something very agreeable, tho' not very easy to describe. His face was round and plump; his nose small, not flat, like the Negroes, a very good mouth, thin lips, and his fine teeth well set, and as white as ivory (Defoe 2003: 162).

This well-known passage, in essence, emphatically states that Friday is different in every possible way from a racialized derogative understanding of Blackness that is reserved for the term "Negroe": He has straight hair, lighter olive-colored skin, and a small nose; and the fact that the opposite attributes are considered to be negative, attests to emerging racist classifications, which increasingly frame Blackness as aesthetically displeasing and intellectually inferior.[10] Like Aphra Behn's Oroonoko, his features are even favorably compared to those of a European. In the Americas, there is a spectrum of non-whiteness including the "tawny" complexion of the "*Brasilians*, and *Virginians*, and other natives of *America*." This also shows that slavery is not yet linked to an idea of hereditary chattel slavery based on race as a fixed category. Both Friday's skin color, which is "not very easy to describe," and his masculinity, wavering between manliness and softness, cannot be classified in a straightforward binary manner thus, linking racial ambiguity to a certain degree of gender trouble. Nonetheless, there is a clear hierarchy from darker to lighter complexions.

Interestingly, Crusoe, in turn, is described as "white"[11] mostly from Friday's point of view: Friday reports of the Spanish who are "white bearded men" like Crusoe:

10 Despite the explicit characterization in the text, there is a persistent "Africanization" (Wheeler 1995: 847) of Friday in the cultural imaginary. This trend shapes contemporary film adaptations of *Robinson Crusoe* to this day (while films like *Man Friday*, 1975, directed by Jack Gold, or *Crusoe*, 1988, directed by Caleb Deschanel, to varying degrees challenge the politics of race of the original, they visually contrast a white Crusoe with Black masculinity; for an explicit focus on filmic *Robinsonades*, cf. Mayer 2002). But this tendency can already be witnessed in the 18th century with visual representations of Friday in book illustrations, for example, which either follow said Africanization, or, alternatively, resort to images closer to the myth of the noble savage. This visual ambiguity of Friday also points to the complicated colonial constellation of the diminishing indigenous and the quickly growing African enslaved populations in the Caribbean.

11 The term "white" was used mostly in the colonies to describe all Europeans, as does Friday. Wheeler argues that the British at that time did not consider themselves a "white people" rather, "they believed themselves to be Christians or denizens of a civil society who possessed a white complexion" (Wheeler 2000: 272).

> He told me, that [...] west from their country, there dwelt white bearded men, like me, and pointed to my great whiskers, which I mention'd before; and that they had kill'd *much mans*, that was his word: by all which I understood, he meant the *Spaniards*, whose cruelties in *America* had been spread over the whole countries, and were remember'd by all the nations from father to son (Defoe 2003: 170).

Lighter skin color and beardedness then are considered signs of Europeanness and uncontested masculinity. Friday constructs a similarity between Crusoe and the Spanish on account of their looks, but Crusoe immediately distinguishes his English identity again from the cruel "Black" Spanish. So, while to a certain degree, Friday's views are incorporated into the narrative, they are almost always conveyed through Crusoe's speech, except for the supposedly amusing emphasis on the faulty "much mans," which "was his word," and is quickly corrected by his 'Master.' Whereas the Spanish and the English could indeed be conflated as looking the same from a non-European point of view, the discursive construction of a racially unifying white identity is undermined by an emphasis on national distinction and Spanish 'barbarity.' In other words, color-codings of Friday as darker and naturally inferior to the lighter Englishman are considered a given in the logic of the text, but race is not yet ossified, and it seems that throughout the story, Crusoe's nationality and his religion are the most important identity forming elements, especially in relation to other Europeans. Hence, the debate about Friday's status as a "grateful slave" (cf. Boulukos 2008) in *Robinson Crusoe* I believe needs to be sutured to reflections on racial multiplicity. On a metatextual level, one can also link the master-slave debate to the notion of dialogicity and how Crusoe's relationship to Friday affects his self-understanding. Critical opinion, as stated, is divided: while postcolonial scholars like Peter Hulme stress the muting of the native voice and Friday's status as enslaved,[12] others, like John Richetti and Daniel Carey (cf. 2009: 121), highlight reciprocity. Richetti sees the modernity of the novel specifically in Crusoe's lengthy reflections on cannibalism which he reads as "pure dialogism" (2000: 344) in the Bakhtinian sense. Following Mikhail Bakhtin's influential framing of novelization (1994), the novel, in contrast to the monologic world view of the epic, is polyphonic. Different voices, for example, the speech of the narrator and the characters, are in a dialogic relationship, embodying different world views in different languages. In this logic, the foundational tone of modernity that is

12 Contemporary postcolonial rewritings, like J.M. Coetzee's *Foe* (1986) and Derek Walcott's *Pantomime* (1980), begin from a similar critique of the Friday character and offer creative re-arrangements of the power dynamics in the story.

expressed in novelistic discourse then is a subject in dialogue with itself. And indeed, this can be related directly to Defoe's depiction of Robinson Crusoe.

As explained, Crusoe has to cope with an increasingly challenged understanding of his cultural identity. In the process of travelling, this English subject recognizes the tension between similarity and difference, which can also quickly change as in his relation to Xury that is first understood as similar and then hierarchically structured as different (not only through religion, but also age). Even before the mentioned first encounters with Friday, it is the often-discussed and for Richetti central passage on cannibalism that exemplifies Crusoe's capacity for dialogue. While at first it is outright disgust that he feels when he speculates about the anthropophagic natives motivating his initial plan to attack the invaders (cf. Defoe 2003: 131), Crusoe slowly interrogates his own truth and becomes much more relative in his views on cannibalism (cf. Defoe 2003: 135). He comes to the realization that he cannot adopt the position of the judge of these men if God himself does not punish them, and apparently in their moral universe, a form of cannibalism is permissible. Following this introspection, Crusoe, Richetti argues,

> not only thinks but dramatizes the conditions of thought, narrates the function of thinking within his personal development, and defines himself as a mind making its way through a series of positions, each of which has a distinct validity and personal rightness for him at different points in his experience. Or in Bakhtin's terms we might want to say that he locates his personality at the intersection of competing explanations, rational, emotional, historical, political, of cannibalism, with his own personal situation as the lone European inhabitant of the island (2000: 341).

From this he concludes that "Crusoe's reflections are cross-cultural, for him a revelation of tolerant relativity" (Richetti 2000: 342). While I share Richetti's classification of Crusoe as self-reflexive about his status as a European and Christian in a foreign setting, I would draw a slightly different conclusion. His self-interrogation does not necessarily amount to "tolerant relativity" since his morally superior position is never really challenged; he simply learns humility in the sense that he relegates the authority to judge the cannibals to *his* God. He accepts that it is not for him to judge in relation to the higher divine power. Later, with regard to *his* subjects on *his* island, he is more than happy to assume the position of the uncontested monarch and ruler. Richetti, in fact, too, admits that in the novel we find a "dialogue with himself" about how to come to terms with cultural alterity (2000: 344). Literally, Crusoe "debates with himself." Moreover, shortly after his display of moral tolerance towards the cannibals,

Crusoe immediately contrasts his insight once more with the backward "conduct of the *Spaniards* in all their barbarities practis'd in *America*" (Defoe 2003: 136). The abhorrence against these barbarous Europeans, as which Protestant Crusoe marks the Catholics recurrently, is also clearly then a new form of inner-European and Christian distinction, that differentiates in color-coded language the 'whiter' Europeans of the North from the 'Blacker' ones in the South. In Defoe's writing we can identify a new form of foundational self-reflexivity that challenges the position of an individual in relation to his God *and* that faces cultural alterity. However, this cultural alterity for the most part functions as a form of gratuitous obstacle that Crusoe, clearly an adventurer more than an ethnographer, simply needs to bring under his control. Therefore, in Crusoe's relationship to Friday, true dialogicity is more difficult to assert: as mentioned, Crusoe uses language mainly to teach Friday to obey his orders, and except for some short direct quotes of Friday's characteristic, faulty English interspersed in Crusoe's summaries of Friday's actions, there is very little direct speech, except in a central dialogue again on the role of cannibalism (cf. Defoe 2003: 169). As throughout the novel, there is never autonomy of Friday's speech, it cannot stand for itself or be regarded as communicating meaning without the focalizer and interlocutor Crusoe explaining what Friday means. This finally brings me back to the critical debate about Friday's contested status as a slave. Carey stresses Friday's agency in voluntarily submitting to Crusoe and urges critics to grant predominance to the text itself. But, as there is no internal focalization, Carey, I would argue, to a certain degree here falls prey to the solipsistic perspective of the text, limiting epistemic authority to Crusoe's interpretation of Friday, who is reduced to a mirroring function and what Hulme calls the fiction of voluntary servitude:

> Friday is certainly a slave inasmuch as he has no will of his own; and Crusoe, unwilling as he may be ever to call Friday 'slave,' has no qualms about adopting the other half of the dialectic [...]. Yet within the fiction the term 'slave' can be avoided because Friday's servitude is voluntary, not forced (Hulme 1992: 205).

Hulme convincingly links the text to the unrealistic wish-fulfilment of the romance rather than the realism of the truly modern novel thus emphasizing the hybridity of *Robinson Crusoe* as a text still wavering between older and newer forms of narrative fiction.

Crusoe might be the first psychological hero of the novel, but he is not yet truly a social subject, as Helga Schwalm has argued as well.[13] Apart from his inner struggles, there is no challenging interpersonal interaction with Friday or even his own family, which is completely marginal in this tale of literally insular male individualism. Interrogating 18th century literature, a postcolonial reading should not hastily gloss over the historical specificities of race and slavery, but it also cannot stop at the literal to make sense of the mutual dependence in the master-slave dialectic. With this historically further removed example in mind, I now turn to an early 20th century ordering of modern racialization. In response to the more rigidly binary US-American color line, Harlem Renaissance author Nella Larsen employs a form of irritated focalization of bi-racial subjectivity.

Nella Larsen: *Quicksand* (1928)

In contrast to the financial success story of the *homo economicus* Crusoe, the mixed-race heroine of *Quicksand*, Helga Crane, daughter of a white Danish woman and an African American man, ultimately fails in a racialized logic of labor and success and was often read in the framework of the so-called 'tragic mulatta' formula.[14] Aspiring to middle-classness, Crane experiences the affective relationality of racism across a number of settings in this short novel set in the 1920s: from the alienation as a teacher in an assimilationist Southern College and her financial struggles in Chicago to the growing sense of entrapment in the initially exciting Harlem, to her white relatives in Copenhagen, who exploit her as an exotic spectacle, back to the eventually fatal domesticity in the Southern United States as a preacher's wife, dying during childbirth of her fifth child in rapid progression.

Crane clearly does not fit the aspirational "Du Boisian notion of uplift" and Black respectability, which Judith Butler reads as a "moral notion of 'race' which, […] requires the idealization of bourgeois family life in which women

13 Schwalm calls this "prioritization of the economic drive for autonomy" (2007: 243; my translation) and later argues: "Crusoe's self-fashioning resembles autonomy that escalates into solipsism which needs to subjugate, or rather destroy the other or completely demarcate oneself from the other" (2007: 248; my translation).

14 Jeanne Scheper too notes this predominant reading of Larsen's work and characterises the formula as "a melodramatic form in which the mixed-race character is seen as a split subject, tragically flawed by 'nature' and trapped in a narrative trajectory inevitably leading to rejection or death" (2008: 680). Cf. also Walker (2016) for a critique of reducing interpretations of Larsen's work to this formula.

retain their place in the family" (1993: 178). There is no space for a racial in-between in what Werner Sollors (2000) calls an American exceptionalism in the policing of so-called miscegenation, of which Crane is a visible product. In Jim Crow America, the logic of the *one-drop rule* undermines the supposed visual evidentiality of race up to the point where you can pass as white, and still be considered Black. Crane's illicitness in both white and Black company is an effect of the regulation of race as a product of heterosexual reproduction that also makes Crane herself fear being read as overly sexual (cf. Carby 1987: 174). Linking introspection on color and affect, Larsen offers a prism of race in reaction to the ill-fitting Black and white binary, describing objects and colors vividly (cf. Hostetler 1990). Right at the beginning of the text, Crane is sitting in her room where "a single reading lamp, dimmed by a great black and red shade, made a pool of light on the blue Chinese carpet, on the bright covers of the books [...], on the shining brass bowl crowded with many-colored nasturtiums" (Larsen 2009: 1). In opposition to the decorum at the college that ascribed dark colors as appropriate dress, Crane's "loyalty to the inherent racial need for gorgeousness told her that bright colors *were* fitting and that dark-complexioned people *should* wear yellow, green, and red" (Larsen 2009: 18).

Crane experiences race not as given, but as always relational. In the U.S., she does not feel at ease with the narrative of progress and the prospect of a decent marriage to colleague James Vayle; even Harlem becomes suffocating, as she is to keep her mixed heritage a secret. Her lack of belonging is not mended by her opportunist European family in Copenhagen either, and tellingly, Denmark's own colonial history of slave trading with the West Indies is not present in the story as Lunde and Stenport contend (2008). In Europe, she refuses marriage again, this time to the white artist Axel Olsen. She explains "It isn't just you, not just personal [...]. It's deeper, broader than that. It's racial. [...] [I]f we were married, you might come to be ashamed of me, to hate me, to hate all dark people. My mother did that" (Larsen 2009: 88). Crane suddenly longs for connectivity with Black people and, at the same time, feels deeply alienated from the United States when considering a return, as the following quote underlines.

> Go back to America, where they hated Negroes! To America, where Negroes were not people. To America, where Negroes were allowed to be beggars only, of life, of happiness, of security. To America, where everything had been taken from those dark ones, liberty, respect, even the labor of their hands. To America, where if one had Negro blood, one mustn't expect money, education, or, sometimes, even work whereby one might earn bread. Perhaps she was wrong to bother about it now that she was so far away. Helga couldn't, however, help it. Never could she recall the shames and often the absolute

horrors of the black man's existence in America without the quickening of her heart's beating and a sensation of disturbing nausea. It was too awful (Larsen 2009: 82).

The passage runs through the adversarial effects of racism in five different variants, all beginning with "To America": even far away in Denmark, she recalls the "shames" – note the unusual plural here – and "absolute horrors of the black man's existence," interestingly using the universalized masculine form for what seems to shake her personally to the core. Typical of passages of free indirect discourse, there is a blurring between the voice of the extradiegetic narrator and Crane's perceptions as the internal focalizer, describing her physical reaction as "a sensation of disturbing nausea" (Larsen 2009: 82).

This modernist interrogation of the color line explicitly links racialization and affect as visceral. In her insightful work on *Ugly Feelings* Sianne Ngai identifies "irritation" as the primary tone of the novel offering us not psychological explanations, the predominant form of realist novelistic discourse that Defoe helped shape, but irritated affective surface reactions – and that is despite the title's association with depth – wavering between identification and disidentification with Blackness (cf. 2005: 190). Ngai resists a reading of *Quicksand* as a failure of racial authenticity – the mentioned stereotypical tragedy of mixed-racedness – but argues instead that there is no essence attached to race in the text. It is not that Helga Crane is not *really* Black or *really* white; she experiences color as increasingly disorienting. Erotically, she is attracted both to the "queer" gray-eyed Robert Anderson (Larsen 2009: 20), and to Audrey Denney with the "golden" skin[15] (Larsen 2009: 60), who is ostracized for passing in white company. After kissing the married Dr. Anderson, the regretful Crane converts to Christianity. Thus, Crane's suppressed sexuality heralds her downfall: She enters the ill-fated marriage to the Reverend Mr. Pleasant Green who, despite his name, is described as a "rattish yellow man" (Larsen 2009: 118). Exchanging the queer appeal of Harlem's "goldenness" for the yellowness of her husband and Alabama domesticity, *Quicksand* paints in vivid colors how race is always co-articulated with gender, class and sexuality.

While Robinson Crusoe becomes part of the emerging white English middle class by accumulating wealth abroad, Helga Crane flirts with whiteness/lightness as class mobility, but is reduced to the (fatal) gendered labors of reproduction. In *Black-ish*, my final and briefest example from contemporary TV, class and race are also convoluted. Protagonist Andre Johnson constantly wants to prove his 'realness' and fears his family's disconnection from African American culture

15 Dawahare sees this "association of gold with whiteness and thus with value" (2006: 33) as a recurring theme in Larsen's writing.

precisely because of their financially elevated status in an all-white upper-class suburban neighborhood.

Black-ish: TV Series (*ABC*, 2014-)

The US sitcom *Black-ish*, created by Kenya Barris, in its fifth season in 2018, focuses on the Johnson family which unites mixed-race mother and doctor Rainbow (Tracee Ellis Ross), who embodies affluent Hippie privilege, and advertising executive Dre (Andre) (Anthony Anderson), who worked his way up from LA's notorious neighborhood of Compton to live the 'American Dream.' Beginning in the America of Barack Obama's second presidency, the narrative of tragic bi-raciality is supposedly overcome for good and even turned into a success story. Thus, the show addresses the question how much race still matters in 21st century America. Repeatedly, *Black-ish* emphasizes that the four children, stylish 15-year-old Zoey (Yara Shahidi), 14-year-old nerd Junior (Marcus Scribner), the six-year-old twins Jack (Miles Brown), and Diane (Marsai Martin), do not see race in the same way as their parents or grandparents do. As any six-year-old, Jack's reaction to Dre's emphasis that Obama is the *first* Black president, is telling: Obama is the *only* president he is aware of. While there is a longing for African Americanness as familiarity, it is also decreasingly naturalized as simply given but rather seen as a cultural heritage that needs to be passed on (and that Dre's white colleagues in the advertising industry misunderstand or try to appropriate).

Nevertheless, *Black-ish* is, of course, not post-race[16] in the sense that race no longer matters, or that there is no racism. Quite the contrary, as the title of the show indicates, it constantly refers to the insecurity of what it means to be Black or rather Black-*ish* in the United States today, addressing issues such as the rise of the Black Lives Matter movement in reaction to continuing police brutality

16 Nana Adusei-Poku, for example, rejects post-race as a label, but discusses post-Black as an aesthetic strategy of artists who try to delink Blackness from a representational logic in art. Cf. "I would argue instead that the term [post-race] has to be modified to 'post-racism.' Post-racism is a necessary distinction as it expresses a desire to [...] [live] in this world without being affected by the consequences of racism. The utopic assumption that we have entered a post-race era shows how contested and politically charged the discussions about Blackness and race are. Some individuals feel offended, others don't care, many are already bored, most don't really know what to think about it and some claim that we are already post post-black" (Adusei-Poku 2012). Cf. also Paul (2014).

and the consequences for parents raising African American children. What is more, in the third season, episode 12, aired in January 2017, the election of Donald Trump as the 45th president of the United States is played out as a traumatic event for the family that ends with a solemn reconsideration of Martin Luther King's "I have a Dream" speech during the March on Washington more than fifty years ago in August 1963, thus disrupting ideas of linear progress.

The choice of genre, especially in the context of the issue of race, constitutes a powerful statement. The family sitcom is a notoriously conservative and appeasing format that traditionally builds on heteronormative assimilation. This is self-reflexively played out in *Black-ish*, specifically with references to the most successful aspirational African American TV family ever: The Huxtables of *The Cosby Show*, which aired on *NBC* from 1984 to 1992, and whose legacy has become even more complicated given the multiple rape allegations against actor Bill Cosby, once labelled "America's funniest father" on the cover of *Life* magazine (June 1985). In the episode "The Johnson Show" (season 2, episode 21), for example, the family recreates the famous Cosby opening titles dancing around Dre, covering his face with their hands and then revealing his big smile in close-up. At another occasion for Father's Day when the Johnson kids lack an idea for a present, Diane suggests lip-syncing a Ray Charles song, citing her TV predecessor Rudy Huxtable, then quickly dismissing the idea by noting that "I think this has been done better before." While acknowledging the tainted legacy of *The Cosby Show* explicitly in that episode, the Johnsons also express nostalgia for the representation of an upwardly mobile African American family and specifically highlight the character of the mother Clair (Phylicia Rashad). This is a TV family reminiscing about another TV family.[17]

As a TV show of the 21st century, *Black-ish* clearly exhibits postmodern self-reflexivity in how genre and race interact. Blackness in some ways seems less restrictive. After all, the family can dress up as the First Family for Halloween and in this day and age, the black *paterfamilias* is also ok with a gay sister;[18] the format, however, pushes towards didacticism, with episodes trying hard to

17 Another parallel is the recent 2018 spin-off of the series called *Grown-ish* which puts a younger female perspective at the center and focuses on Zoey's experiences in college and thus clearly echoes the *Cosby Show* universe and their spin-off, *A Different World* (1987-1993), which very successfully told the stories of Denise Huxtable (Lisa Bonet) and her friends at college.

18 Dre's sister Rhonda fittingly is played by former teen star Raven-Symoné who had her breakthrough as a child actor on the *The Cosby Show* and while not endorsing the identity label 'gay' has publicly acknowledged being in a relationship with a woman in 2014.

resolve issues, such as who can use the n-word, or gun violence. This is reinforced in the framing device of the overly explanatory voice-over of the male head of the family at the beginning and end of most episodes. Blackness here is no longer racial in the sense that it is pre-defined, it needs to be claimed and remade, within a society that is built on the legacy of slavery: the codes of the Black vernacular, such as the competitive insulting game of The Dozens, need to be taught to Junior (season 1, episode 15), and, tellingly, it is the Harlem Renaissance which is construed as heritage in Granddad's (Laurence Fishburne's) fake family history to compensate for the fact that his grand-children cannot trace their lineage back for several generations like their white class mates (season 1, episode 24). Hence, how we imagine racial difference then, as chromatic, intercultural or nationally specific, is not irrelevant in an increasingly mediatized postcolonial world.

Conclusion: Postcolonial Theory as Intersectionality

As a way of closing, I want to return to my proposal of a non-binary and non-static framing of categories of difference as a postcolonial contribution to an interdisciplinary understanding of intersectionality. By now the term "intersectionality" has come to be used almost synonymously with any research agenda that addresses difference and/or diversity as co-constructed by more than one axis of social stratification.[19] However, this has been predominantly a sociological framing of difference. Given its roots in Kimberlé Crenshaw's pioneering work on radical justice activism, intersectionality has often become shorthand for describing processes of Othering rather than being understood as a way of critiquing all identity formations, including hegemonic ones. Moreover, and more fundamentally, the all too clear-cut use of categories does not match the multi-layered and conflicting process of subjectification.[20] A postcolonial

19 In particular, Kimberlé Williams Crenshaw's diagnosis of the intersection of race and gender discrimination in US legal frameworks (1997) and Patricia Hill Collins' Black feminist standpoint theory (1999) have proven influential. In Germany particularly – but by no means solely – in the social sciences, intersectionality is often framed as an import from U.S.-American critical race studies which thus problematically positions 'race' too as exterior to Germany. Kerner (2012) offers an overview of the German debate on intersectionality.

20 This section draws on ideas that I have developed together with Gabriele Dietze and Beatrice Michaelis. A more extensive elaboration on these potential pitfalls of

perspective on literature and media can act as an important supplement to this framing of difference as a crossing or intersecting of identity categories, such as race, class, and gender. It helps expose the instability of each of these assumed categories along different temporalities and spaces, on their own and in co-articulation with each other. Such a postcolonial tracing of racial landscapes can unearth the specific effects of racisms and help defamiliarize ontologizing notions of race as biological difference. In reading cultural texts, we cannot simply apply a matrix of race, class, gender. We have to interpret the contradictory mechanisms of racialization in co-articulation with other markers of difference and we need to embed this process of meaning-making within specific socio-historic contexts. To be clear, I am not proposing that there is a linear progression of a proto-racial understanding in the 18th century that is not quite race yet, followed by 'proper' race as the result of scientific racism in the 19th century and that we now have 'magically' surpassed all of this in an era of 'post-race.' I am arguing that racialization might at certain times seem discursively more flexible than at others. It can also be situated in different generic and aesthetic contexts: Defoe is championing the modern novel as foundational for white dialogical self-reflexivity, while Larsen offers modernist descriptions of the affective and gendered labors of the color binary, *Black-ish* finally struggles to establish both a popular, postmodern understanding of race and the family sitcom.

In sum, postcolonial literary and cultural studies can contribute to intersectionality research historically situated close readings, transnational comparative insights and foster a productive understanding of the instability of categories of difference, which is precisely not to underestimate the violent effects of world-making that these entail. In such a conception of the *longue durée* of European and American modernity, postcolonial theory should not only be understood as the temporal after colonialism, it also signifies the epistemological beyond.

intersectionality, especially from a queer point of view was published as Dietze, Haschemi Yekani and Michaelis (2018).

Works Cited

Adusei-Poku, Nana (2012): "The Challenge to Conceptualise the Multiplicity of Multiplicities – Post-Black Art and Its Intricacies." In: Dark Matter November 29 (http://www.darkmatter101.org/site/2012/11/29/the-multiplicity-of-multiplicities-%E2%80%93-post-black-art-and-its-intricacies).

Aravamudan, Srinivas (1999): Tropicopolitans. Colonialism and Agency, 1688-1804, Durham: Duke University Press.

Bakhtin, Mikhail Mikhaĭlovich (1994[1981]): The Dialogic Imagination. Four Essays. Michael Holquist (ed.), Caryl Emerson/Michael Holquist (trans.), Austin and London: University of Texas Press.

Barrett, Lindon (2014): Racial Blackness and the Discontinuity of Western Modernity. Justin A. Joyce/Dwight A. McBride/John Carlos Rowe (eds.), Urbana: University of Illinois Press.

Black-ish (2014): ABC, September 24, created by Kenya Barris.

Boulukos, George (2008): The Grateful Slave. The Emergence of Race in Eighteenth-Century British and American Culture, Cambridge: Cambridge University Press.

Butler, Judith (1993): "Passing, Queering: Nella Larsen's Psychoanalytical Challenge." In: Judith Butler, Bodies That Matter. On the Discursive Limits of 'Sex', New York and London: Routledge, pp. 167-185.

Carby, Hazel V. (1987): "The Quicksands of Representation." In: Hazel V. Carby (ed.), Reconstructing Womanhood: The Emergence of the Afro-American Woman Novelist, New York and Oxford: Oxford University Press, pp. 163-176.

Carey, Daniel/Festa, Lynn (eds.) (2009): The Postcolonial Enlightenment: Eighteenth-century Colonialism and Postcolonial Theory, Oxford: Oxford University Press.

Carey, Daniel (2009): "Reading Contrapuntally. Robinson Crusoe, Slavery, and Postcolonial Theory." In: Daniel Carey/Lynn Festa (eds.). The Postcolonial Enlightenment: Eighteenth-century Colonialism and Postcolonial Theory, Oxford: Oxford University Press, pp. 105-135.

Cohen, Jeffrey Jerome (2000): "Introduction Midcolonial." In: Jeffrey Jerome Cohen (ed.), The Postcolonial Middle Ages, New York: St. Martin's Press, pp. 1-18.

Collins, Patricia Hill (1999): "Moving beyond Gender. Intersectionality and Scientific Knowledge." In: Myra Marx Ferree/Judith Lorber/Beth B. Hess (eds.), Revisioning Gender, Thousand Oaks: Sage, pp. 261-284.

Crenshaw, Kimberlé Williams (1997): "Mapping the Margins. Intersectionality and Identity Politics. Learning from Violence Against Women of Color." In: Mary Lyndon Shanley/Uma Narayan (eds.), Reconstructing Political Theory. Feminist Perspectives, London: Pennsylania State University Press, pp. 178-193.

Dawahare, Anthony (2006): "The Gold Standard of Racial Identity in Nella Larsen's Quicksand and Passing." In: Twentieth Century Literature 52/1, pp. 22-31.

Defoe, Daniel (2003[1719]): Robinson Crusoe. John Richetti (ed.), London: Penguin Books.

Dietze, Gabriele/Haschemi Yekani, Elahe/Michaelis, Beatrice (2018): "Modes of Being vs. Categories – Revising the Tools of Queer Intersectionality." In: Greta Olson/Mirijam Horn-Schott/Daniel Hartley/Regina Leonie Schmidt (eds.), Beyond Gender: An Advanced Introduction to Futures of Feminist and Sexuality Studies, Abingdon and Oxon: Routledge, pp. 117-136.

El-Tayeb, Fatima (2011): European Others. Queering Ethnicity in Postnational Europe, Minneapolis: University of Minnesota Press.

Ganguly, Keya (2004): "Temporality and Postcolonial Critique." In: Neil Lazarus (ed.), The Cambridge Companion to Postcolonial Literary Studies, Cambridge: Cambridge University Press, pp. 162-180.

Gilroy, Paul (2001): Against Race. Imagining Political Culture Beyond the Color Line. 4th edition, Cambridge: Belknap Press of Harvard University Press.

Hartman, Saidiya (2008): Lose Your Mother. A Journey along the Atlantic Slave Route, New York: Farrar.

Heng, Geraldine (2011a): "The Invention of Race in the European Middle Ages I: Race Studies, Modernity, and the Middle Ages." In: Literature Compass 8/5, pp. 315-331.

Heng, Geraldine (2011b): "The Invention of Race in the European Middle Ages II: Locations of Medieval Race." In: Literature Compass 8/5, pp. 332-350.

Heng, Geraldine (2015): "Reinventing Race, Colonization, and Globalisms across Deep Time: Lessons from the Longue Durée." In: Publications of the Modern Language Association of America 130/2, pp. 358-366.

Hostetler, Ann E. (1990): "The Aesthetics of Race and Gender in Nella Larsen's Quicksand." In: Publications of the Modern Language Association of America 105/1, pp. 35-46.

Hulme, Peter (1992): Colonial Encounters. Europe and the Native Caribbean, 1492-1797, London: Routledge.

Kerner, Ina (2012): "Questions of Intersectionality: Reflections on the Current Debate in German Gender Studies." In: European Journal of Women's Studies 19/2, pp. 203-218.

Larsen, Nella (2009): Quicksand and Passing. 19th edition, Deborah E. MacDowell (ed.), New Brunswick: Rutgers University Press.

Lunde, Arne/Anna Westerstahl Stenport (2008): "Helga Crane's Copenhagen: Denmark, Colonialism, and Transnational Identity in Nella Larsen's Quicksand." In: Comparative Literature 60/3, pp. 228-243.

Mayer, Robert (2002): "Three Cinematic Robinsonades." In: Robert Mayer (ed.), Eighteenth-Century Fiction on Screen, Cambridge: Cambridge University Press, pp. 35-51.

McClintock, Anne (1995): Imperial Leather. Race, Gender and Sexuality in the Colonial Contest, New York and London: Routledge.

Ngai, Sianne (2005): Ugly Feelings, Cambridge: Harvard University Press.

Paul, Heike (2014): "'Race,' Racism, and Tacit Knowing." In: Winfried Fluck/Eric Redling/Sabina Sielke/Hubert Zapf (eds.), American Studies Today: New Research Agendas, Heidelberg: Winter, pp. 263-287.

Richetti, John (2000): "Ideas and Voices: The New Novel in Eighteenth-Century England." In: Eighteenth-Century Fiction 12/2-3, pp. 327-344.

Said, Edward W. (2003[1978]): Orientalism. 25th anniversary edition, New York: Vintage.

Scheper, Jeanne (2008): "The New Negro Flâneuse in Nella Larsen's Quicksand." In: African American Review 42/3-4, pp. 679-695.

Schwalm, Helga (2007): Das eigene und das fremde Leben. Biographische Identitätsentwürfe in der englischen Literatur des 18. Jahrhunderts, Würzburg: Königshausen & Neumann.

Sollors, Werner (2000): "Introduction." In: Werner Sollors (ed.), Interracialism. Black-White Intermarriage in American History, Literature, and Law, New York: Oxford University Press, pp. 3-22.

Stam, Robert/Ella Shohat (2012): Race in Translation. Culture Wars around the Postcolonial Atlantic, New York: New York University Press.

Walker, Rafael (2016): "Nella Larsen Reconsidered: The Trouble with Desire in Quicksand and Passing." In: Multi-Ethnic Literature of the United States 41/1, pp. 165-192.

Watt, Ian P. (2000[1957]): The Rise of the Novel. Studies in Defoe, Richardson and Fielding, London: Pimlico.

Wheeler, Roxann (1995): "'My Savage,' 'My Man': Racial Multiplicity in Robinson Crusoe." In: ELH. A Journal of English Literary History 62/4, pp. 821-861.

Wheeler, Roxann (2000): The Complexion of Race. Categories of Difference in Eighteenth-Century British Culture, Philadelphia: University of Pennsylvania Press.

Passing Tone/Note[1]

Marius Henderson

A want to abandon fidgeting about in "ruses of analogy,"[2]
to obstruct the reels of the perpetually turning disavowing fetishization –
oh, so very real – epistemological onslaughts and assaults tend to preclude
(critical) self-implication.

No more sneering puns.
You turn out indigestible
as you regurgitate,
back and forth and back,
way back again.

Fear no more, of feeding on liquefied food for thought,
unfounded lines of transmogrification,
disemboweling dehiscence –
an about-face, in the light of those moments when
racialized timidity turns into ignorance.

being pitched & then dropped
in cold water –
never quite sure about one's safety in sitting,
in occupying "intellectual" space –

---

1 This text emerged out of embodied notational inquiries into (tonalities of) "racial (non-)passing," particularly in the context of educational settings, in Germany (and elsewhere).

2 Cf. Frank B. Wilderson III, *Red, White & Black: Cinema and the Structure of U.S. Antagonisms* (Durham: Duke UP, 2010), 35-54.

the danger of losing ground and
falling – a relation of subordination.
These instances when a seminar session or a panel discussion turns into a county fair –
half-standing in front of a class: will I pass, will I pass?
The question detonates,
in anticipation & afterthoughts.

And when the balance shifts,
you become
a specimen,
from inquiring agent
to object of inquiry
and fear, as interlocutors watch their tongues.

You come to be a blotting book, engraved & gutted –
fence sitting on paper-thin palisades,
sharp edges, wired encasement,
and you find yourself,
objectified matter in a case study,
or stretched across a screen, to reflect a tragic projection –
What's one more "tragic m******"?
From tragedy to farce & back to tragedy again.
Alas, does the glorification of the underdog
keep the underdogging conditions in place?

From a pedagogy of disgust & detestation
to a pedagogy of the detested.
A congregated conflagration and the fragrance of
revulsion & abjection.
I want that mucus, that hawking up, hemming, harrumph.

What transpires?
A clearing of the throat & the inhalation of particles.
What do you hear? Sores, soars, lumps that leak.
Not noticing, but checking.
Not noticing, but observing.
You're being checked.
You're being observed.

"Do you see yourself? Do you see yourself being seen?"[3]

Constant constipation – holding:
"outness."
Our vocabularies smash and shatter, and we don't care.
The difference between running for and running from.
And then that's precisely the point,
to be imprecise when it comes to pointing,
to be beside or beyond the point.
Or "that point after apprehension but before abstraction."[4]

It has become a custom to stretch out & then latch out at what's
surreptitiously and intuitively perceived as an intrusion –
this thin line of historical felt, between inclusion & intrusion, or
the thin line between proposed inclusion & felt intrusion,
felt, as gates are kept in check.
"And whiteness is a violently upheld dream that safety is real."[5]

Thinking of this nexus of "passing" & loss,
and absence,
like the absence of a sense of control.

An attempt to careen past those incredulous looks,
a condition that knows no rest,
distrust weighs down,
wears us out,
and perspires between us.
How did _ become so suspicious? How did we turn out so inquisitive?
We simply aspire to clarity.
We like to know and worry
who's who.
But don't worry,

3 Rozena Maart (2015): "Decolonizing Gender, Decolonizing Philosophy: An Existential Philosophical Account of Narratives from the Colonized." In: Radical Philosophy Review 18/1, p. 77.

4 Denise Ferreira da Silva (2013): "To Be Announced: Radical Praxis or Knowing (at) the Limits of Justice." In: Social Text 31/1, p. 59.

5 Hannah Black, "New World Disorder," *artforum.com*, February 27, 2017.

"We don't check pedigrees."[6]

And still, there are these moments of intense study,
when one aspirates: "I am with you."
Albeit not resorting to a form of solidarity that presupposes or strives for
a "unity of feeling," perpetuates "people-of-color-blindness,"[7] and bulldozes over differences in relational positionality. But solidarity of a different kind,
as the seemingly paradoxical attempt to simultaneously stand with and provide space for the most vulnerable and most abjected.[8]

How much can you even take?
Blisters & blueberries in the morning.
Unbelonging, uncovering
the icing of the white cubic cake.
To leave the register of vision, specularity,
not running for:
office, visibility, and what have you –
what some have access to –
but perpetually running from,
or dropping out
of the spectacular disciplinary passe-partout,
which feigns free accessibility,
free "passing" (*passer*) for all, everywhere and through everything (*partout*).
What's your favorite frozen ivory on ivory frame?

The perpetual reproductive labor
of rinsing the bubbles forming around one's gums,
as labor spent in sticky speech acts from way back hums.
They will not inter-mingle with you,
but interrogate & inter.

And "passing" & its entwinement
with lack.
Writing from –

6 Nell Zink, *Mislaid: A Novel* (New York: Harper Collins, 2015), 63.

7 Cf. Jared Sexton (2010): "People-of-Color-Blindness: Notes on the Afterlife of Slavery." In: Social Text 28/2, pp. 31-56.

8 I am indebted and grateful to Dont Rhine from Ultra-red for this distinction between different conceptualizations and practices of solidarity.

a lack of fitting terms for "this" in German,
apart from further insults,
and thus also un-speaking of "it" in the following sense:
"To be beyond the range or compass of (a faculty or expression)"[9]

Alabaster and pumice covered lungs,
enunciation, emission, ejection,
heavy clay breath,
clouded glances at
Archibald J. Motley, Jr.'s painting *The Octoroon Girl*,
and how the fingers of her left hand seem to fade into each other.
Is she losing grip?
Is this her writing hand?
Is loss enacted in contortion?
Is loss enacted in composition?
Is it possible to cultivate a molten hapticality,
or a haptic plasticity between bodies that pass each other by,
even if only "as some combination of flight and submergence?"[10]

When being rendered in a register of clairvoyance,
does your body become a cloudy obstruction?
A stain on the shoes of
progress marching onward?

Hush, hush, pant, pant,
garlanding our slanted & skewed perspectives – what we inhale
till exhaustion.
Mutation & miraculation.
Clunky shivers in honeyed suspension:
"We care about each other so militantly, with such softness, that we exhaust ourselves, and then record, in the resonance of our slightly opened mouths, the sound of that."[11]

Shiv-err-ring, not shying away from erring.

---

9 "pass, *v.*" *oed.com.*

10 Cf. Fred Moten, "Manic Depression: A Poetics of Hesitant Sociology," Lecture at the University of Toronto, April 4, 2017.

11 Fred Moten, "hand up to your ear," in *The Little Edges* (Middletown: Wesleyan UP, 2015), 12.

These algae covered lips
won’t wait any longer,
we never know, just wonder.

Contributors

Maria Alexopoulou, Postdoctoral Scholar and Lecturer at the Chair for Contemporary History at the University of Mannheim, is working on her second-book project (*Habilitation*) on "Racial knowledge and migration in Germany since 1945" and is teaching migration history and postcolonial studies. She is a member of the migrant organization "Die Unmündingen e.V.," whose antiracist and emancipatory work ranges from political activism to historical exhibitions and the production of documentaries.

Mahmoud Arghavan completed his PhD in American Studies at Free University of Berlin in 2013 with a dissertation entitled "Iranian American Literature: from Collective Memory to Cultural Identity." From 2014 to 2016 he worked as Adjunct Lecturer at the American Studies Department of the University of Tübingen. Currently he is working as coordinator for the Unit of Aid for Refugees, Migration, and Integration at *Innere Mission* in Munich. His research interests include Iranian American Literature, Diaspora Studies, Postcolonial Studies, Critical Race Theory, and Border Studies. He has recently authored the article "Postcolonial Orientalism: A Study of the Anti-Imperialist Rhetoric of Middle Eastern Intellectuals in Diaspora" in the edited collection *Postcolonial Justice* (Brill, 2017), and co-written the article "Writing against Neocolonial Necropolitics: Literary Responses by Iraqi/Arab Writers to the US 'War on Terror'," which appeared in a special issue on "Global Responses to the 'War on Terror'" in the *European Journal of English Studies* (2018).

Amina Grunewald attended Humboldt University of Berlin and Goldsmiths College/University of London. She received an M.A. degree in English/ American Studies, French Studies, and Educational Studies, and obtained her PhD with a dissertation on Native self-designs in the works of Sherman Alexie at Humboldt University of Berlin. Related research projects during her time as a

PhD candidate encompassed a research stay in Vancouver/Canada on community-based representations of First Nations in Canadian museum and gallery spaces. She was integrated as a PhD research fellow at CEREV/ Concordia University, Montréal, to investigate Indigenous trauma narratives, Native epistemologies, and potential transferability to (non-)Native audiences. Her postdoctoral studies focus on representations that thematize deep ecologies, alterNative ethics, and ontologies in American literatures and cultures, while broadening her horizon on critical animal studies, and contemporary Native North American visual arts. Currently, she is an associated fellow at Forum Transregionale Studien/Art Histories and Aesthetic Practices and a teaching assistant at Humboldt University of Berlin's Department of American Studies.

Elahe Haschemi Yekani is Professor of English and American Literature and Culture with a focus on Postcolonial Studies at the Department of English and American Studies at Humboldt University of Berlin. Her research interests include the Anglophone novel, Postcolonial Studies, and Queer/Gender Studies. She currently works on her second book *Familial Feeling: Entangled Tonalities in the Rise of the British Novel and Early Black Atlantic Writing*, in which she traces an entangled literary history of canonical bourgeois novels of the late 18th and 19th century with the earliest written testimonies of Black British writers. Other publications include: *The Privilege of Crisis. Narratives of Masculinities in Colonial and Postcolonial Literature, Photography and Film* (Campus 2011) and the edited volume (together with Eveline Kilian and Beatrice Michaelis) *Queer Futures: Reconsidering Ethics, Activism, and the Political* (Ashgate/ Routledge 2013; paperback 2017).

Marius Henderson studied English and American Studies as well as Gender and Queer Studies at the University of Hamburg and Johns Hopkins University. Currently, he is a research and teaching assistant at the Chair of American Studies: Culture and Literature at Friedrich-Alexander University of Erlangen-Nuremberg. He is also a PhD candidate at the University of Hamburg and has just submitted his PhD thesis. His PhD project scrutinizes modes of rendering suffering poetically and theoretically, in the context of contemporary experimental poetry. In recent conference papers, essays, and poetic texts he has explored (im-)possible enmeshments between queer-feminist theory, Afro-Pessimism, experimental poetics, and black metal theory. He regularly cooperates with other writers and artists, and is interested in collaboratively bridging gaps between artistic, scholarly, and activist practice from a position of critical self-implication. He is a founding member of the PhD student network,

“Perspektiven in der Kulturanalyse: Black Diaspora, Dekolonialität und Transkulturalität,” which is based at the University of Bremen. This network consists of differently positioned members, fosters the critical reflection of sociopolitical positionalities within and beyond academia, and regularly organizes workshops, e.g., on Black Diasporic theory, empowerment, and critical whiteness. Recent publications include the edited volume, *Entangled Memories: Remembering the Holocaust in a Global Age* (co-edited with Julia Lange, Heidelberg: Universitätsverlag Winter 2017), and a contribution to the edited collection *The Failed Individual: Amid Exclusion, Resistance, and the Pleasure of Non-Conformity* (edited by Katharina Motyl and Regina Schober, Frankfurt: Campus Verlag 2017).

Saskia Hertlein is a non-tenured Assistant Professor in American Literary and Media Studies and Inter-American Studies at the University of Duisburg-Essen, Germany, where she has recently served a term on the *Committee for Diversity Management*. After having covered *Tales of Transformation. Emerging Adulthood, Migration, and Ethnicity in Contemporary American Literature* extensively in her doctoral research, her current research interests include critical diversity literacy from a literary and cultural studies perspective, modes of conviviality, and the early U.S. city. In classes preparing future English teachers for their mandatory stay abroad as well as in her literary and cultural studies classes at the University of Duisburg-Essen, which prides itself with its awareness of its diversity, e.g. in its student population, she continually works on reflecting “Who Speaks and Who Is Heard/Hurt?” both in class content and teaching practice.

Nicole Hirschfelder is Associate Professor of American Studies at Eberhard Karls University of Tübingen, Germany, and a member of the Collaborative Research Center 923 “Threatened Order – Societies under Stress.” She studied in Frankfurt and at the University of Wisconsin, Madison and conducted research in the course of Yale University’s PhD research scholar program. Her main interests include inequality, oppression, the Civil Rights Movement, new social movements, and ways of seeing and representing disasters. In 2014, she published her first book, *Oppression as Process: The Case of Bayard Rustin*. In January 2016, she was one of the organizers of the workshop, “Who Can Speak and Who Is Heard/Hurt?: Ethnic Diversity, Race, and Racism in American Studies in Germany” in Tübingen. In fall 2016, she was a visiting professor at the University of Maryland on a Fulbright scholarship and is returning to UMD for the spring term of 2019.

Luvena Kopp is a doctoral candidate at the Collaborative Research Center (CRC) 923 "Threatened Order – Societies under Stress" at the University of Tübingen in Germany. She earned her Magistra Artium in American Studies and Theater, Film and Media Studies from Goethe University Frankfurt. Her research interests include figurational sociology, power relations and violence, and African American history, literature and culture. Kopp, a member of the Norbert Elias Figurational Research Network, has been a Fulbright Fellow at New York University. She also participated in the 2016 Fulbright American Studies Institute entitled "Why Black Lives Matter: Race and Politics in the U.S." organized by San Francisco State University. Her publications include "Satirizing Satire: Symbolic Violence and Subversion in Spike Lee's *Bamboozled*," in *Post-Soul Satire: Black Identity after Civil Rights* (eds. Derek C. Maus and James J. Donahue, University Press of Mississippi, 2014), "Understanding Ferguson: Suburban Marginality and Racialized Penality in the Age of Neoliberalism," in *Power Relations in Black Lives: Reading African American Literature and Culture with Bourdieu and Elias* (ed. Christa Buschendorf, transcript, 2018), and the forthcoming essay "Towards a Black Prophetic Critique of Neoliberal State Violence: Spike Lee's *Do the Right Thing* and the Death of Eric Garner," in *Violence from Slavery to #Black Lives Matter: African American History and Representation* (eds. Andrew Dix and Peter Templeton, Routledge, 2019). In her dissertation, Kopp employs the concepts of figurational sociology to investigate representations of hidden mechanisms of power in the films of Spike Lee.

Priscilla Layne is Associate Professor of German and Adjunct Associate Professor of African, African American, and Diaspora Studies at the University of North Carolina at Chapel Hill. Her publications address topics like representations of Blackness in German film, postwar rebellion, and Turkish German culture. She has published essays in the journals *German Studies Review*, *Colloquia Germanica*, and *Women in German Yearbook.* She is the author of *White Rebels in Black: German Appropriation of African American Culture*, published by the University of Michigan Press in 2018. She was a Mellon Mays Undergraduate Fellow, and since 2013 has been a mentor for the Moore Undergraduate Research Apprenticeship Program, which provides underrepresented undergraduate students who would like to pursue a PhD with faculty mentors for a summer research project.

Kai Linke recently finished his PhD in American Studies at Humboldt University of Berlin. His thesis is titled *Good White Queers? Tracing the Limits of (Self-)Critical Engagements with Racism and Whiteness in Queer Comics from the U.S.* He is now in training to become a high school teacher in English and Geography.

Derek C. Maus is Professor of English and Communication at the State University of New York at Potsdam, where he teaches a wide range of courses on contemporary literature from around the world. He is the author of *Jesting in Earnest: Percival Everett and Menippean Satire* (South Carolina, 2019), *Understanding Colson Whitehead* (South Carolina, 2014), and *Unvarnishing Reality: Subversive Russian and American Cold War Fiction* (South Carolina, 2011). He is also the editor or co-editor of *Conversations with Colson Whitehead* (Mississippi, 2019), *Post-Soul Satire: Black Identity after Civil Rights* (Mississippi, 2014), and *Finding a Way: Critical Essays on Walter Mosley* (Mississippi, 2008). Additionally, he has taught at Karl-Franzens University in Graz, Austria as a Fulbright Scholar and at McGill University in Montréal, Canada. He is currently working on a book project comparing representations of black identity in contemporary African American and African Canadian fiction. His full CV is available at: https://potsdam.academia.edu/DerekCMaus.

Courtney Moffett-Bateau is a black woman scholar and lecturer based in Germany. She has taught at the Humboldt University of Berlin, Camp Afflerbaugh-Paige, an LA County juvenile probation camp located in La Verne, California, as well as worked in cooperation with the project "Förderunterricht für Kinder mit Migrationshintergrund," which offers free tutoring for German and non-German identifying POC pupils. Her dissertation surveys the institutional and literary mechanisms that allow American texts to be racialized through categorization. Currently, she is the recipient of the Konrad Adenauer Foundation's scholarship for "*Excellent and Talented*" international students. She is the recipient of the University of Duisburg-Essen's 2015 award for innovative teaching and of a Fulbright Grant (2008-09). She was also an Urban Fellow for the Center for California Cultural and Social Issues in 2007.

Katharina Motyl is Assistant Professor at the American Studies Department of the University of Mannheim. She obtained her PhD in American Studies from Free University of Berlin and has held appointments as a visiting scholar at the University of California, Berkeley, and the University of Southern California. In her second-book project, she investigates the loops of interaction between cultural, legal, and medical discourses on drug use/substance dependence among social minorities from the Early Republic to the "War on Drugs." Her publications include the monograph *With the Face of the Enemy – Arab American Literature since 9/11* (Campus, 2019) as well as the edited volumes *The Failed Individual – Amid Exclusion, Resistance, and the Pleasure of Non-Conformity* (with Regina Schober, Campus, 2017) and *States of Emergency – States of Crisis* (with Winfried Fluck, Donald E. Pease, and Christoph Raetzsch, Narr, 2011). Most recently, she has co-authored the article "Writing against Neocolonial Necropolitics: Literary Responses by Iraqi/Arab Writers to the US 'War on Terror'," which appeared in a special issue on "Global Responses to the 'War on Terror'" in the *European Journal of English Studies* (2018). Further research interests include African American expressive culture, Indigenous cultures and U.S. settler imperialism, issues of gender in U.S. imperial and cultural history, as well as critical posthumanism.

Anthony Obute is currently a Research Assistant at the American Studies Program, Eberhard Karls University of Tübingen. He also holds a German Academic Exchange Service (*DAAD*) scholarship for his PhD research, which examines "Toxic Ecologies of the Global South." His research interest includes Ecocriticism, Environmental Humanities, African American Literatures, Post-colonialism, Diaspora Narratives, and the Global South. His current project examines the material destruction of the human body, precipitating from anthropocentric aggression on the environment, through the model of accelerated global capitalism. His research also seeks to initiate new models for narrating environmental disasters, which have assumed normative dimensions, and also lost sense of the precarious.

Lili Rebstock lives in Leipzig and is a PhD candidate at the Department of Educational Science at Martin-Luther-University of Halle. Her research interests include racism and anti-racism in theory and praxis, gender studies and feminism, as well as youth studies and youth culture. Besides, she acts as a trainer for civic education in the field of anti-racism and anti-discrimination.

Annika Rosbach is a Research Assistant in the American Studies Department at the Faculty of Translation Studies, Linguistics, and Cultural Studies of the University of Mainz, Germany. She also works as a freelance translator and mainly translates legal texts from English and Dutch into German. Annika Rosbach is currently working on a PhD project which examines and analyzes literary representations and German translations of African American English, spanning a time period from the publication of *Uncle Tom's Cabin* in 1852 until today. Her research interests include literary translations, African American literature, and U.S. Christian literature.

Kimberly Alecia Singletary received her PhD in Rhetoric and Public Culture at Northwestern University in 2013. She received her MA from Georgetown University's Communication, Culture, Technology program, and her BJ from the Missouri School of Journalism. She has received German Academic Exchange Service (*DAAD*) and Fulbright fellowships to Germany and Austria, respectively. Her work on blackness in the global public sphere has been published and featured in various journals and online forums. After 10 years of teaching in the U.S. Midwest and Pacific Northwest, she left academia for the educational technology field and is now an independent scholar.

Ismahan Wayah is a PhD student in English, Postcolonial, and Media Studies at the University of Münster, Germany. Her research interests are Muslim diasporic fiction, Black thought, intersectionality, postcolonial German history, and decolonial archival studies. She positions herself as an activist academic and works closely with Black and Muslim communities in Germany on various issues, including racism, anti-Muslim racism, educational inequality, empowerment, and documenting marginalized histories. In addition, Ismahan Wayah works as a researcher and consultant on diversity and migration at the Historical Museum Frankfurt am Main. Recent publications include the articles "Muslimische Frauen im Spannungsfeld neo-kolonialer (Dis)Kontinuitäten," which appeared in *Online IslamiQ* in 2017, "Auch Wir Schreiben Geschichte: Dekoloniale Archivarbeit," which came forth in the edited collection *Connecting the Dots: Geschichte(n) zu Unterdrückung und Widerstand. Modellprojekt für historisch-postkoloniales Lernen* (Mangoes and Bullets, 2017), and "Körper (ver)Stimmen Räume," which was published in *Jüdisch-Muslimische Verflechtungen. Konkurrenzen, Allianzen und Empowerment* (edited by Ozan Keskinkilic and Armin Langer, Yilmaz-Günay Verlag 2018).

Sebastian Weier received his PhD in American Studies from the University of Bremen. He is currently at the Academy of Créteil, where he teaches at high schools in the *banlieues* of Paris. Sebastian Weier is engaged in anti-fascist and critical border militancy and an active member of the French Communist Party.